Designed cover image: © 2026 Shutterstock

First edition published 2026
by CRC Press
2385 NW Executive Center Drive, Suite 320, Boca Raton FL 33431

and by CRC Press
4 Park Square, Milton Park, Abingdon, Oxon, OX14 4RN

CRC Press is an imprint of Taylor & Francis Group, LLC

ISBN: 978-1-041-00329-8 (hbk)
ISBN: 978-1-041-00330-4 (pbk)
ISBN: 978-1-003-60927-8 (ebk)

DOI: 10.1201/9781003609278

Typeset in Times
by Apex CoVantage, LLC

Optimising Biogas Plants
Squeezing Profit Out of Your Digester

Mario Alejandro Rosato

CRC Press
Taylor & Francis Group
Boca Raton London New York

CRC Press is an imprint of the
Taylor & Francis Group, an **informa** business

Optimising Biogas Plants

Keeping theory to a minimum without sacrificing scientific rigour, this comprehensive guide offers readers a practical approach to boosting the profitability of their biogas plants. The techniques explained allow for an assessment of a biogas plant throughout its entire life cycle, from critical review of the business plan to the selection of retrofits to recover additional income from the process by-products. This book:

- Details how to apply the scientific method to the review of biogas projects, considering technical flaws induced by marketing and human factors like cognitive bias.
- Explains a method for the quick startup of digesters.
- Proposes a thermodynamic approach to the selection of biogas upgrading solutions, making the choice independent of marketing claims or biased information.
- Provides a summary of possible marginal incomes from biogenic CO_2, residual heat, and digestate.
- Includes the latest technological developments in laboratory measurement instruments and techniques beyond the mere biochemical methane potential (BMP) test.
- Includes a guideline to perform a technical due diligence, with examples of the most frequent errors in biogas plant business plans.

Building upon the laboratory test techniques published in the author's complementary book, *Managing Biogas Plants*, this text helps readers investing in or managing existing industrial biomass plants optimise processes and profit.

Contents

Preface

Many things have happened since the publication of *Managing Biogas Plants: A Practical Guide* in 2018:

1. I received dozens of acknowledgement emails from readers who found answers to their concrete problems, both in the research lab and in the day-to-day operation of anaerobic digesters.
2. In March 2022, the title was ranked 76th in the list of the '100 best books on chemical engineering of all times' by BookAuthority. By November 2023, it had risen to the 33rd position, and currently (July 2024) it has reached the 20th position.
3. An American company developed an artificial intelligence system to optimise industrial biogas plants based on it and improved the plants' productivity by more than 20%.
4. In the last 5 years, I have lectured for the World Biogas Association (WBA) and other private institutes and companies, training more than a hundred professionals and learning from the practical problems they face every day.
5. The biogas market is evolving from electric generation or combined heat and power (CHP) plants to more complex biogas upgrading systems.
6. The diffusion of social networks has created a new challenge to biogas plant engineers and owners: fake news and conspiracy theories around anaerobic digestion and digestate. Following the diffusion of false information over the Internet, some ideological positions against biogas have even gained the status of law in Italy, despite the scientific evidence showing that anaerobic digestion is one of the most efficient technologies for decarbonising the economy. I found myself involved in debunking misinformation and 'half-truths' in public debates, which is not always an easy task because of generalised cognitive bias.
7. Measuring gas flow at the laboratory scale has followed the same pace of industrial development, leading to third-generation instruments with higher accuracy, precision, and sensitivity. A growing interest in bioupgrading (aka Power to Gas (P2G) or hydrogenotrophic methanation) requires the development of a sound protocol that allows upscaling the laboratory results to the industrial plant. This argument is still the subject of academic research and deserves some detailed consideration and further debate.

Considering all the above-mentioned arguments – and that my retirement age is approaching inexorably – I thought it was time to condense my last 5 years of experience into a new book, for young professionals to build on. If you enjoyed the first volume, I'm sure the present update will not disappoint you.

Disclaimer

Since my work is based mainly in Italy and to a lesser extent in Spain and other countries, most of the case studies in this volume refer to the legislation and peculiarities of the biogas markets in the said territories. In no way does this imply that generalisation to other countries is possible, nor does it mean that European legislation or techniques are better than others. On the contrary, I openly declare myself sceptical about the sustainability of the so-called European model or 'biogasdoneright®' because there is no point in cultivating cover crops to employ them as digester feedstock while millions of tons of organic waste are still left to rot in landfills and open ponds or lagoons throughout the world, emitting methane into the atmosphere. I publicly disagree with the EU's Green Deal policy on hydrogen because it is based on ideological preconceptions, and a simple exergy analysis demonstrates that it is less efficient than other solutions. The said positions do not mean any engagement in favour or against any political party: they are just the expression of facts supported by *unbiased* thermodynamic calculations.

All the critics towards the current (2024) EU legislation, the scientific literature, and widely diffused practices or beliefs contained in this book are supported by my personal hands-on experience and results of laboratory and field tests. I deliberately avoided ideological positions and cognitive bias by strictly applying the Cartesian method. Hence, the spirit of all critics contained in this book is to contribute to an academic debate, with the final goal of improving the overall quality of Science (with uppercase S) in the anaerobic digestion field.

I am aware that being one of the oldest distributors of BPC Instruments AB (formerly Bioprocess Control AB) and using their products in my laboratory may sound like a 'conflict of interest' to some researchers and politicians. There is no conflict at all. Since I only publish claims that I can support with experimental evidence and since the evidence published here results from measurements performed with the said products, dedicating a chapter to describe how they work was necessary to explain the origin of the said results and their uncertainty margin. By including a chapter about BPC Instrument's technology, I did not mean to advertise any product, nor does this constitute any misuse of commercial brands. The reason is that many readers will employ other types of instruments for their tests, so their results may differ much from those published here because of different measurement conditions and error margins. Discretion is advised to those equipped with other laboratory instruments, or employing different test methods, because comparing results may be tricky. In such cases, a thorough analysis of both *error propagation* and *replicate dispersion* is mandatory before drawing any conclusion, as well as including a positive control to check the inoculum's correct methanogenic and hydrolytic activities. Hence, I considered it necessary to provide detailed enough information on how I performed my tests to allow a fair comparison between eventual contradictory measurements. On the other hand, readers who already employ BPC products will find tweaks and tricks that will improve their user experience and facilitate interpreting any 'strange' results they may obtain.

About the Author

Mario Alejandro Rosato is an electric-electronic and environmental engineer. He is the CEO of Sustainable Technologies srl, an Italian company. A member of the World Scientific Journalists Association (WSJA), he has been in charge of *AgroNotizie*'s column on bioenergy since 2014 and has published over 300 articles in other journals. He has been running an anaerobic digestion laboratory in Italy since 2012, providing consultancy services to biogas plant owners and builders, and also serving as a technical expert in litigation cases about underperforming biogas plants. Since then, he has been providing technical training to hundreds of anaerobic digestion professionals, in both academia and industry, throughout the world. He has been providing technical training to the World Biogas Association (WBA) associates since 2023. In 2015, he was a member of the technical commission in charge of redacting the Italian norm on the biochemical methane potential (BMP) test protocol, and his contribution focused on error propagation analysis and improving accuracy and precision. His first book on anaerobic digestion, *Managing Biogas Plants: A Practical Guide*, was published by CRC Press in 2017 and has been in BookAuthority's list of the best 100 books on chemical engineering since 2022.

1 Nullius in Verba!

1.1 OVERVIEW

The Latin locution *Nullius in verba* means 'on the word of no one' or 'take nobody's word for it'. It is the motto of the Royal Society and represents the essence of the scientific method, which is based on experimental proof and counterproof. The content of this chapter is an extension of the concept broadly treated in the first volume *Managing Biogas Plants: A Practical Guide*: the publication of an experiment or case study and the conclusions derived from it in a peer-reviewed journal does not prove universal applicability of the principles to any biogas plant. The present volume proposes a critical analysis of some case studies from the scientific literature on anaerobic digestion (AD). The aim of the study is to drive the attention of engineers, investors, and technicians in the biogas industry, who are not necessarily familiar with the rules of scientific publications nor with some trends that affect the quality of the scientific production of the last 10–20 years [1]. Applying data or concepts from the scientific literature to concrete operational situations in biogas plants can lead to underperformance and loss of profit for the following reasons:

1. The quantity of published information increases exponentially but the general quality tends to be lower. This phenomenon results from a series of pressure factors on researchers known as 'Publish or die!' [2].
2. Despite being a legitimate – and desirable – praxis, securing funding for the research from private companies may somehow condition the publication of the results, sometimes biasing or hyping the effectiveness of a given solution.
3. Open-access publications are the first ones that technicians, stakeholders, and public officers in search of information will consult, often ignoring the fact that the Internet is plagued with scam scientific journals. Such unscrupulous editors will publish anything against the payment of a fee, after a superficial review or sometimes with no review at all [3]. Not all readers have the expertise to assess the quality of a publication, the time or ability to track if the same has been retracted or reviewed, nor the capacity to replicate the experiment in their laboratories.
4. Researchers are humans, so they sometimes draw wrong conclusions from an experiment or generalise a special case, publishing in good faith what they believe to be a discovery with universal validity.
5. Reviewers are human too and, as such, subject to cultural bias and heuristics [4]. In other words, they may reject the publication of a paper containing valuable discoveries or practical information, or approve the publication of research results 'in line with the literature', without questioning the validity of 'the literature' itself. The researcher's age and affiliation often play a role,

DOI: 10.1201/9781003609278-1

although the fame and prestige of a given university or an individual do not necessarily warrant the quality of the research.

The present volume includes both physical and biological aspects of biogas plant optimisation. Physical aspects, such as stirring and heating of the digester during its startup, may indeed influence the biological process. However, the main scope of physical optimisation is to save energy and thereby reduce the operation expenditure (OPEX), without hampering productivity. Following the same approach adopted for the redaction of the first volume, the author proposes some empirical methods, derived from his professional experience, for optimising the biological process. Such methods rely mainly on laboratory batch tests, despite the general opinion of researchers and biogas plant builders that batch tests do not give complete and accurate information. The reason for the author's dissent is partly philosophic, partly mathematic, and mostly pragmatic:

1. From a philosophical point of view, the author argues that most of the scientific literature on AD lacks a correct analysis of the test uncertainties. In the best case, publications only contain the usual analysis of variance (ANOVA) among replicates of the same test but no analysis of metrological error propagation. When assessing the BMP of a feedstock, the result is assumed (wrongly) to be reliable if the dispersion of the replicates is low. When assessing the influence of some factors on the BMP (e.g., adding a booster or pretreating the feedstock), the factor under study is assumed (wrongly) to have a significant influence if $p >$ 5% [5], without considering that sometimes the mere metrological error may be much higher than 5%. Hence, differences greater than 5% are not necessarily significant. No wonder that the attempt to scale up the results of such tests to the plant usually fails: false hypotheses lead to wrong conclusions.
2. From a mathematical point of view, the theory of control systems calls the outcome of a batch digestion test the transfer function, which is the response of any system – in our case, the bacterial ecosystem contained in the anaerobic inoculum – to an impulse, represented by the quantity of feedstock placed in the batch reactor at the start of the test. The Laplace transform of the resulting flow vs. time curve allows us to calculate the inoculum's response to any feeding frequency and amplitude (the mass of volatile solids (or COD) of feedstock per hour or day that must be loaded into the digester to obtain the desired gas flow).
3. From a pragmatic point of view, batch tests are the cheapest and simplest to perform, and can be performed even by personnel with basic skills. Interpreting their outcomes is not always straightforward because the mathematics involved in calculating the Laplace transform and its convolution integral is beyond most biogas plant managers' capabilities, but this problem can be overcome with a simplified approach. Hence, batch tests are the simplest and cheapest way to optimise the OPEX of a biogas plant with high accuracy.

All that said, it must be clear that despite all the effort and goodwill put into writing this book, the Cartesian doubt '*nullius in verba*' applies to its content too. Replicating and counter-proofing are the essence of scientific progress, so the author welcomes any sound experimental evidence that challenges the methods, views, and opinions published in this volume. Carl Sagan's principle 'Extraordinary claims require extraordinary evidence' will apply, so any challenge or criticism must include all the information and experimental data necessary for its analysis.

The next sections of the present chapter will provide examples of how to apply the Cartesian method in two particular cases: critically reviewing scientific publications and business plans, and debunking conspiracy theories and NIMBY arguments against biogas plants. Readers living in happy countries where the population and local authorities accept biogas plants as a normal component of industrial and economic development can simply overlook Section 1.3 and jump to the next chapter.

The last chapter will summarise all the concepts in this book, resulting in a systematic method to perform the technical due diligence of biogas projects. A critical review of a biogas project's business plan and the plant's design and operation strategy before starting the construction will minimise financial risk and ensure that the investor obtains the foreseen profitability. It is also possible to apply the same method to an existing biogas plant in case of underperformance. The outcomes can be of two types: reaching acceptable profitability is feasible – usually at a given cost – or if the project was badly conceived or the initial market conditions have drastically changed, the method will allow one to decide the project's temporary suspension or termination on a rational basis.

1.2 READING SCIENTIFIC PUBLICATIONS AND BUSINESS PLANS CRITICALLY

The author's goal when writing this section was to benefit industrial and academic readers alike. Industrial readers will learn how to

a) discern good-quality publications;
b) decide if and to what extent they can apply the published information to their specific cases;
c) design lab tests to double-check the validity of the published results; and
d) pinpoint cases of cognitive bias, false hypotheses in business plans, and hypes about the performance of machinery and products for biogas plants.

Academic readers could be either students, young researchers, or peer reviewers. The first ones will avoid common pitfalls typical of juvenile enthusiasm: adopting 'creative' (i.e., nonstandard) protocols, discarding potentially interesting results because they are 'not in line with the literature', blindly retrieving numbers from ANOVA software, and triumphally announcing 'ground-breaking results' without having considered all variables. Mature researchers and peer reviewers will improve the overall quality of scientific publications by pinpointing cherry-picking, oversimplification, and confusion of cause with effect; avoiding the $p > 5\%$ criterion; and correctly reporting measurements with their associated uncertainty.

All the examples in the next subsections will be anonymous. The author does not intend to blame any researcher or research group but to point out common defects in the redaction and peer-reviewing process, suggesting a general method to avoid them.

1.2.1 Cognitive Bias and How it Influences the Biogas Market and Research

Cognitive bias is a systematic thought process caused by the tendency of the human brain to simplify information through a filter of personal experience and preferences. The filtering process is a mechanism that enables the brain to prioritise and elaborate large amounts of information quickly. Although an advantage at the early stages of human evolution, cognitive bias is a hurdle to rational thinking. All humans are subject to cognitive bias because it is 'hard-wired' in our brains. The only way to overcome this hurdle is to be humble enough to admit that it exists and that our judgements and decisions are more prone to emotional impulse and simplified intuition than rationality. The notion of cognitive bias was introduced by Amos Tversky and Daniel Kahneman in 1972 and grew out of their experience of people's innumeracy, i.e., the inability to reason intuitively with higher orders of magnitude. Tversky, Kahneman, and colleagues demonstrated several replicable ways when human judgements and decisions differ from the rational choice theory. Kahneman received the Nobel Prize in Economic Sciences in 2002 for his studies about how investors base their decisions on irrational perceptions or logical fallacies. To date, more than 200 types of cognitive biases have been identified. The reader can find abundance of freely accessible information on cognitive biases, both as text articles and as educational videos on YouTube. The present discussion will be limited to those biases and logical fallacies that the author identified more frequently in the biogas industry and technical literature:

1. **Unconscious bias**: It is the result of our limited cognitive capacity. We don't see things as they are: we see things as we are. Emotions and beliefs play a fundamental role in unconscious bias. Activists opposing the construction of biogas plants described in Section 1.3 are a good example of people under unconscious bias, acting under purely irrational and emotional convictions. Another classical example is the wrong solution to a simple arithmetic problem: a bar of soap and a toothpaste tube costing \$1.10. If the price of the toothpaste is \$1 higher than the soap's price, what is the cost of the soap alone? Most of us will answer immediately '10 cents', representing unconscious mental shortcut, a wrong answer. If we were perfectly rational beings, we would refrain from answering immediately and solve the following two equations with two unknowns first (t is the price of the toothpaste and s the price of the soap):

   $s + t = \$1.10$
   $t = \$1 + s$, and replacing this one in the first equation:
   $s + \$1 + s = \$1.10 \Rightarrow 2\,s = \$0.10 \Rightarrow s = \$0.05$

2. **Anchoring bias**: A cognitive bias that causes us to rely too heavily on the first piece of information we are given about a topic. When we set plans or make estimates about something, we interpret newer information from

the reference point of our anchor, instead of seeing it objectively. The most common examples found in the biogas industry are the almost religious belief of most operators and researchers that a digester will not perform if the volatile fatty acids to total alkalinity ratio (VFA/TA, aka FOS/TAC in the European literature) is higher than 0.3 and the belief that only full-scale tests are representative and reliable.

3. **Overconfidence bias**: People tend to think they are better than others. In general, the lower the level of knowledge an individual has on a certain argument, the higher his/her conviction of being better informed than expert professionals. It is very frequent among investors, who believe they will succeed where others already failed; young researchers, who believe to be more brilliant than their elders; and old researchers, who believe to be infallible.
4. **Confirmation bias**: The tendency to search for, interpret, favour, and recall information in a way that confirms or supports one's prior beliefs or values. People taking part in a debate display this bias when they select information supporting their views while ignoring contrary information, or when they interpret ambiguous evidence as supporting their beliefs. The effect is strongest for desired outcomes, emotionally charged issues, and deeply entrenched beliefs. Confirmation bias is insuperable for most people, but one can manage it, e.g., by education and training in critical thinking skills. In science, the worst case of confirmation bias is called *cherry-picking*, i.e., the act of selectively choosing or highlighting only the information that supports a particular viewpoint or argument, while ignoring or disregarding any conflicting data or evidence. The usual sentence found in scientific publications 'Our results are in line with XY et al.' could be an indicator of cherry-picking, although not necessarily fraudulent. In business, some salespeople showcase the success stories of some customers and label the unsatisfied ones as 'too blunt to understand the benefits', 'technically uncultured', or 'too lazy to work correctly'. The author found a clear propensity to confirmation bias among manufacturers of additives for anaerobic digesters, biogas plant builders, and pre- and posttreatment equipment suppliers.
5. **Survivorship or survival bias**: It is the logical error of concentrating on entities that passed a selection process while overlooking those that did not. This can lead to incorrect conclusions because of incomplete data. It is different from cherry-picking because there is no deliberate intention to exclude information to confirm a predefined hypothesis but only a logical flaw in assuming that only 'positive' information is useful. The author noticed this bias in investors building a biogas plant in Sicily. At that time, there were already several hundreds of biogas plants running on corn silage in Northern Italy and Central Europe, so the investors decided to replicate 'the successful business model' in Sicily without considering that corn is not a suitable crop for the Sicilian pedoclimatic environment.
6. **Group thinking or groupthink**: The term 'groupthink' was first used by social psychologist Irving L. Janis in 1972. It refers to a cognitive bias that encourages people to desire harmony or conformity within a group. In many

cases, people will set aside their own beliefs to adopt the opinion of the rest of the group. The author observed this phenomenon among some researchers who uncritically adhere to the convictions of their groups – university, national association, or board – on a given method or laboratory protocol. Activists of 'no-biogas' political groups and 'only electric' NGOs are other perfect examples of group thinking.

7. **Heuristic bias or oversimplification**: Kahneman defines it as 'attempting to solve complex problems in an intuitive or oversimplified manner'. Trying to manage the complexity of an anaerobic digestion process with a table and a few cheap tests, like FOS/TAC and electric conductivity, is the most common example of oversimplification bias in the biogas industry.
8. **Implicit bias**: It is a negative attitude, of which one is not consciously aware, against a specific social group. It is the unconscious root of thinking that leads to explicit racism. Of course, no European politician or economist would admit that the European biogas policies and industry are implicitly racist. There are several millions of household digesters providing biogas for self-consumption in China and India, and the model has proved efficient and less dependent on state subsidies for at least 50 years. Despite factual evidence gathered from the International Energy Agency's statistics [6], all the European analysts and policymakers the author has met so far affirm that small biogas plants and direct use of biogas at the household or farm level are 'Third World solutions, not profitable here because they lack scale economy'. Such a biased conception hampers the diffusion of small biogas plants in Europe. For instance, in Italy, household composters are not only legal but some local authorities also promote their use. On the other hand, a citizen installing a micro-digester with the same treatment capacity as a composter in his garden or farm could be liable for a double offence: illegal waste disposal and fiscal fraud for not paying tax on the produced methane. European cooperation agencies and industrialists try to export the European *biogasdoneright*® model, i.e., big, complex, centralised biogas plants fed with cover crops, to Africa, India, and Latin America. The said model is profitable in Europe only because of strong state subsidies and as a subterfuge to build biogas plants without being subject to the European overregulation about the concept of 'waste' [7]. Strong subsidies are necessary to compensate for the unnecessary cost burden arising from European ideological attitudes towards a resource generally considered as 'stinking waste'. No matter how much evidence they are presented with, economists and politicians honestly believe that the European business model is superior. The implicit bias behind such a belief is that the model originated in 'developed countries' and is supported by 'high-tech companies', while small digesters are 'homemade' by 'semi-illiterate farmers in underdeveloped countries' (textual comment of a European professional).

Now that we have a general idea of what cognitive bias is, let's have a look at other sources of misinformation and irrational misconceptions that hamper the deployment of biogas plants and their profitability.

1.2.2 Logical Fallacies

In Chapter 6 of his first book [8], the author discussed the issue of logical fallacies, examining it within the context of critically reviewing and extrapolating data from the scientific literature to the routine operations of an already established biogas plant. In this section, we will discuss how cognitive bias relates to logical fallacies in a plant's engineering and business plan, ultimately hampering the biogas plant's profitability and environmental benefits. Chapter 10 will summarise the concepts presented here, within the context of case studies.

Classical philosophy relied on two methods of thinking: induction and deduction.

a) **Induction:** The inference of a general law from particular instances. In other words, it is a bottom-up thinking method, starting from a single case – or a limited number of cases – and reasoning to derive logical conclusions widely applicable – ideally universally applicable – to other situations.
b) **Deduction:** The inference of a particular instance from general laws or principles. In other words, it is a top-down reasoning method, starting from a statement widely believed to be true, or self-evident (axiom) and reasoning towards a logical inference applicable to a particular case.

The scope of induction and deduction is creating a mental model, i.e., taking a subject and objectifying it so we can manipulate it in our minds. Induction cannot guarantee reaching final true conclusions because it is based on the axiom that 'what is true for the many is true for the whole'. This axiom reflects the belief that nature is governed by uniform and immutable laws, which is true for deterministic processes but not applicable to random phenomena. Induction, at best, can guarantee that a conclusion is 'likely to be true'. Deduction usually leads to true conclusions if it is based on true assumptions. Syllogism is a particular case of the more general deduction method, based on analysing only two propositions or premises at a time and leading to a single logical conclusion. Basing our reasoning on true assumptions is a necessary condition, but it is not sufficient as the two premises must also be logically sound. Conversely, sometimes false or logically unsound propositions may lead to true (or partially true) conclusions by pure chance, e.g., the use of the VFA/TA (aka FOS/TAC) method to determine the stability of the anaerobic digestion process, which is already discussed in Chapter 6, Section 6.2.2 of the first book, reference 8 already cited.

Inductive reasoning based on strong evidence leads to conclusions with a high probability of truth, while superficial, weak, or poor-quality experimental evidence will likely produce questionable or false conclusions. Using a deductive method allows us to check the validity of the results derived by inductive reasoning: if we reach a false conclusion, then the premises must be false. For instance, the author demonstrated in the first volume that the FOS/TAC method is based on a logical fallacy. Indeed, the FOS/TAC recommended values result from inductive reasoning (an extrapolation from a few empirical cases). Since the publication of the first volume, the author has received many messages from biogas plant managers confirming that their plants worked more efficiently with FOS/TAC values well above the established 'optimal' ranges. This process of observing, formulating hypotheses, and then checking the validity of the hypotheses with further observations is the basis of the scientific method sketched in Figure 1.1. It is also called the Cartesian method in honour of the French philosopher and mathematician René Descartes (Cartesius in Latin), who was the first to define it formally.

American philosopher Charles Sanders Peirce developed a third method, called abductive reasoning, in the nineteenth century. Abductive reasoning does not necessarily lead to true conclusions, and sometimes it may even lead to a logical fallacy. Its scope encompasses formulating an 'educated guess', i.e., reaching a *plausible* (not necessarily probable) conclusion without proving it. Abductive reasoning is useful when we have insufficient information for formal deductive reasoning or lack the time to gather particular observations leading to a 'probable' hypothesis with inductive reasoning. A typical example of abductive reasoning in the biogas industry is using trace elements or 'booster' additives when the plant manager observes a systematic decrease in productivity over a certain time. There are hundreds of reasons why a biogas plant may decrease its productivity, most of them having nothing to do with the biological process itself: rainy or moist weather, calibration drifts of the weighing platforms, hidden or tiny leaks in the gas dome, low water level of the overpressure valves (hydraulic guards), low atmospheric pressure causing lower efficiency of the motor (especially small motors without turbo compressor), etc. Nevertheless, managers who do not systematically check the specific methanogenic activity (SMA) of the inoculum, the BMP of the feedstock, and the efficiency of digestion experience panic when observing a systematic fall in productivity and begin to search for a quick solution. They usually assume a problem with the biological process and, more

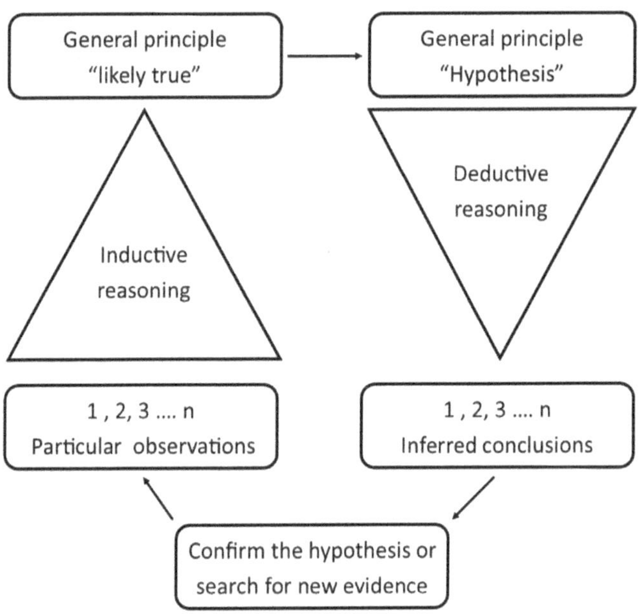

FIGURE 1.1 Circular Process of Inductive and Deductive Reasoning.

specifically, a deficit in trace elements as 'the most plausible cause' of the unexpected productivity drop. It is possible to avoid falling into the said wrong abductive reasoning (or oversimplification bias) with the application of another thinking tool: the philosophical razor.

A philosophical razor is a tool that helps us make decisions when faced with competing hypotheses or explanations. It allows us to focus on the minimum instances necessary for logical analysis by eliminating redundant or irrelevant information. The main idea behind using a philosophical razor is to choose the simplest explanation that adequately accounts for the available experience, rather than resorting to unnecessary complexity or too many assumptions. To be effective, a philosophical razor should be expressed in a short sentence that is easy to remember and apply. It should eliminate improbable, needlessly complex, or unsupported pieces of evidence in favour of simple and probable ones. Additionally, there should be an argument explaining why the simpler hypotheses are true, or else we risk falling victim to the oversimplification bias. Some classical philosophical razors are as follows:

Ockham's razor (sometimes spelt in its Latinised form **Occam's razor**, aka **parsimony principle**): Its original form (in Latin) translates into English as 'Entities should not be multiplied unnecessarily'. In modern terms, we could state: 'When you have a set of competing theories that make the same predictions, the simplest one is the best'. Albert Einstein is credited as the author of a shorter version: 'Everything should be made as simple as possible, but not simpler'.

Hanlon's razor: 'Never attribute to malice what is adequately explained with carelessness'. The term 'carelessness' can be interchanged with words such as 'ignorance', 'incompetence', or 'negligence'. Scammers exist in all industrial sectors, and the biogas industry is no exception. Serving as an expert witness in legal cases involving biogas plants, the author has sometimes encountered situations where one party acted maliciously. However, in some cases, Hanlon's razor was a useful tool in resolving disputes. The issues often stemmed from the incompetence of one party or a combination of negligence and superficiality from both parties.

Hitchen's razor: 'What can be asserted without evidence can be dismissed without evidence'. It is also called 'the burden of evidence': if somebody makes a statement without providing evidence or proof, we don't necessarily need to embark on a discussion to prove it wrong. Of course, debunking fakes will improve our credibility, but it can be extremely time-consuming because of the so-called Brandolini's law: 'The amount of energy needed to refute bullshit is an order of magnitude bigger than that needed to produce it'. Despite being ironical and academically unorthodox, Brandolini's law thoroughly describes one of the biggest problems of our society: social networks diffusing false information on a global scale (see Section 1.3 for some examples in the (Italian) biogas industry).

Feynman's razor: 'If you can't explain something simply, then you don't understand what you are talking about'. Many politicians and technical vendors resort to abstruse bureaucratic or technical jargon to appear knowledgeable. Politicians' goal is to get the votes of citizens who oppose projects of biogas plants for purely ideological reasons or conspiracy theories (see Section 1.3). Salespeople's goal is, of course, to sell a given technical solution, product, or consultancy service.

1.2.3 Methods for Assessing the Robustness of a Decision-making Process

Modern problem-solving techniques resort to combinations of the Cartesian method, philosophical razors, and mental models. Such techniques generate and test hypotheses about a situation, or assess the potential causes behind certain effects. The following list is not exhaustive and is included here to stimulate the reader to hone his/her reasoning skills. Using the title of each of the following paragraphs as a keyword for an Internet search will yield hundreds of freely downloadable papers and articles from the academic, military, and business domains.

1.2.3.1 Strengths, Weaknesses, Opportunities, Threats (SWOT) Analysis

This technique is widely diffused among investors in assessing business plans.

It consists of dividing a sheet of paper into four quadrants and reasoning to identify the information to fill each of them. The upper quadrants, strengths and weaknesses, are related to the internal factors, i.e., the characteristics and resources within your organisation that directly influence its operations and performance. These factors are completely within your organisation's control, so they can be modified, improved, or capitalised upon. In the concrete case of an industrial biogas business, the internal factors to assess as strengths or weaknesses could be, for instance, technical skills of the personnel, control of the feedstock supply chain, dependence on third parties' proprietary technology or components, availability of manure or digestate to inoculate and stabilise the process along time, financial capacity, proven technology, or business model.

The lower quadrants, opportunities and threats, represent the external factors, i.e., elements outside the organisation's control that affect its operations, market position, and success. These factors arise from the industry climate and the broader business environment. You typically have no control over external factors but can respond to them. Examples of external factors in a biogas business plan are the volatility of energy prices, political stability, NIMBY groups (see Section 1.3), changing regulations, disruptive technologies, and extreme climatic events.

1.2.3.2 Plus, Minus, Interesting (PMI), or Determining the Pros and Cons

PMI is a brainstorming, decision-making, and critical thinking tool used to encourage the examination of ideas, concepts, and experiences from more than one perspective. It can help brainstorm ideas, make decisions quickly by analysing and weighing the pros and cons, reflect upon or evaluate a product or process after a failure or fact, and identify strengths and weaknesses for future improvement.

1.2.3.3 Key Assumptions Identification (KAI)

This involves a systematic process to identify the assumptions that guide a decision maker's interpretation of evidence and reasoning about a particular problem, and then judge which ones are key. It is like a philosophic razor but constructed *ad hoc* for a given problem. This technique is especially useful when analysing a new biogas plant construction project or the revamping of an old one. The line of thinking is a four-step process:

a. Review the current line of thinking/reasoning on an issue. Just write it down as a list of the assumptions you base the current reasoning upon.
b. Identify and articulate all the premises and assumptions – stated and unstated – that are accepted as being true for the line of thinking to be valid.
c. Identify which of the assumptions are key. Challenge each premise and assumption, asking whether it 'must' be true for the line of reasoning to be valid and whether it remains true under all conditions. The assumptions that 'must' be true become the key assumptions. You can now focus your further research on this narrowed list of key assumptions to conserve resources. In addition, you will consider under what conditions or in the face of what information these might not hold, and analyse the implications for the line of thinking.
d. Evaluate key assumptions. Conduct further research and analyse the implications if not true. Once you have the refined list, you can consider the following questions: how much confidence exists that the assumption is correct? What explains the degree of confidence in the assumption? What circumstances or information might undermine the assumption? Is a key assumption more likely a key uncertainty or key factor? Could the assumption have been true in the past but less so now? If the assumption proves to be wrong, how would the plan need to be altered? Has the process identified new factors that need further analysis?

1.2.3.4 Quality of Information Check

The fundamental principle of deductive logic is that the premises must be true for the conclusions also to be true. Information tends to be overabundant in our era, but its quality tends to be poorer or applicable only to some particular cases. The method consists of making an inventory of the sources of information we rely on for making our decisions. For instance, consider the BMP of the feedstock. Is it a value obtained from the literature? What method and instrument were employed for its measurement? Was it measured with our substrate and inoculum, or are we assuming measurements performed in other plants or countries to be applicable to our case? Who performed the measurement? What is the uncertainty of the measurement?

The notions contained in our short overview of logical and critical thinking allow us to address the key problem that often hampers the profitability of biogas plants: logical fallacies deriving from cognitive bias or incorrect reasoning.

a. *Post hoc* logical fallacy and real-scale tests

   There is a misconception among biogas plant engineers, and even among some academicians, that only experimental data obtained at full scale are reliable because laboratory tests are performed under 'ideal, unrealistic conditions'. It is easy to prove such a misconception wrong by employing a simple uncertainty analysis. Laboratory instruments and methods are designed to directly measure the values we need to base our decisions on. Such values can be, for instance, the net amount of methane produced by the combination of a given feedstock and inoculum, the reactor's efficiency, or the process. If used correctly, laboratory instruments usually ensure the minimum error margin technically possible. Furthermore, laboratory tests can be replicated to check consistency over time or varying conditions. Hence, accurate and precise laboratory measurements provide in relatively short times a large number of strong instances that, by inductive deduction, will provide strong hypotheses, i.e., a high probability of our conclusions being true. On the other hand, full-scale tests generate a limited number of weak instances (sometimes only one!) because the instruments are of industrial grade (usually having high error margin, sometimes they have not been calibrated in years) and/or measurements are not taken in the optimal range that guarantees the minimum absolute error. In all cases, the constants assumed to link the productivity of a given feedstock with its energy yield are not constants. For instance, think of the common practice of 'saving' the cost of a lab test and checking the quality of a feedstock lot (or the efficacy of a given pretreatment of the feedstock) using a 'sample' of several hundred tons for some weeks. The output variable – electricity injected into the grid – is the result of the methane yield (the variable we want to assess) multiplied by the lower heating value of methane (a universal constant free of uncertainty) multiplied by the electrical efficiency of the generator (which is a 'constant' given in the technical specifications of the generator but valid only under certain reference conditions). But the total amount of methane produced during the test period results from a combination of four factors: the biological activity of the inoculum, the feedstock composition, and feeding rate of the last month; the real BMP of the feedstock; the accuracy of the weighing system; and to a certain extent, the technology and state of maintenance of the digester. The efficiency of the generator is not a constant: it depends on the hours it has been running and its state of maintenance; the ambient temperature, pressure, and moisture; and the electric load percentage (the efficiency of an electric generator is maximum at nominal load, in some models at 105% load). So, the 'pseudo-BMP' obtained by a test at full scale is useless because it is easy to apply Ockham's razor and demonstrate that such a figure is the result of too many assumptions that cannot be proven true. With such a measurement method, it is impossible to discern if an eventual under- or overperformance in comparison to the assumed benchmark derives from the feedstock quality or a combination of all other factors. Usually, it is impossible to have at least three identical plants and feed them with the same lot of feedstock to check the standard deviation (noise or random factors that influence the measurement). It is even more impossible to have three additional plants identical to the ones under test, fed with a benchmark feedstock. To overcome this practical impossibility, advocates of full-scale tests resort to another logical fallacy: the *post hoc*.

This fallacy was already known to ancient thinkers, and its Latin form is '*Post-hoc, ergo propter hoc*', which translated into English means 'If something happened after a given event, then the said event is the cause'. Unless it is possible to prove a causality nexus between the first and the second observation beyond any reasonable doubt, both events are independent, and hence, the first does not necessarily imply the second. Think of somebody who goes to the casino, bets $X, and loses all. Some days later, this person gets a rabbit paw, returns to the casino, bets $X again, and wins $Y. The flawed logical conclusion is that bringing a rabbit's paw to the casino causes the carrier to be lucky. Now think of a biogas plant fed with X tons/day of a given feedstock, producing Y MWh electricity in one month. The following month the plant is fed with X tons/day of the same feedstock pretreated in some manner, or with an alternative feedstock. The electricity production during the second month results to be Z MWh. The flawed conclusion in many peer-reviewed papers, commercial articles, and congress posters is that the difference (either positive or negative) between Y and Z is the result of change in the feedstock or pretreating it. It may be so sometimes by pure chance, or perhaps the company selling the pretreatment or the alternative feedstock just chose the best results among many such tests (cherry-picking), or the difference is not relevant (it is of the same order of the method uncertainty, roughly 10%). From a logical point of view, the full-scale test performed in the way such tests are usually performed is as sound as the theory of the rabbit paw.

b. The gambler's fallacy

The gambler's fallacy is the incorrect belief that if a particular event occurs more frequently than normal during the past, it is less likely to happen in the future (or vice versa), when it has otherwise been established that the probability of such events does not depend on what has happened in the past. Such events, having the quality of historical independence, are referred to as statistically independent. The fallacy is commonly associated with gambling, where it may be believed, e.g., that the next dice roll is more than usually likely to be six because there have recently been fewer than the expected number of sixes. The author has observed this fallacy among municipal waste plant stakeholders. Everybody can understand that municipal waste is heterogeneous and its composition variable in time for a given locality. After some years of operation within design parameters, the plant continues to receive the same quantity of waste every day, but the production is much lower than expected from historical records and remains so for some months. Considering that the plant had performed within expected limits in the past, the stakeholders wrongly assume that it should perform likely now and start searching for a culprit: the feedstock's supplier, the engineer, the new worker, etc. Suppose they were familiar with Hanlon's razor. In that case, they should conduct a technical survey to assess other (more probable) causes; for instance, some transient biological problem, a thick layer of sediment on the digester's bottom, or corrosion leading to gas leaks. Or they should check if the reason is demographic: perhaps the population is changing its feeding habits because of cultural influences or economic crisis, wasting less food, or producing waste with a lower BMP.

1.2.4 Hypes and 'Groundbreaking Results'

In his excellent book on hype, bias, and fraud in science (ref. 1), Stuart Ritchie quotes:

> A study plotted the proportion of abstracts that contained particular terms, starting in 1974. The use of the words 'innovative', 'promising' and 'robust' increased exponentially; 'unique' and 'unprecedented' (perhaps paradoxically) became vastly more common; 'favourable' rose steadily. 'Groundbreaking' trundled along near zero until around 1999, when for some reason it suddenly shot upward. On average, the use of positive words in abstracts increased by nearly nine-fold over the analysed period: just 2 per cent of abstracts in 1974 had contained such self-praise, compared to 17.5 per cent in 2014.

and concludes:

> To paraphrase the biologist Ottoline Leyser, the point of breaking ground is to begin to build something; if all you do is groundbreaking, you end up with a lot of holes in the ground but no buildings.

Hypes in catalogues and commercial presentations are common and not a surprise. We all know that companies need to sell, so some rounding in excess or some bias in sampling is to be expected.

Hypes in the anaerobic digestion scientific literature may have many causes, and Hanlon's razor can dig out most of them. Deliberate hypes may hide fraudulent scopes, but, fortunately, this is seldom the case in the biogas academic world, although it is more common in other fields of science, like psychology or medicine (ref. 1 already cited). Some degree of hype may arise from the need to publish a large number of papers to escalate the ranks of bibliographic metrics and advance career, the so-called publish or perish effect [9], and hence a text containing 'groundbreaking results' is most likely to be accepted than the one containing negative results. Accepting only positive results seems to be a frequent flaw of some peer reviewers (survival bias), although making negative results publicly available helps science progress as much as publishing successful experiments. Public negative results are useful to dismiss false hypotheses, thereby avoiding useless experiment repetition and wasting resources in dead-end lines of research. The following are some good examples of pure superficiality, juvenile enthusiasm, and overconfidence bias.

1.2.4.1 Example 1: The Case of the Self-heating Digester

In a peer-reviewed paper freely available on the Internet, a group of researchers from a European university discuss the issue of cooling biogas plants during summer to avoid negative impacts on bacterial ecosystems. They measured what they considered 'all incoming and outgoing energy flows' in a full-scale digester and found a component of heat that they could not explain. The researchers hypothesised that a group of anaerobic bacteria that have not yet been discovered produces metabolic heat under anaerobic conditions, thus explaining the 'discrepancy' with their heat balance. They supported their hypothesis with the finding of a direct correlation between the use of fresh silage as feedstock and the self-heating phenomenon although this assumed fact was not universal and could not be reproduced in laboratory conditions for smaller digesters. Occam's razor should be enough for dismissing the hypothesis of the existence of an unknown group of exothermic anaerobic bacteria: the simplest theory to explain why the temperature of a digester increases in summer and not in winter, or under laboratory conditions, is just because of a higher average environment temperature and solar radiation. Nevertheless, the idea of the digesters self-heating because of the fresh silage is an example of what Nobel Laureate Irving Langmuir called 'pathological science' [10], so Occam's razor will not be enough to debunk it.

We can use the critical thinking tools described in the former section to analyse if this 'groundbreaking discovery' is such or if it is just a hype resulting from incomplete observations and abductive reasoning. The Cartesian method requires us to doubt conclusions that cannot be proved true, and we must formulate alternative hypotheses to arrive at an absolute truth. In this case, we can employ deductive reasoning to demonstrate that the temperature rise of a digester during summer is not caused by an exotic and mysterious phenomenon of metabolic heat but just because it is summer.

Let us compare the facts reported in the paper to our existing knowledge of anaerobic digestion:

a. Digesters tend to overheat only in the summer due to increased solar exposure and thick insulation if they lack cooling systems. This is a common-sense explanation, and technical standards of civil engineering allow the accurate calculation of the temperature flows in summer and winter. The common practice of bioclimatic architecture shows that even at latitudes as high as that of Central Europe, the energy consumption in summer air conditioning can be higher than the heat demand in winter if the building has thick insulation, which is usually the case for industrial digesters.
b. The paper states that the alleged self-heating is more common in larger digesters, which indicates that thermal capacity could be the main cause of the temperature increase.
c. The thermal model described in the paper is static, and the description is somehow vague. Apparently, the authors did not consider important factors such as solar radiation and heat capacity.
d. Anaerobic digestion is an endothermic phenomenon that requires heat to proceed. This is an absolute truth, demonstrated in thousands of experiments (see, for instance, [11]).
e. The correlation between the increase in the digester's temperature and silage-based diets is an effect, not the cause of the phenomenon. Or, more probably, there is no correlation at all. Most European biogas plants run on silage, but this does not necessarily mean that digesters fed with other substrates will not suffer temperature increases during spring or early summer. Conversely, the paper states that many plants fed with silage do not show a 'self-heating' effect.

f. The researchers were unable to replicate the phenomenon in the laboratory. In other words, they carried an abductive reasoning process, reaching a hypothesis and publishing it without any proof by deductive reasoning.

The reader will find a simple, yet rigorous mathematical model of heat transfer between the digester and the environment in Chapter 6 – Quick Startup and Thermal Stability of the Digester. The model employs an analogy with a passive electrical grid, where the formulas that regulate heat flows are similar to those that describe electric currents. Solving thermal transfer problems with the help of electrical models is a sound and well-established method, widely employed for assessing the thermal performance of buildings with the finite elements method (FEM). The FEM consists in modelling the digester as a passive electrical circuit: the casing of the digester (walls and insulation) is represented as resistances, and the thermal mass of the digester and walls as capacity. Heat input flows – solar radiation and heating coil – are equivalent to current generators: a pseudo-random variable current in the first case and a constant current in the second. The external air temperatures are equivalent to a pseudo-random variable voltage generator, and the temperature of the soil is assumed as a constant voltage generator. The imbalance between the heat input during the day and heat dissipation during the night over several days or weeks accumulates in the water heat capacity, increasing the digester's temperature during summer.

The thermal mass of a digester is mainly composed of water, a substance that requires large quantities of energy to vary its temperature. The strong insulation with which the digesters are equipped equates to a large series resistance, which in mathematical terms translates into a high 'time constant' or the equivalent capacitive-resistive system.

System designers, especially those from cold countries, mistakenly assume that heat flows are only winter heat losses, overlooking the overheating during summer. During the day, the sun shining on the external cladding of the digester can increase the surface temperature above 50°C, and this is true even at high latitudes. Therefore, heat flows from the outside to the inside, which is commonly at 38–40°C. The large thermal mass of the digester stores this heat throughout the day, increasing the temperature by a fraction of a degree. During summer nights, the outside temperature drops, but it can stay above 25°C. Since it is equipped with an insulating coat, the digester disperses little heat, and its temperature drops by a fraction of a degree smaller than the heating. The heat balance is therefore positive, and the large liquid mass inside the digester acts as an accumulator, slowly but steadily increasing its temperature as we move from spring to summer. At a certain point, the plant manager turns the heating system off, but the temperature continues to increase in an apparent 'self-heating'. Counterintuitively, this effect could be more conspicuous at higher latitudes because in summer the day is much longer than the night. This cannot be explained by the model employed for the redaction of the paper because it didn't consider the thermal inertia. The introduction of thermal capacity and solar radiation in the bioclimatic model of the digester also explains why no 'self-heating' has ever been found in the laboratory: the temperature of a laboratory is usually around 20°C all day long, so the heat flow of the laboratory reactor is always from the interior to the exterior. If heating is interrupted, the reactor will cool down until it reaches room temperature.

1.2.4.2 Example 2: The New Digester Technology that Reduces Nitrogen

During an AD congress, a young researcher presented a poster about a digester packed with inert support to which anaerobic bacteria adhere. The 'discovery' was that this 'novel' technology had the non-buoyant packing near the top of the digester instead of the classical neutrally buoyant plastic rings, and it was capable of reducing chemical oxygen demand (COD) and N simultaneously. We know that AD reduces COD and converts most organic N into NH_3 or NH_4^+. The young researcher was a victim of overconfidence bias and failed to apply Ockham's razor. Let's assume that all the measurements were performed correctly, and the concentration of total N at the exit was lower than at the entry. There are many competing explanations for the said observation, so the supposition that a new geometrical configuration of an anaerobic filter would favour the growth of a new kind of bacteria able to simultaneously eliminate carbon and nitrogen is simply the most artificial. The simplest explanation is just that bacteria are using the incoming N and C to grow on the packing. Incidentally, the young researcher did not check the growth of the biomass film on the packing; hence, this term was missing in the balance. The wrong assumption was that operating the 'novel' digester for N days would bring it to a 'steady state', so bacteria would cease to grow.

1.2.4.3 Example 3: The New Electro-digester Technology that Doubles the BMP

The paper described a metallic digester with a coaxial electrode that, according to the authors, 'more than doubled' the production of CH_4. The rationale was that the electric current produced H_2, which was converted to CH_4 by hydrogenotrophic archaea. So, all carbon would become CH_4. The researchers performed a test with a stainless steel cylindrical reactor having its body connected as the cathode and a coaxial electrode as the anode. They claim that during an assay with sodium acetate, the efficiency of the process was 64% and that the BMP was 230% of the theoretical value. The percentage of CH_4 in biogas was 98%. A dose of healthy scepticism about the said claims is advisable because of the following:

a. It is not clear how they define 'efficiency'. The potential energy of the additional CH_4, intended as its lower heating value (LHV), is thermal energy, while the process consumes electricity, i.e., pure exergy.
b. As usual in the AD literature, they do not analyse the measurement uncertainty. Strangely, the increase in the BMP of acetate is 230%. We must recall that the stoichiometry of digestion produces 1 mole of CH_4 and 1 mole of CO_2 per mole of acetic acid. In the best case, one would have expected the conversion of all the C into methane, i.e., doubling the BMP. Therefore,

which is the origin of the remaining 30% of C being converted into CH_4? This leads to three logical explanations: either the measurement was inaccurate (plausible because the paper contains no analysis of the error margin) or the electrolysis is killing part of the bacteria in the inoculum, and other bacteria are converting the carbon into CH_4 or, finally, the hydrogenotrophic archaea are taking the CO_2 and the H_2, producing CH_4 and water. If the first supposition is true, then the experiment should be repeated, ideally by other research groups with more attention to metrological aspects. If the second supposition proves to be the explanation, then such a digester is impractical because its operation would be unstable over time, requiring frequent re-inoculations. If the third explanation is true, then there will be a further question to answer: what is the fate of the sodium? How could the hydrogenotrophic archaea be quicker than sodium in capturing the CO_2 resulting from the digestion of acetate?

c. Producing biogas with 98% CH_4 during a batch test with sodium acetate is a usual result, even in a standard reactor. The reason is that the stoichiometry of sodium acetate yields methane and sodium carbonate (see Chapter 5, Section 5.2 example b). In other words, claiming that the biogas produced during the electromethanogenesis of sodium acetate has 98% CH_4 does not prove anything.
d. What is the fate of O_2?

The author made a test with a different reactor, composed of a glass body with two stainless steel electrodes. The electrodes of the control reactors were left open, i.e., with no current at all. One control was a blank and the other two were fed with cellulose. Another set of three reactors was connected in series so that the current across them was the same: one was blank, and the other two were fed with the same quantity of cellulose as the control. All were connected to an AMPTS II via CO_2 traps. The reactors connected to the current produced larger amounts of CH_4 (or perhaps a mixture of CH_4 and H_2. The author has no gas chromatographer to check this possibility.). However, the BMP of cellulose calculated with the set of electro-reactors was 95% of that of the control set, which was within the norm. The overall uncertainty of both sets was around 5%, so despite a tendency to a lower (average) BMP obtained from the electro-reactors, the difference is not relevant. The only remarkable difference is just a higher brute production of gas from the three electro-reactors compared to their corresponding controls, but since the current was the same in all of them, the difference could be explained as gas deriving from electrolysis. Remarkably, the stainless-steel electrodes that carried the current (just a few mA) were severely corroded and the inoculum had lost more VS solids than the control, so probably, the current or the O_2 produced killed part of the bacterial population, resulting in part of the CH_4 derived from the degradation of the dead bacterial biomass.

The claim of a 'novel disruptive technology' should be quarantined in this case, until more convincing experimental evidence is gathered with several reference substrates (e.g., acetic acid instead of sodium acetate, or cellulose or gelatine), and the economic and operational convenience of this technology is demonstrated.

1.3 DEBUNKING MISINFORMATION: THE CASE OF 'NO BIOGAS' ACTIVISTS

The opposition by residents to proposed developments in their local area can be understandable –and even legitimate – if the planned works are unnecessarily polluting; if the cost–benefit analysis is weak, biased, or inexistent; or if there is sound evidence of corruption behind the project. But the systematic opposition to any project, under any circumstances, against all reasonable explanations is an irrational attitude widely diffused throughout the world, usually labelled with the acronym NIMBY (not in my backyard). Each country seems to have a particular sensitivity to some type of project: highways, waste treatment facilities, wind farms, and energy infrastructures are those that most frequently raise popular opposition. Italy is probably the country with the highest frequency of the NIMBY syndrome against biogas plants. Abuses of the public administration, failure to educate the people – very often aggravated by false information, deliberately disseminated with purely electoral purposes – and even petty local baronial logic or disputes between neighbours have contributed to demonising the anaerobic digestion technology. The Italian trend is the opposite of other countries and international institutions' policies:

1. In Sweden, the energy gas usage in 2021 was 21.8 TWh, of which approximately 79% was fossil gases (natural gas, LPG, and LNG) and 21% was biogas/biomethane. Biogas and LNG use has increased, and natural gas use has decreased over the last 10 years. Most of the biogas is upgraded and used in the transport sector. The average biomethane share in 2021 was 96% in the CNG/CBG used for transport and 65% in the LNG/LBG market [12].
2. The Joint Research Council (JRC) demonstrated in its well-to-wheel analysis that, depending on the feedstock, biomethane can ensure the lowest emissions in transport [13].
3. The International Panel for Climate Change (IPCC) states that emissions of greenhouse gases from energy produced with biogas are among the lowest ever, depending on feedstock and land use [14].

Despite all the scientific evidence and recommendations of the international bodies, the opposition to biogas in Italy has been fierce. Local committees have founded superregional associations and published a manual on how to stop biogas and biomass projects: *The Biomass & Biogas Vademecum*. The said document became viral among conspiracy bloggers and social networks, with additional promotion from some 'antisystem' mass media.

Italy is an extreme case of 'biogas hysteria', but the author has received notice that similar situations are beginning to rise in other countries. If the reader is facing anti-biogas activists, the following analysis will surely provide insights to debunk some of the most recurring hoaxes, hypes, and fallacies against biogas plants, searching for the locally applicable legislation when necessary. Italian readers will find the complete article in [15].

1.3.1 Example 1: Misleading Interpretation of Local Laws and Regulations

'A biogas plant authorised to run on dedicated crops can also legally resort to the OFMSW (organic fraction of municipal solid waste) based on Legislative Decree no. 387 of 12/29/2003 and the ruling of the Council of State Sec. V n°5333 of 07/29/2004'.

The words 'municipal waste' recall in every population – not only Italians – the idea of bad smell, heavy traffic, and nuisance for the citizens. In this case, the activists were diffusing incomplete information with intent on scaremongering. In the Italian bureaucratic-administrative context, the use of OFMSW in an agricultural plant implies that the digestate no longer has the legal status of a by-product, becoming 'waste'. Therefore, the disposal costs make this option unfeasible for an agricultural company, whose interest is in adding value to its residues and replacing fertilisers with digestate to lower the production costs. From a purely scientific point of view, there is no difference between OFMSW and any other fermentable organic matter: it is proven that anaerobic digestion is the best technique for reducing the polluting load and containing the greenhouse effect. The legal complications of digestate management in Italy arise from the illogical interpretations of the European definition of 'waste' and 'end of waste' [16].

1.3.2 Example 2: False Information to Scaremonger about the 'Dangerousness' of Biogas Plants and Build the First Impression Bias in the Population

'Digestate is waste (CER codes: 190600–03; -04; -05; -06)'.

Behavioural scientists call it 'the first impression bias': a limitation in human information processing that causes us to make quick and incomplete observations about others based on the first piece of information we perceive. First impressions are often crucial as they lead to quick assumptions and judgements and tend to persist in time, even if further evidence proves them wrong. In this case, the activists are trying to manipulate the common citizen, who is unfamiliar with the European Waste Catalogue (EWC, CER in Italian) and the EU Waste Framework Directive. The EWC is a set of codes that allows waste to be classified according to its industry and/or origin, regardless of whether the waste is being processed or not, and does not contain any judgement on its eventual dangerousness. In this particular case, it is easy to debunk the activists' argument. A search on the Internet provides dozens of entries, from Wikipedia to the official European Gazette [17]. The meanings of the codes are as follows:

19 06 00: waste resulting from anaerobic treatment of waste;
19 06 03: liquor from anaerobic treatment of municipal waste;
19 06 04: digestate from anaerobic treatment of municipal waste;
19 06 05: liquor from anaerobic treatment of animal and vegetable waste;
19 06 06: digestate from anaerobic treatment of animal and vegetable waste.

The activists' argument is partially true and limited to the case of biogas plant projects for the treatment of waste. Most of the argument is misleading: code 19 06 00 refers to 'wastes' such as plastics, metal, and silt, resulting from the process but not necessarily to the digestate itself. The activists also omit an important piece of information: in a waste treatment plant, digestate and liquor (the liquid fraction of the digestate) will never leave the plant because they are intermediate products of a more complex process. Waste treatment plants usually produce compost; i.e., the 'end of waste' directive applies because the composting process radically changes the features of the digestate.

The argument is false in the case of agricultural plants, at least in Italy [18]. Agricultural facilities, by definition, are authorised to digest only agricultural by-products, and therefore, the digestate is not 'waste' but a 'by-product'; hence, it does not require further processing.

In any case, there are two implicit logical fallacies in the activists' intent to present digestate as 'waste':

a. The fact that a given substance has the legal status of 'waste' does not mean it is a 'pollutant' or 'dangerous'.
b. The fact that an industry proposes the construction of a waste treatment facility does not mean that there is an 'eco-mafia' conspiring with corrupt governments to pollute a given territory and make people sick.

1.3.3 Example 3: Misleading and Conspiracy-mongering Information

'These power plants are used by entrepreneurs who invest just to benefit from generous State incentives provided for renewable sources. Without State subsidies, the main economic reason for this activity would disappear. In any case, it is possible to believe that the generalised propensity for biomass and biogas plants is also part of a more general perspective of the reuse of these plants for waste treatment. The organic fraction of solid urban waste (OFMSW) is equated to biomass by ministerial decree. It is easy to foresee that once these power plants have been built, instead of being fuelled with agricultural biomass – which Italy does not have and which will have an ever-increasing cost – they will be fuelled with OFMSW, whose disposal cost is already a primary source of profitability. The income from OFMSW collection and treatment is worth between €80 and €110/t, green waste around €60/t and sewage sludge around €90/t'.

The 'NIMBY committees' insinuate that the state is corrupt to the point of authorising biogas plants as agricultural (because of a 'shorter' authorisation procedure) and then using them for waste treatment purposes, for the benefit of the 'entrepreneurs'. As explained in example 1, admitting OFMSW into a plant means that the digestate automatically has the legal status of 'waste', so the owner of the plant would have to pay to dispose of the digestate rather than use it as an organic fertiliser. Furthermore, the incentives for plants that use waste are decidedly lower than those for agricultural plants, without considering that the taxation on agricultural activities is lower than that for industrial activities, such as waste treatment. An agricultural entrepreneur therefore has neither convenience nor interest in feeding the biogas plant with waste. And the state, however corrupt it may be, needs to 'raise cash', so it is indeed more interested in encouraging OFMSW plants to collect taxes, which in Italy are higher for industrial activities than for agricultural ones.

Regarding the alleged ministerial decree that equates OFMSW to biomass, the '*Vademecum*' does not provide any reference, so it is impossible to check if such a decree exists and if it is still in vigour. It is true that Directive 2001/77/EC opened the door to fraud and favoured existing incineration plants, assuming at a flat rate that 50% of the waste was biomass, and rewarding the energy produced therein accordingly. This directive, however, was repealed in Italy by Legislative Decree 3 March 2011 n. 28 [19]. In summary, the Italian legislation states:

> *biomass:* the biodegradable fraction of products, waste and residues of biological origin coming from agriculture (including plant and animal substances), forestry and related industries, including fishing and aquaculture, grass cuttings and pruning from public and private green areas, as well as the biodegradable part of industrial and urban waste.

The decree 28/2011 and the whole body of subsequent financial laws make a clear distinction between the incentives provided to biogas plants fed with 'by-products' and those that use 'waste' instead, and also between plants put into operation before and after the said decree. The authorisation procedures are also very different between both types of plants. The statement tends to create alarmism and also win popular consensus, taking advantage of the people's growing distrust in the institutions, on the extreme and sometimes artful complexity that the same institutions introduce in the legislation they produce.

1.3.4 Example 4: False Statement

'Biogas is more polluting than natural gas because it contains only 55/60% methane'.

Apart from the fact that the *Vademecum* does not specify which calculations support this statement, or which component they deem 'polluting', common sense tells us that a mixture containing an alleged contaminant at 55 or 60% cannot be more polluting than the same substance at 100%. Let's clarify things a bit: biogas is a mixture composed of 55–60% CH_4, while the rest is CO_2. Its combustion emits CO_2 and water vapour, exactly like that of natural gas (which is almost pure CH_4). Page 66 of the aforementioned IPCC study clearly shows that the greenhouse effect potential of biogas combustion is much lower than that of natural gas because the latter is a fossil fuel and requires transport over very long distances. Biogas, on the other hand, is essentially a 'zero-kilometre' resource and also 'neutral' – because the emitted CO_2 is the same that will be recaptured by the plants, producing the biomass of subsequent years. The alleged 'polluting power' of methane is, once again, the citation of a fact in a manipulative and alarmistic way. The IPCC has studied the 'greenhouse effect power' of various gases, so it is scientifically indisputable that the release of 1 ton of CH_4 into the atmosphere causes the 100-year effect of 29 tons of CO_2. The 'no biogas committees' fraudulently overlook the fact that the goal of a biogas plant is precisely to prevent the release of CH_4 into the atmosphere and to exploit it rationally. Objectively, the CO_2 and water vapour resulting from the combustion of biogas in a single large plant – equipped among other things with a catalyst and sophisticated control systems – are decidedly less harmful and easier to subject to strict controls than gas exhaust from cars, domestic heating boilers, and wood-burning fireplaces, which are checked only on a random basis.

1.3.5 Example 5: A Hoax

'Anaerobic digestion systems are unable to completely neutralise the bacteria present in the waste, in particular, Clostridia which are heat-resistant bacteria (the bacteria that cause botulism and tetanus belong to this family). In Germany, some researchers have suggested that the Escherichia coli epidemic that hit Germany in the summer of 2011, causing 18 deaths and the thousands of cases of botulism observed in animals between the summer of 2011 and early 2012, would have been caused by the presence of biogas power plants'.

A search on the EC official site shows that the investigations of sanitary authorities demonstrated that the *E. coli* epidemic of 2011 was caused by fenugreek seeds imported from Egypt, germinated in Germany, and consumed raw in salads [20].

The author conducted a study of the peer-reviewed literature on the possible diffusion of botulism via the digestate [21], finding that the said hypothesis is false. Not only is there no relationship between digestate and botulism, but the experimental evidence published in the literature indicates that the anaerobic digestion process eliminates, or at least substantially reduces, the presence of *C. botulinum* and other pathogens in digestates (see, for instance, [22]). The NIMBY committees provide some newspapers and recordings of radio/TV interviews as 'proofs' of the nexus between botulism and biogas. A thorough fact-checking of the same (presented in ref. 21 already cited) shows that the information is false, incomplete, or lacks sound scientific evidence. For instance, in the case of botulism in a dairy farm in Padua, local health authorities report that 'probably during the production phase of the hay bales, a wild animal (a hedgehog, a rat or a pheasant) was caught by the machinery and incorporated into one of them. The carcass caused the proliferation of the bacterium. Subsequently, the infected bale was placed in the mixer, where all the concentrated foods and flours were inserted, and the fodder was distributed to the entire herd'. The 'no-biogas handbook' and the arguments of the 'NIMBY committees' regularly overlook or pretend to ignore the official communication of the local sanitary authority and insist on the newspaper article on 'contamination caused by the nearby biogas plant', without however providing any concrete evidence.

BIBLIOGRAPHY

[1] Stuart Ritchie, Science fictions – Exposing fraud, bias, negligence and hype in science, *Vintage Digital* (July 16, 2020).

[2] Morten Sivertsvik, Publish or die, *Journal of Aquatic Food Product Technology*, 28:5, 451 (2019). https://doi.org/10.1080/10498850.2019.1616277

[3] Holly Else & Richard Van Noorden, The fight against fake-paper factories that churn out sham science, *Nature* 591, 516–519 (2021). https://doi.org/10.1038/d41586-021-00733-5

[4] M Horgan, Biased decision making in materials science: Where does it originate and can it be avoided?. *MRS Bulletin*, 46, 361–367 (2021). https://doi.org/10.1557/s43577-021-00104-5

[5] Ronald L Wasserstein & Nicole A Lazar, The ASA statement on p-values: Context, process, and purpose, *The American Statistician*, 70:2, 129–133 (2016). https://doi.org/10.1080/00031305.2016.1154108

[6] IEA, *Outlook for biogas and biomethane: Prospects for organic growth*, IEA, Paris (2020). https://www.iea.org/reports/outlook-for-biogas-and-biomethane-prospects-for-organic-growth, License: CC BY 4.0.

[7] Directive 2008/98/EC of the European Parliament and of the Council of 19 November 2008 on waste and repealing certain Directives https://eur-lex.europa.eu/legal-content/EN/TXT/?uri=CELEX%3A02008L0098-20240218

[8] MA Rosato, *Managing Biogas Plants – A Practical Guide*; CRC Press, 2017, ISBN 9781138626614.

[9] S Rawat & S Meena, Publish or perish: Where are we heading?, *Journal of Research in Medical Sciences*, 19:2, 87–89 (February 2014). https://www.ncbi.nlm.nih.gov/pmc/articles/PMC3999612/. PMID: 24778659; PMCID: PMC3999612.

[10] Irving Langmuir, *Transcription of a lecture on pathological science* (December 18, 1957). https://galileo.phys.virginia.edu/~rjh2j/misc/Langmuir.pdf

[11] Sung T Oh & Alastair D Martin, Thermodynamic study equilibrium model in anaerobic digestion process, *Biochemical Engineering Journal*, 34, 256–266 (2007). https://doi.org/10.1016/j.bej.2006.12.011

[12] Linus Klackenberg, *Biomethane in Sweden – Market overview and policies*, Swedish Gas Association (January 12, 2023). Downloadable from https://www.energigas.se/media/amrji21q/biomethane-in-sweden-230313.pdf

[13] M Prussi, M Yugo, L De Prada, M Padella & R Edwards, *JEC Well-To-Wheels report v5*, EUR 30284 EN, Publications Office of the European Union, Luxembourg (2020). https://doi.org/10.2760/100379, JRC121213. https://op.europa.eu/en/publication-detail/-/publication/7a2ecdc8-fed8-11ea-b44f-01aa75ed71a1/language-en. ISBN 978-92-76-20109-0

[14] Renewable Energy Sources and Climate Change Mitigation, *Special report of the IPCC, summary for policymakers and technical summary*, IPCC (2012). Downloadable from https://www.ipcc.ch/site/assets/uploads/2018/03/SRREN_FD_SPM_final-1.pdf (see file pages 32, 63, 64, and Figure TS.2.5 on page 66).

[15] MA Rosato, I 'comitati del no' ed il vademecum biogas e biomasse, *AgroNotizie* (April 28, 2017). https://agronotizie.imagelinenetwork.com/bio-energie-rinnovabili/2017/04/28/i-quotcomitati-del-noquot-ed-il-vademecum-biogas-e-biomasse/53738

[16] MA Rosato, La giungla normativa europea sul digestato, *AgroNotizie* (February 22, 2022). http://agronotizie.imagelinenetwork.com/bio-energie-rinnovabili/2016/02/22/la-giungla-normativa-europea-sul-digestato/47577

[17] https://ec.europa.eu/eurostat/documents/342366/351806/Guidance-on-EWCStat-categories-2010.pdf/0e7cd3fc-c05c-47a7-818f-1c2421e55604

[18] 27-12-2014 Supplemento ordinario n. 97 alla GAZZETTA UFFICIALE Serie generale – n. 299 http://www.gazzettaufficio.it/do/atto/serie_generale/caricaPdf?cdimg=14A09910005000101110001&dgu=2014-12-27&art.dataPubblicazioneGazzetta=2014-12-27&art.codiceRedazionale=14A09910&art.num=1 &art. Tiposerie=SG

[19] Ministry of Development, *Complete text of the Decree*. http://unmig.sviluppoeconomico.gov.it/unmig/norme/28dlg11.htm

[20] Parliamentary question – E-001755/2019(ASW), European Parliament, answer given by Mr Andriukaitis on behalf of the European Commission. https://www.europarl.europa.eu/doceo/document/E-8-2019-001755-ASW_IT.html

[21] MA Rosato, Botulismo e rifiuti: esiste davvero un nesso?; Ingegno Ambientale, *Journal of the Italian Association of Environmental Engineers and the Territory*, 1, 4–6 (May 2014). https://www.ingegneriambientali.it/docs/docs_bacheca/IngegnoAmbientale1-2014.pdf

[22] B Fröschle, U Messelhäusser, C Höller & M Lebuhn, Fate of *Clostridium botulinum* and incidence of pathogenic clostridia in biogas processes, *Journal of Applied Microbiology*, 119, 936–947 (2015). https://doi.org/10.1111/jam.12909

2 The Cartesian Approach to Maximise Profit

2.1 CHAPTER OVERVIEW

The design of commercial biogas plants is always suboptimal because it is the outcome of a series of trade-offs. The reasons are multiple, and the following list includes the most frequent ones in the author's personal experience:

1. Overestimated biochemical methane potential (BMP) during the engineering phase. The plant owner must use more feedstock than foreseen for a given productivity.
2. Mix of substrates: The designer did not consider that the resulting C/N/P is not optimal.
3. Assuming that the BMP of a mixture of feedstocks equals the proportional sum of the individual BMPs.
4. Wrong inoculum chosen during the startup phase.
5. Long-term inhibition: A common issue in plants fed with a single substrate (e.g., sewage sludge, cereal silage, or swine manure).
6. The designer estimated the BMP using a theoretical method instead of testing the feedstock in the laboratory for eventual inhibition problems.
7. Suboptimal digester choice, usually resulting from the plant builder's 'One-type-fits-all' industrial concept, e.g., proposing a 22-m-diameter continuously stirred reactor tank (CSRT) for a plant with 100 kW electric nominal power. The oversized digester will require at least 20 kW for stirring and pumping.
8. The "standard" solution in a country may not be suitable or economically convenient somewhere else. For example, steam explosion systems are known to improve the AD of OFMSW but are not widespread in Italy because steam generators must comply with many national and local laws and the bureaucratic burden is too expensive.
9. 'Scale economies' that lead to under- or oversizing some plant components. Such 'scale economies' increase the profit of the plant builder but often also increase the operation expenditure (OPEX) for the owner much more.

Increasing the productivity of a biogas plant requires three improvements, which are often contradictory (Figure 2.1):

a. Introducing modifications to the process and sometimes re-engineering the plant.
b. Increasing the biological activity of the inoculum and allowing it to digest more biomass with the same digester volume. Reaching this goal requires

DOI: 10.1201/9781003609278-2

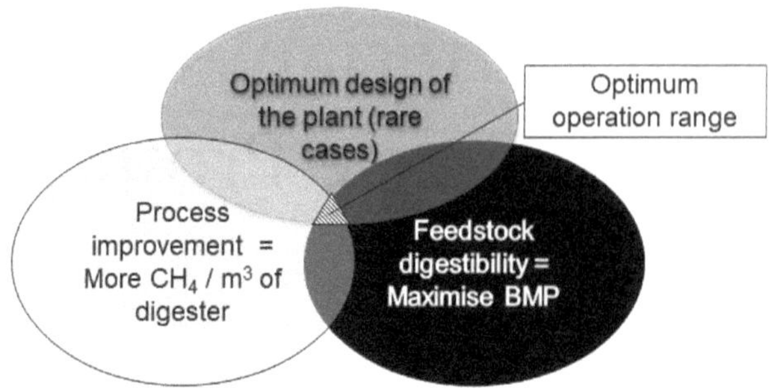

FIGURE 2.1 The Optimum Operation Range Is Usually Narrow Because of Design Trade-offs.

increasing the organic loading rate (OLR) and unavoidably reducing the hydraulic retention time (HRT) unless it is possible to increase the volatile solids (VS) concentration of the feedstock. Reducing the HRT translates into lowering the digestion efficiency. Digesting more biomass in less time without losing the conversion efficiency requires increasing the metabolic rate of the bacterial flora, i.e., increasing the digester's temperature. It is also possible to increase the bacterial population, i.e., the cellular density of the inoculum, or both strategies at the same time. Regardless of choosing either, the practical problem is maintaining the ecological equilibrium among thousands of microorganisms that compose the anaerobic sludge's flora.

c. Increasing digestibility, i.e., producing more methane with the same quantity of feedstock. Several strategies exist for the said scope: enzymatic pretreatments, physical pretreatments (fine milling, sonication, cavitation, etc.), chemical pretreatments (using acid, alkaline, and organic solvents), and finally, thermal pretreatments (pasteurisation, boiling, steam explosion, and thermal hydrolysis). The efficacy of the different technologies is difficult to foresee because it depends on the plant design and the features of the feedstock. Efficacy is often loosely related to efficiency because the increase in BMP always comes at the expense of high energy or product consumption.

The literature provides some insights on possible techniques that may be worth considering to improve a biogas plant's performance. However, it also abounds in hypes, generalisations of specific cases, and even conceptual mistakes. Developing the analytical capacity to discern potentially useful information for a given concrete case represents a competitive advantage for the biogas plant manager or owner. This chapter proposes a critical review of some papers in the literature and some case studies from the author's personal experience.

2.2 REVIEWING THE PLANT'S ENGINEERING AND SETTINGS

2.2.1 Mesophilic or Thermophilic Operation?

Anaerobic digestion allows the production of methane, a high-density energy vector, by investing just heat from a low-temperature source, often residual heat. It is a common belief that increasing the digester's temperature will increase a plant's productivity. The rationale is that bacterial activity will be faster at higher temperatures, thus reducing the time to digest a unit mass of feedstock, so it is possible to produce more methane with a unit volume of the digester. The literature on thermophilic digestion is abundant, as the reader can check in a recent study [1]. In general, thermophilic digestion (50°C—55°C) tends to produce biogas with a higher percentage of CH_4 than mesophilic digestion, and the digestion time reduces from the usual 30 days to 20 days. The residual BMP of digestate is lower, which means that the conversion efficiency of the thermophilic digester is higher for a given OLR. However, comparing thermophilic to mesophilic batch tests, the differences in BMP are null or irrelevant. In other words, the differences are of the same order of magnitude as the measurement uncertainty. Table 2.1 (from the paper already cited above) shows the BMP of the same samples, measured with the same AMPTS instrument at two different temperatures.

N.B.: The author rounded the values published in the paper in compliance with the norm UNI CEI 70098–3:2016, ref. 2 (the Italian translation of the ISO/IEC GUIDE 98–1:2009, ref. 3) because, as usual in the biogas literature, the paper's authors calculated the BMP with two decimals. Decimals are irrelevant since the uncertainty of the result is in the order of units or even tens.

The eventual higher CH_4 percentage in the biogas is irrelevant too because, in a batch test, part of the CO_2 produced will never leave the reactor. The maximum CH_4 percentage registered during a batch test is a misleading indicator of biogas quality that a continuous digester will produce. This argument will be treated in detail in Chapter 9, Section 9.2.

An important advantage of thermophilic over mesophilic digestion, especially for plants treating waste in the EU territory [4], is the ability to sanitise the waste. The said benefit is officially recognised in the Regulation (EU) 2019/1009 of the European Parliament and of the Council of 5 June 2019 [5]. The two cases in which thermophilic digestion alone is sufficient to sanitise organic waste are

a. thermophilic anaerobic digestion at 55°C for at least 24 hours followed by a HRT of at least 20 days and
b. thermophilic anaerobic digestion at 55°C with a treatment process including pasteurisation as described in point 1 of Section 1 of Chapter I of Annex V to Commission Regulation (EU) Nr. 142/2011 (applicable to the digestion of non-edible animal by-products and slaughterhouse waste, see 6).

TABLE 2.1
BMP and Uncertainty Ranges Measured at Different Temperatures

	BMP @ 35°C (Ncm^3/g VS)	BMP @ 50°C (Ncm^3/g VS)	Difference between Averages (Ncm^3/g VS)
Sample 1	255 ± 18	266 ± 35	11
Sample 2	325 ± 17	325 ± 22	0
Sample 3	325 ± 17	325 ± 22	0

The combination of anaerobic digestion – regardless of its temperature – and composting is treated in detail in Chapter 8, Section 8.2.

Unfortunately, making a mesophilic digester to run at thermophilic temperature is not always possible. In some cases, thermal dispersion of the digester, in particular those with plastic membrane gas domes, or an undersized heat exchanger prevents the temperature from exceeding 45°C, especially in places with harsh winters.

In combined heat and power (CHP) biogas plants, there is always plenty of heat available, so increasing the temperature of the digester is not a problem as long as the thermal dispersions and external temperature allow it. In plants designed for producing biomethane, it is necessary to use part of the product to heat the digester or to employ some external heat source (e.g., lignocellulosic biomass, geothermal energy, or solar energy). In such plants, keeping the temperature at a minimum, compatible with enough methanogenic activity, is crucial to minimise the OPEX.

Another disadvantage of thermophilic digesters is their less biodiverse bacterial ecosystem that makes the digester more sensitive to sudden changes in its temperature or the composition of its diet.

Changing the design temperature from mesophilic to thermophilic or vice versa requires pondering all the pros and cons. It is highly advisable to conduct a laboratory simulation, simultaneously comparing at least two identical continuous reactors, one thermophilic and the other mesophilic, fed with the same OLR and the same daily sample divided into two equal parts. The simultaneity of both tests is a necessary condition to avoid the *post hoc* logical fallacy. Before starting the test, running both digesters at their chosen temperatures for at least 45 days is necessary to reach a stable composition of the bacterial ecosystem. The goal of the test is to compare the net production of CH_4 at each temperature, so placing a CO_2 trap before the flow meter and frequently changing the alkaline solution are necessary conditions to obtain reliable results. Collecting biogas in a gas bag and analysing its composition daily are acceptable, but this adds uncertainty to the results (the metrological errors of the flow meter and the gas analyser add up). The risk of adopting the latter solution is that the differences in CH_4 production between both reactors will be of the same order of magnitude as the total measurement uncertainty.

2.2.2 Temperature Control

Chapter 6 is entirely dedicated to modelling and assessing thermal flows in the digester. In general, industrial biogas plants benefit from an enormous thermal inertia, so the temperature is intrinsically stable. On the other hand, the biological activity of mesophilic bacteria and archaea does not change much in the range 35°C–45°C.

2.2.3 The Correct Setting of the Stirrers

Most of the existing biogas plants are of the continuously stirred tank reactor (CSTR) type, operating wet digestion, i.e., the total solids concentration of the slurry is less than 12%. Stirring the slurry is necessary to facilitate bacteria access to nutrients. Too little stirring will slow down the digestion rate because the microorganisms will consume the nutrients around them but cannot move further to continue feeding.

Biogas will remain trapped in bubbles, and foam will form. Finally, insufficient stirring will favour the sedimentation of solids. In the long term, a thick layer of sediment will accumulate at the bottom, reducing the active digestion volume, which means shorter HRT and reduced biomass conversion into biomethane. Such a condition is called a 'hydraulic short circuit' (Figure 2.2) and is frequent in digesters that have been operating for more than 5 years. Abundant sediment formation is frequent in plants fed with chicken manure because it contains high levels of calcium and ammonia that cause insoluble hydroxyapatite and struvite crystals to precipitate. Sand and silt mixed with the feedstock entering the digester are also common sediments in agricultural biogas plants.

A certain stirring intensity will consume more energy but ensure adequate heat and nutrient exchange between the microorganisms and their surrounding medium. The mechanical action of stirring will disrupt bubbles, letting the gas escape from the slurry and preventing foaming. Figure 2.3 shows the instant gas flow in a laboratory

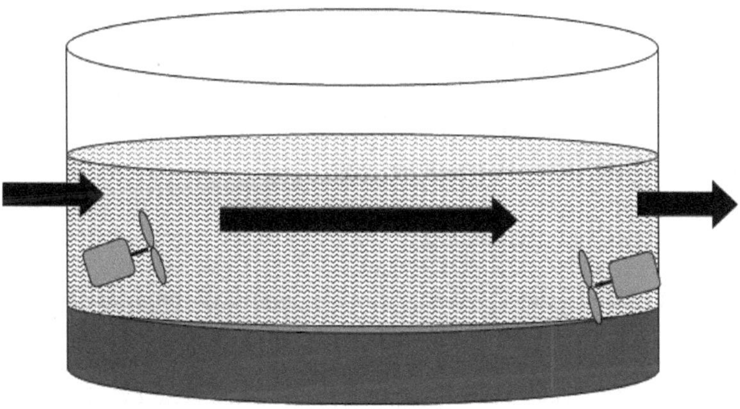

FIGURE 2.2 Thick Sediment Layers Substantially Reduce the HRT and Hence the Digestion Efficiency.

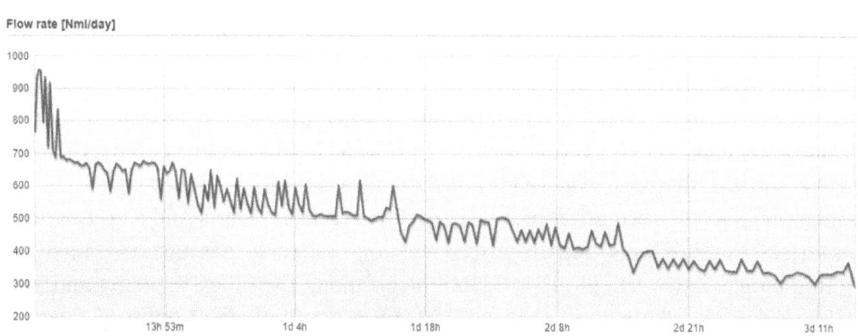

FIGURE 2.3 Flow Rate vs. Time in a Laboratory Reactor.

reactor with intermittent stirring, where the gas flow peaks correspond to the 'on' cycle.

Too much stirring will consume more energy and does not necessarily increase the digestion rate; on the contrary, phenomenon such as cavitation and shear forces from the speed gradient can damage the plasmatic membranes of the microorganisms. The dilemma every biogas plant manager faces is then: which is the optimum stirring intensity?

The anaerobic slurry is a non-Newtonian fluid; i.e., its viscosity is a non-linear function of temperature and speed gradient. Defining the analytical formulas for calculating the power necessary to ensure optimum mass and thermal exchange coefficients is difficult, if not impossible. Applying empirical criteria is then the quickest method that provides acceptable results. Depending on the type of stirring, three different empirical criteria apply.

2.2.3.1 Turnover Rate (TR)

Some plants do not possess internal stirrers and serpentine heaters. Instead, one or more pumps aspire the slurry from the bottom, force it through an external heat exchanger, and send it back to the top of the digester. The stirring is adequate when the recirculation flow is between two and three times the volume of the digester. There are three ways of controlling the total flow:

a. **Switching the pumps on and off**: This control is inefficient because the start current is five to six times the nominal current, leading to increased ohmic losses and overheating of the motor if the start and stop cycle is too short.
b. **Partially closing a valve placed on the aspiration or the impulsion tube**: This control is inefficient because it dissipates energy as friction in the partially choked valve. In the case of centrifugal pumps, the rotor runs with a combination of flow and water head, which may reduce the overall pumping efficiency.
c. **Varying rotation speed with an inverter**: This is the most modern and efficient way of controlling any electric centrifugal pump. The flow is proportional to the square of the rotation speed, and the inverter keeps the torque constant throughout the operating range, so the power effectively absorbed is proportional to the flow.

2.2.3.2 Unit Gas Flow

Wastewater treatment plants sometimes have a set of diffusers placed at the bottom. A compressor aspires biogas from the headspace and bubbles it through the diffusers, creating vertical sludge streams (Figure 2.4). This solution reduces sedimentation, but it can lead to the formation of a floating layer of light debris and scum, with the same effect as sediments. An improved version, called a 'gas piston', consists in injecting biogas into vertical tubes. Bubbles rise along the tubes, aspiring and pushing the sludge up, like pistons (Figure 2.5). From the biological point of view, gas sparging is a mild stirring method that does not damage the sludge's bacterial flora.

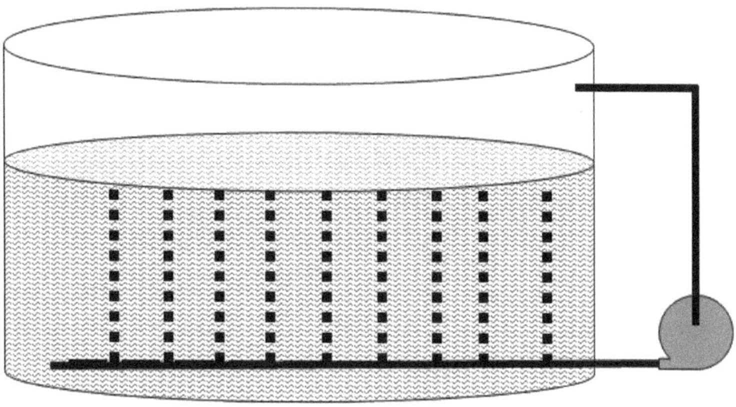

FIGURE 2.4 Operation Principle of Gas Bubbling Stirring.

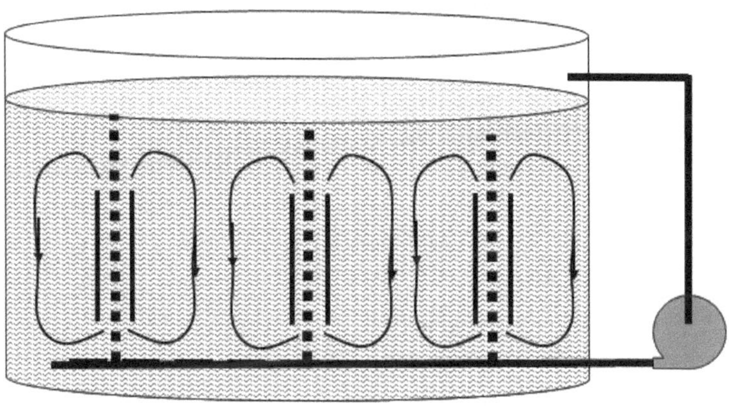

FIGURE 2.5 Operation Principle of Gas Piston Stirring.

From the energy point of view, gas compression is an inefficient process because of the thermodynamic irreversibility resulting in heat losses. Furthermore, gas bubbling works well with very thin sludge, in the order of 5% total solids or less.

The criterion for an acceptable stirring intensity with gas bubbling systems is checking an empirical coefficient known as unit gas flow (UGF):

$$UGF = 1{,}000\ Q_{gas}/V_{dig},$$

where

Q_{gas} = gas flow of the compressor [m^3/min];
V_{dig} = volume effectively occupied with sludge [m^3].

The acceptable values are as follows:

UGF = 4.5–5 m^3/min./m^3 of digester (spargers or free bubbling diffusers),
UGF = 5–7 m^3/min./m^3 of digester (gas pistons).

The methods for controlling gas flow are the same as already described in the case of pump stirring. The same considerations apply.

2.2.3.3 Unit Power

Most of the existing CSRT and plug flow reactor (PFR) biogas plants have either propeller or paddle wheel stirrers. In this case, the empirical coefficient for assessing stirring intensity is called unit power (UP):

$$\mathrm{UP} = P_{\mathrm{m}}/V_{\mathrm{dig}},$$

where
P_{m} = mechanical power at the shaft of the stirrer [W],
V_{dig} = volume effectively occupied with sludge [m^3].
Empirical rule: The stirring intensity is adequate when 6 W/m^3 < UP < 8 W/m^3.

Onsite measurement of the mechanical power at the shaft is almost impossible. Therefore, it is necessary to employ the nominal values shown on the motor's name-plate to calculate mechanical power with the following formula:

$$P_{\mathrm{m}} = \eta_{\mathrm{el}}\, \eta_{\mathrm{mec}}\, P_{\mathrm{el}},$$

where
η_{el} = electrical efficiency (resulting from the motor's rated IEC class),
η_{mec} = mechanical efficiency (assume 1 for direct drives, 0.95 for stirrers driven through reduction gears),
P_{el} = nominal electric power on the nameplate.

The international standard IEC 60034–30–1:2014 [7] Rotating electrical machines—Part 30–1: Efficiency classes of line operated AC motors (IE code) (formerly IEC 60034–30:2008) defines the efficiency classes for two-pole and four-pole asynchronous three-phase induction motors (aka squirrel cage motors or just cage motors). In the United States, the applicable standard is NEMA GM1. The equivalence between both standards (July 2024) is as follows:

IEC	NEMA
IE1	Standard Efficiency
IE2	High Efficiency
IE3	Premium Efficiency
IE4	Super Premium Efficiency
IE5	No standard yet (probable name will be Ultra-Premium Efficiency)

In all motors, the IE or NEMA class is stated in the nameplate (Figure 2.6 shows a conceptual example, and the layout may vary depending on the manufacturer). In

Brand				NEMA or IE class	
Type		Serial N.		Year	
kW		RPM		Phases	
V		A		Hz	
cos Φ		Duty		Poles	
Insul.type		IP		Serv. Factor	
I_a/I_n		M_a/M_n		Weight kg	
Labels (e.g.: CE, UL, NEMA, RoHS, TÜV, etc.)					

FIGURE 2.6 Facsimile of an Electric Motor Nameplate. The Actual Layout Varies from Brand to Brand.

older motors, the efficiency class may be stated as 'EFF1' or 'EFF2', equivalent to IE1 and IE2.

Efficiency is a function of the motor's nominal power. If the nameplate only states the efficiency class but not the efficiency percentage, the latter can be retrieved from Tables 2.2 and 2.3.

If neither the IEC class nor the efficiency percentage is stated on the nameplate, it is probably a very old motor, so assume IE1. If your plant is based in Europe, it is possible to estimate the IE class from the plant's construction date. EC Regulation 640/2009 and successive EC Regulation 4/2014 define the progressive adoption of the IEC standard 60034–30–1:2014 as follows (practical guidelines, examples of nameplates, and efficiency curves publicly available in ref. 8):

- All three-phase induction motors produced from 16 June 2011 must be at least IE2 class.
- From 1 January 2015, all motors having $7.5 < P_{el} < 375$ kW must be at least IE3 class (single speed), or at least IE2 if they have an electronic speed controller.
- From 1 January 2017, all motors having $7.5 < P_{el} < 375$ kW must be at least IE3 class (single speed), or at least IE2 if they have an electronic speed controller based on frequency variation, in such a way that the power is proportional to the rotation speed.

The regulation currently in force is Commission Regulation (EU) 2019/1781 of 1 October 2019 (text amended in 2023, available for download in all EU official languages 9). The text contains tables with rated efficiencies of each IE class as a function of the number of poles and nominal power.

In most cases, the installed power of the stirrers is higher than necessary, and an effective stirring power can be obtained by two different control strategies:

1. **Power controlled by an inverter**: In this case, the formula for calculating P_m has an additional factor:

$$P_m = \eta_{el}\, \eta_{mec}\, P_{el}\, f/f_n,$$

TABLE 2.2
Efficiency of 50-Hz Induction Motors according to IEC 60034–30–1:2014

Efficiency % for 50-Hz Induction Motors according To IEC 60034-30-1:2014																
	IE1				IE2				IE3				IE4			
Output	Poles				Poles				Poles				Poles			
kW	2	4	6	8	2	4	6	8	2	4	6	8	2	4	6	8
0.12	45.0	50.0	38.3	31.0	53.6	59.1	50.6	39.8	60.8	64.8	57.7	50.7	66.5	69.8	64.9	62.3
0.18	52.8	57.0	45.5	38.0	60.4	64.7	56.6	45.9	65.9	69.9	63.9	58.7	70.8	74.7	70.1	67.2
0.20	54.6	58.5	47.6	39.7	61.9	65.9	58.2	47.4	67.2	71.1	65.4	60.6	71.9	75.8	71.4	68.4
0.25	58.2	61.5	52.1	43.4	64.8	68.5	61.6	50.6	69.7	73.5	68.6	64.1	74.3	77.9	74.1	70.8
0.37	63.9	66.0	59.7	49.7	69.5	72.7	67.6	56.1	73.8	77.3	73.5	69.3	78.1	81.1	78.0	74.3
0.40	64.9	66.8	61.1	50.9	70.4	73.5	68.8	57.2	74.6	78.0	74.4	70.1	78.9	81.7	78.7	74.9
0.55	69.0	70.0	65.8	56.1	74.1	77.1	73.1	61.7	77.8	80.8	77.2	73.0	81.5	83.9	80.9	77.0
0.75	72.1	72.1	70.0	61.2	77.4	79.6	75.9	66.2	80.7	82.5	78.9	75.0	83.5	85.7	82.7	78.4
1.1	75.0	75.0	72.9	66.5	79.6	81.4	78.1	70.8	82.7	84.1	81.0	77.7	85.2	87.2	84.5	80.8
1.5	77.2	77.2	75.2	70.2	81.3	82.8	79.8	74.1	84.2	85.3	82.5	79.7	86.5	88.2	85.9	82.6
2.2	79.7	79.7	77.7	74.2	83.2	84.3	81.8	77.6	85.9	86.7	84.3	81.9	88.0	89.5	87.4	84.5
3	81.5	81.5	79.7	77.0	84.6	85.5	83.3	80.0	87.1	87.7	85.6	83.5	89.1	90.4	88.6	85.9
4	83.1	83.1	81.4	79.2	85.8	86.6	84.6	81.9	88.1	88.6	86.8	84.8	90.0	91.1	89.5	87.1
5.5	84.7	84.7	83.1	81.4	87.0	87.7	86.0	83.8	89.2	89.6	88.0	86.2	90.9	91.9	90.5	88.3
7.5	86.0	86.0	84.7	83.1	88.1	88.7	87.2	85.3	90.1	90.4	89.1	87.3	91.7	92.6	91.3	89.3
11	87.6	87.6	86.4	85.0	89.4	89.8	88.7	86.9	91.2	91.4	90.3	88.6	92.6	93.3	92.3	90.4
15	88.7	88.7	87.7	86.2	90.3	90.6	89.7	88.0	91.9	92.1	91.2	89.6	93.3	93.9	92.9	91.2
18.5	89.3	89.3	88.6	86.9	90.9	91.2	90.4	88.6	92.4	92.6	91.7	90.1	93.7	94.2	93.4	91.7

22	89.9	89.9	89.2	87.4	91.3	91.6	90.9	89.1	92.7	93.0	92.2	90.6	94.0	94.5	93.7	92.1
30	90.7	90.7	90.2	88.3	92.0	92.3	91.7	89.8	93.3	93.6	92.9	91.3	94.5	94.9	94.2	92.7
37	91.2	91.2	90.8	88.8	92.5	92.7	92.2	90.3	93.7	93.9	93.3	91.8	94.8	95.2	94.5	93.1
45	91.7	91.7	91.4	89.2	92.9	93.1	92.7	90.7	94.0	94.2	93.7	92.2	95.0	95.4	94.8	93.4
55	92.1	92.1	91.9	89.7	93.2	93.5	93.1	91.0	94.3	94.6	94.1	92.5	95.3	95.7	95.1	93.7
75	92.7	92.7	92.6	90.3	93.8	94.0	93.7	91.6	94.7	95.0	94.6	93.1	95.6	96.0	95.4	94.2
90	93.0	93.0	92.9	90.7	94.1	94.2	94.0	91.9	95.0	95.2	94.9	93.4	95.8	96.1	95.6	94.4
110	93.3	93.3	93.3	91.1	94.3	94.5	94.3	92.3	95.2	95.4	95.1	93.7	96.0	96.3	95.8	94.7
132	93.5	93.5	93.5	91.5	94.6	94.7	94.6	92.6	95.4	95.6	95.4	94.0	96.2	96.4	96.0	94.9
160	93.7	93.8	93.8	91.5	94.8	94.9	94.8	93.0	95.6	95.8	95.6	94.3	96.3	96.6	96.2	95.1
200-1000	94.0	94.0	94.0	92.5	95.0	95.1	95.0	93.5	95.8	96.0	95.8	94.6	96.5	96.7	96.3	95.4

TABLE 2.3
Efficiency of 60-Hz Induction Motors according to IEC 60034–30–1:2014

	Efficiency % for 60 hz Induction Motors according to IEC 60034-30-1:2014															
	IE1				IE2				IE3				IE4			
Output	Poles				Poles				Poles				Poles			
kW	2	4	6	8	2	4	6	8	2	4	6	8	2	4	6	8
0.12	57.5	62.0	48.0	36.0	59.5	64.0	50.5	40.0	62.0	66.0	64.0	59.5	66.0	70.0	68.0	64.0
0.18	62.0	66.0	52.5	40.0	64.0	68.0	55.0	46.0	65.6	69.5	67.5	64.0	70.0	74.0	72.0	68.0
0.25	64.0	68.0	57.5	50.5	68.0	70.0	59.5	52.0	69.5	73.4	71.4	68.0	74.0	77.0	75.5	72.0
0.37	70.0	70.0	62.0	57.5	72.0	72.0	64.0	58.0	73.4	78.2	75.3	72.0	77.0	81.5	78.5	75.5
0.55	72.0	74.0	66.0	59.5	74.0	75.5	68.0	62.0	76.8	81.1	81.7	74.0	80.0	84.0	82.5	77.0
0.75	74.0	77.0	72.0	64.0	75.5	78.0	73.0	66.0	77.0	83.5	82.5	75.5	82.5	85.5	84.0	78.5
1.1	78.5	79.0	75.0	73.5	82.5	84.0	85.5	75.5	84.0	86.5	87.5	78.5	85.5	87.5	88.5	81.5
1.5	81.0	81.5	77.8	77.0	84.0	84.0	86.5	82.5	85.5	86.5	88.5	84.0	86.5	88.5	89.5	85.5
2.2	81.5	83.0	78.5	78.0	85.5	87.5	87.5	84.0	86.5	89.5	89.5	85.5	88.5	91.0	90.2	87.5
3.7	84.5	85.0	83.5	80.0	87.5	87.5	87.5	85.5	88.5	89.5	89.5	86.5	89.5	91.0	90.2	88.5
5.5	86.0	87.0	85.0	84.0	88.5	89.5	89.5	85.5	89.5	91.7	91.0	86.5	90.2	92.4	91.7	88.5
7.5	87.5	87.5	86.0	85.0	89.5	89.5	89.5	88.5	90.2	91.7	91.0	89.5	91.7	92.4	92.4	91.0
11	87.5	88.5	89.0	87.5	90.2	91.0	90.2	88.5	91.0	92.4	91.7	89.5	92.4	93.6	93.0	91.0
15	88.5	89.5	89.5	88.5	90.2	91.0	90.2	89.5	91.0	93.0	91.7	90.2	92.4	94.1	93.0	91.7
18.5	89.5	90.5	90.2	88.5	91.0	92.4	91.7	89.5	91.7	93.6	93.0	90.2	93.0	94.5	94.1	91.7

22	89.5	91.0	91.0	90.2	91.0	92.4	91.7	91.0	91.7	93.6	93.0	91.7	93.0	94.5	94.1	93.0
30	90.2	91.7	91.7	90.2	91.7	93.0	93.0	91.0	92.4	94.1	94.1	91.7	93.6	95.0	95.0	93.0
37	91.5	92.4	91.7	91.0	92.4	93.0	93.0	91.7	93.0	94.5	94.1	92.4	94.1	95.4	95.0	93.6
45	91.7	93.0	91.7	91.0	93.0	93.6	93.6	91.7	93.6	95.0	94.5	92.4	94.5	95.4	95.4	93.6
55	92.4	93.0	92.1	91.5	93.0	94.1	93.6	93.0	93.6	95.4	94.5	93.6	94.5	95.8	95.4	94.5
75	93.0	93.2	93.0	92.0	93.6	94.5	94.1	93.0	94.1	95.4	95.0	93.6	95.0	96.2	95.8	94.5
90	93.0	93.2	93.0	92.5	94.5	94.5	94.1	93.6	95.0	95.4	95.0	94.1	95.4	96.2	95.8	95.0
110	93.0	93.5	94.1	92.5	94.5	95.0	95.0	93.6	95.0	95.8	95.8	94.1	95.4	96.2	96.2	95.0
150	94.1	94.5	94.1	92.5	95.0	95.0	95.0	93.6	95.4	96.2	95.8	94.5	95.8	96.5	96.2	95.4
185	94.1	94.5	94.1	92.5	95.4	95.4	95.0	93.6	95.8	96.2	95.8	95.0	96.2	96.5	96.2	95.4
220–1000	94.1	94.5	94.1	92.5	95.4	95.4	95.0	93.6	95.8	96.2	95.8	95.0	96.2	96.8	96.5	95.4

where f is the frequency set in the inverter and f_n is the nominal frequency of the motor (50 Hz or 60 Hz, depending on the country).

If your plant has stirrers controlled by inverters, then optimisation of the stirring power requires just adjusting their frequency to obtain an UP within the optimum range.

2. **On/off control**: In this case, the agitation will be intermittent. The formula for calculating P_m includes a correction factor representing the average mechanical power over an arbitrary period t:

$$P_m = \eta_{el}\, \eta_{mec}\, P_{el}\, k,$$

where $k = t_{on} / t$ and $t = t_{on} + t_{off}$.

If your plant has stirrers controlled by timers, then optimisation of the stirring power requires calculating k for $6 < UP < 8$ and adjusting the timers accordingly.

Important: starting an electric motor by just switching it on produces a peak of current, in the order of four to six times the nominal current. Such a current will produce a lot of heat by ohmic losses in the motor's windings, and a certain time is necessary to dissipate it during the normal operation. Very short on/off cycles may lead to overheating the motor, reducing its service life. It is then advisable to choose t arbitrarily and check that the resulting $t_{off} > 4$ hours, to give the motor enough time to cool down. In tropical countries, $t_{off} > 6$ hours may be enough to stay on the safe side.

The following case studies show the stirrer optimisation procedure in detail.

CASE STUDY 2.1: DIGESTER EQUIPPED WITH A VARIABLE-SPEED STIRRER

The plant was built in Italy in 2017 with the scope of digesting cattle and pig slurry. The digester has a gross volume of 1,200 m^3 and a nominal digestion volume of 1,000 m^3. The stirrer is a paddle wheel driven by a two-pole motor of 7.5 kW, with a reduction gear and speed controlled by an inverter (Figure 2.7).

$\eta_{el} = 88.1\%$ (IE class 2, from Table 2.1 in the EU Regulation 2019/1781, or estimated from the corresponding curve in Figure 2.7)

$\eta_{mec} = 95\%$
$P_{el} = 7.5 \text{ kW} = 7{,}500 \text{ W}$
$UP = 7{,}500 \times 0.881 \times 0.95 / 1{,}000 = 6.3 \text{ W/m}^3$

The stirring power is the minimum acceptable for the motor running at its nominal frequency (50 Hz) and the digester at full capacity. Should this stirring be insufficient (for instance, if a floating mat forms after some operation time), the possible alternatives are as follows:

FIGURE 2.7 Paddle-Wheel Stirrer in a Digester under Construction.

Photo by the author.

a. Increase the frequency. Electric motors can admit up to 10% permanent overload without damage. Since the output power is directly proportional to frequency, setting the inverter at 55 Hz will increase the power to 8.25 kW. With this setting, the UP will be 6.9 kW.
b. Reduce the water level of the digester, i.e., the volume of sludge. Keeping the motor at 50 Hz and reducing the volume by 10% have the same effect of increasing the agitation power. This solution saves energy but implies a slight reduction in the HRT from 35 days to 31 days. This is not a problem because 30 days of HRT is enough to digest cattle and pig slurry.

CASE STUDY 2.2: DIGESTER EQUIPPED WITH THREE SINGLE-SPEED STIRRERS, A CIRCULATION PUMP, AND AN ON/OFF CONTROL

The digester runs on cattle slurry and small quantities of straw and other vegetal residual biomass. The digestion volume is 700 m^3. The stirrers are immersed direct-drive propellers and have power of 5 kW and 11 kW, respectively, with no inverter control. A chopper pump is immersed in the digester and has 8 kW power. From the manufacturer's catalogue, with a 6-m (estimated) head, the flow is 115 m^3/h. The plant was built in Italy in 2015, and the IE class of the motors is 3.

Considering the stirrers alone:
$\eta_{el.1} = 89.2\%$ and $\eta_{el.2} = 91.2\%$ (from the table in the already cited regulation);
$\eta_{mec} = 100\%$ because both stirrers have direct drive;

$$p_{mec} = 0.892 \times 5,000\,\text{W} + 0.912 \times 11,000\,\text{W} = 14,492\,\text{W};$$

$$\text{UP} = 14,492\ \text{W} / 700\ \text{m}^3 = 20.7\ \text{W} / \text{m}^3.$$

Considering the pump alone:

$$\text{TR} = 115\ \text{m}^3 / \text{hr} / 700\ \text{m}^3 = 0.16\ \text{volumes per hour}$$

The UP is oversized, and the TR is neglectable (8% of the minimum acceptable TR of twice the digester volume in 1 hour). Controlling the stirrers and the pump with timers will provide the required average power with an on/off cycle.

We can assume turning the pump on only when the bigger stirrer is off, to avoid current peaks. The scope of the pump is chopping the suspended solids as small as possible, so pumping the digestion volume in one day should be enough (arbitrary choice, must be checked on site!). So, the pump will run 6 hours/day, and the said hours must coincide with the off cycle of the bigger stirrer.

The smaller stirrer already provides the minimum stirring when it is on.

$$\text{UP}_1 = 0.892 \times 5000 / 700 = 6.4\ \text{W/m}^3$$

Assuming that the bigger stirrer provides an average UP equal to the maximum in an on/off cycle of 1 hour, the maximum duration of the on state (t) is

$$\text{UP}_2 = 8\ \text{W/m}^3 = 0.912 \times 11,000 \times t / 700 \rightarrow t = 0.56\ \text{hour} = 33\ \text{minutes}$$

A possible stirring strategy could be the following:

Small stirrer on, bigger stirrer and pump off: 12 hours/day
Big stirrer on, pump and small stirrer off: 6 hours/day
Chopper pump on, both stirrers off: 6 hours/day

Note that the individual duty cycle of each motor is 2 hours on and 4 hours off. This ensures enough time for the motor to cool down, reducing machine wear and increasing the life of the insulation. The average daily UP is 6.9 W/m^3, which is a reasonable value.

TABLE 2.4
Possible Timer Setting Strategy for the Stirrers and the Resulting UP

Time	Stirrer 5 kW	Stirrer 11 kW	Pump	UP (W/m^3)
0:00	On	Off	Off	6.4
1:00	On	Off	Off	6.4
2:00	Off	On	Off	14.3
3:00	Off	On	Off	14.3
4:00	Off	Off	On	0
5:00	Off	Off	On	0
6:00	On	Off	Off	6.4
7:00	On	Off	Off	6.4
8:00	Off	On	Off	14.3
9:00	Off	On	Off	14.3
10:00	Off	Off	On	0
11:00	Off	Off	On	0
12:00	On	Off	Off	6.4
13:00	On	Off	Off	6.4
14:00	Off	On	Off	14.3
15:00	Off	On	Off	14.3
16:00	Off	Off	On	0
17:00	Off	Off	On	0
18:00	On	Off	Off	6.4
19:00	On	Off	Off	6.4
20:00	Off	On	Off	14.3
21:00	Off	On	Off	14.3
22:00	Off	Off	On	0
23:00	Off	Off	On	0
Average UP				6.9

The following is the personal opinion of the author about the design of this biogas plant: the efficacy of including a submerged chopper pump inside the digester and an external pump to load the digester is dubious. Since the volumetric loading rate (VLR) of this digester is about 10 m^3/day, it would have been more logical to foresee an external slurry tank of about the same volume, with a chopper pump installed in it and a bypass valve at its output. The same pump can then have a double function: when loading the tank with slurry and biomass, the bypass valve is set to recirculate the slurry; after homogenising the slurry, the bypass pump diverts the feedstock to the digester. Since the volume of the loading tank is much smaller than the nominal flow of the pump, turning it on for about 10 minutes will be more than enough for a thorough homogenisation of the feedstock mixture. Diverting the flow to the digester with the bypass valve allows loading the digester with its daily feedstock portion within 10 minutes. The pump's duty cycle is limited to only 20 minutes/day instead of 6 hours/day. Finally, homogenising the feedstock before pumping it into the digester ensures quicker digestion of the vegetal biomass.

2.3 INCREASING THE INOCULUM'S BIOLOGICAL ACTIVITY

It is generally accepted that the bottleneck of AD processes is the final acetoclastic stage. This is true for the most frequent substrates employed in industrial biogas plants, like energy crops and OFMSW, both mostly composed of polysaccharides and minor quantities of proteins and lipids. When dealing with more complex substrates, such as soapy wastewater or residual soap paste, or with industrial effluents containing potentially inhibiting substances, there are two possibilities:

1. The bottleneck of the AD process is the fermentation stage.
2. There is no bottleneck because the substrate inhibits most, or all, microbial activity.

Figure 2.8 shows the BMP test result of a tannery sludge sample. Figure 2.9 shows the control with sodium acetate of the inoculum employed for the test, sampled from the plant intended to treat the sludge. The control test demonstrates that the inoculum has normal methanogenic activity. The BMP test shows that the sludge's digestion is quick – so there is no bottleneck – but after the 10th day, the curve's descending portion indicates some inhibition of the methanogenesis.

There could be several reasons for the inhibition after a quick digestion, and there is no way to determine them with a single test. For instance:

a. The sludge's digestion releases H_2S or NH_3, which partly inhibits the microbial flora.
b. Some trace element was consumed or chelated during the sludge's digestion, lowering the inoculum's SMA.
c. The sludge contains two diverse qualities or fractions of organic matter: one is easily digestible and the other is more complex, perhaps producing propionic acid.

The strategies to ensure AD treatment of a waste stream with such features without compromising the stability of the digestion plant and its productivity are several, and all of them should be checked in the laboratory before making a decision:

a. Repeat the test with fresh inoculum samples taken from the plant plus a supplement of trace elements solution. This will allow us to decide if there is something in the first 10 days of sludge digestion that is somehow depleting

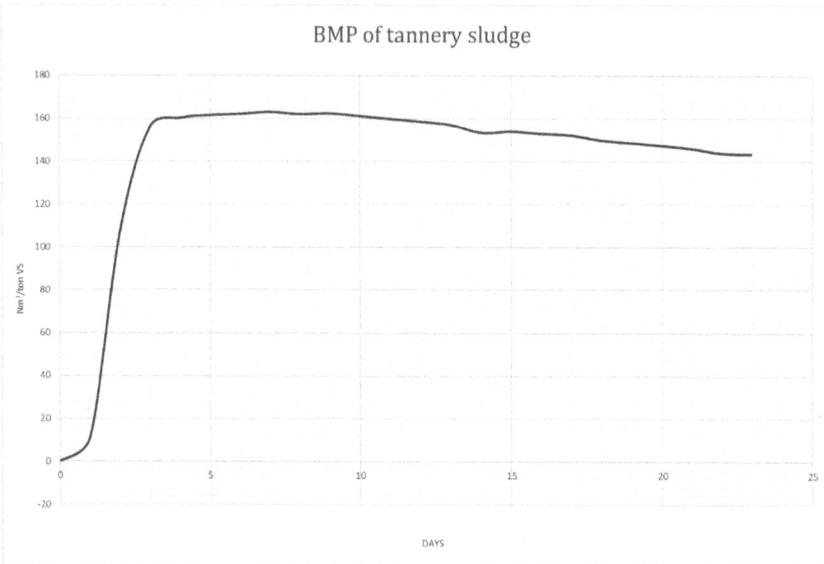

FIGURE 2.8 Negative Slope after the 10th Day of the Specific Methane Production (SMP) Curve of Tannery Sludge Demonstrates That Such a Substrate Causes the Inoculum's Inhibition.

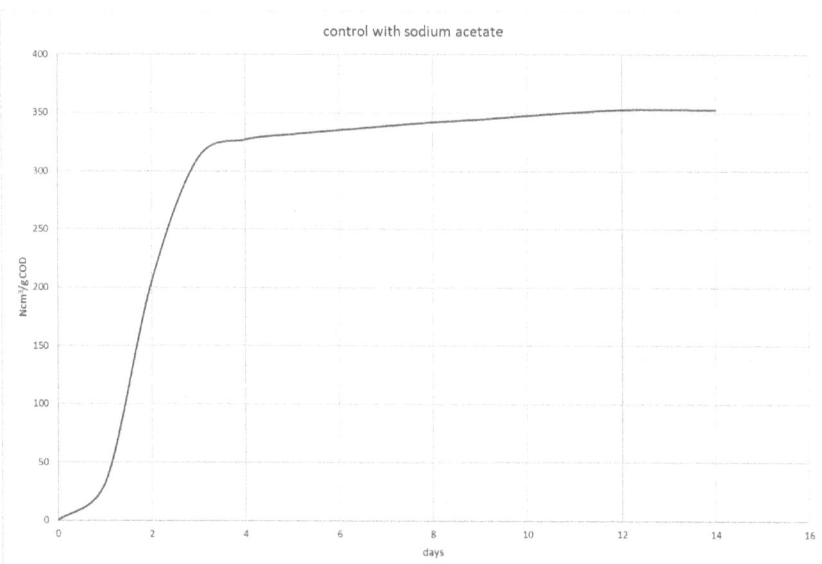

FIGURE 2.9 Normal Sigmoid Shape of the SMP Curve of the Control Reactor (Sodium Acetate) Demonstrates That the Inoculum Has Adequate Methanogenic Activity.

trace elements and compromising the SMA of an initially good inoculum. For instance, a high rate of H_2S production could precipitate Ni, Co, and other trace elements, making them unavailable for the archaea.

b. Mix the content of all the reactors employed in the first run and repeat the test using this mixture as the inoculum. This will allow us to check if the microbial population has somehow adapted to digest the sludge. If the descending portion of the BMP curve disappears, then it means that the digester can be 'accustomed' to digest tannery sludge with time. The practical solution is to start feeding the plant with small quantities of tannery sludge at the beginning and gradually increase the dose over time. It is advisable to test different feeding strategies in the lab by conducting continuous digestion tests with 2- to 10-litre reactors, choosing the most suitable one.
c. Sample fresh inoculum from the plant and mix it with variable percentages of digestate from another plant treating different kinds of waste. This will increase microbial biodiversity and the chances that no inhibition arises. Alternatively, adding mixtures of cattle, swine, and chicken manure to the plant's inoculum could produce the same result.
d. Some companies offer mixtures of lyophilised bacteria, sometimes mixed with enzymes, specifically selected for treating 'difficult' substrates. Since such products are usually expensive, testing their efficacy for a given situation in the laboratory is always the best choice.
e. Adding different percentages of active carbon to the inoculum may adsorb eventual H_2S or NH_3 produced during the AD of tannery sludge.
f. Try co-digesting the sludge with other substrates, ensuring that the C/N ratio of the resulting mixture is as close to 30 as possible.

Some industrial waste streams will always cause inhibition, regardless of the robustness of the microbial flora composing the inoculum, because they contain substances that are biocides, i.e., they will kill most of the microbes. Examples of such substances are

- sludge from macadamia nut processing (contains anacardic acid, a biocide and irritating agent 10)
- residues of *cannabis* (contain several terpenes, in general inhibitors or even bactericides)
- effluents of the pharmaceutical industry (may contain antibiotics or disinfectant agents)
- waste labelled as 'industrial sludge' without any further specification (may contain heavy metals, sodium hypochlorite, sodium chloride, sulphuric acid, hydrocarbons, etc.)

Examples of organic waste that may partially inhibit the inoculum when the latter is not 'accustomed' to digest it are

- olive oil pomace and wastewater
- soap paste and soapy water from the vegetable oil refining industry

- sludge from the vegetable oil refining industry
- slaughterhouse waste (especially in digesters that are usually fed with only corn silage and cattle manure)
- bone and meat meal
- jatropha cake. Figures 2.10 and 2.11 show a BMP test made with inoculum from an agricultural biogas plant usually fed with corn silage and cow manure on two jatropha samples. The first sample (Figure 2.10) was jatropha whole meal, i.e., containing oil. The second sample was jatropha cake, i.e., the meal after oil extraction. The whole meal was expected to reach a higher BMP because oils and fats have a theoretical yield of about 1,000 Ncm^3/g. In practice, the presence of oil partially inhibited the inoculum. Both samples achieved substantially the same BMP. In this case, the strategies to increase the productivity of this substrate are a gradual adaptation of the plant's inoculum to it, inoculation with sludge taken from a plant that already digests fatty substrates, or employing some commercial additive or enzyme. In any of the three cases, checking the validity of the hypothesis in the laboratory is mandatory.

Finally, increasing the operation temperature may – at least theoretically – improve the digestion of complex substrates. It is known that at 30°C–55°C, the biological activity of anaerobic microbes is a direct function of the incubation temperature. In some cases, like the digestion of slaughterhouse waste, choosing a thermophilic

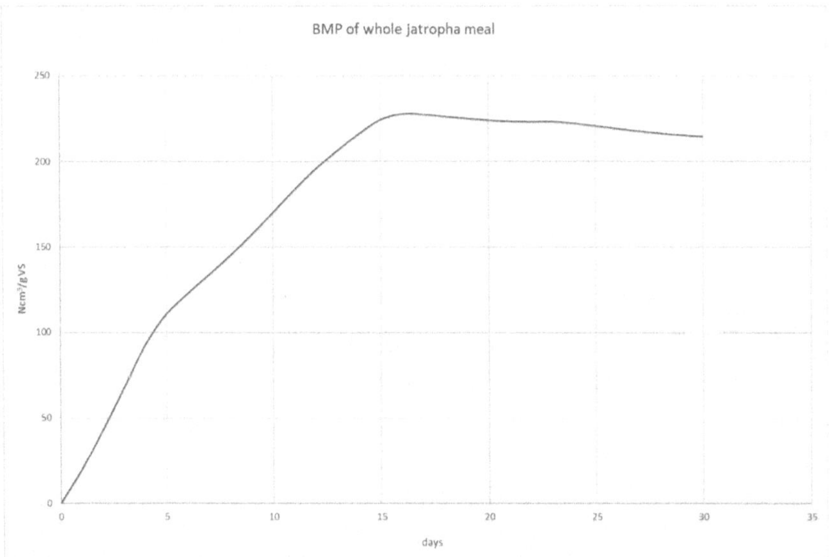

FIGURE 2.10 Negative Slope after the 16th Day in the SMP Curve of Jatropha Seed Digestion Indicates That This Substrate Causes Partial Inhibition of the Inoculum.

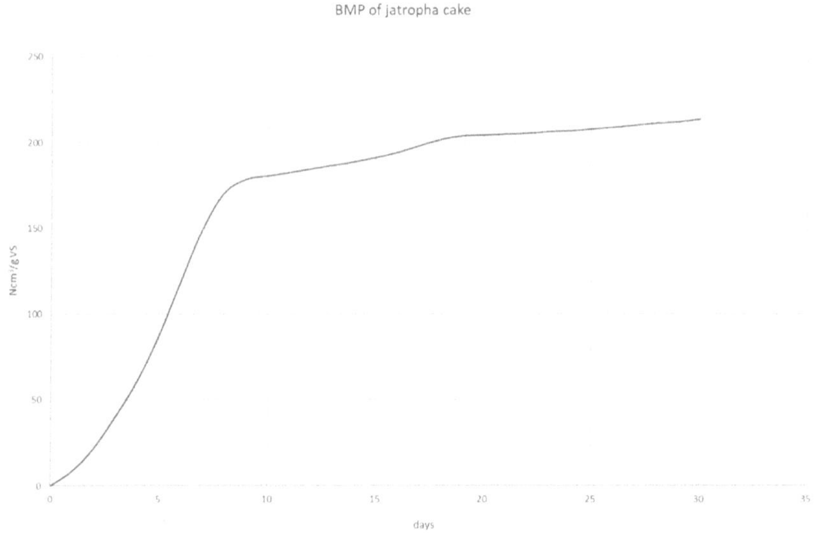

FIGURE 2.11 SMP Curve of the Same Jatropha Seed Sample. On repeating the test with the inoculum employed in the former test, the slope is always positive. This demonstrates that the bacterial flora adapted to digest this substrate despite its toxicity.

process may not necessarily improve the digestion rate but has an additional advantage from the sanitary compliance point of view in the specific case of European anaerobic plants (see 11). In any case, converting a mesophilic AD plant into a thermophilic one requires gradual heating. It is impossible to foresee the optimum heating rate, as it is limited by the available thermal power, the insulation thickness, the external temperature, and the efficiency of the heat exchangers. A dynamic test with 2- to 10-litre continuous reactors is highly advisable to find out the maximum heating rate that prevents process disruption and the optimum operation temperature.

2.4 INCREASING THE DIGESTIBILITY OF THE FEEDSTOCK: EXERGY ANALYSIS OF MECHANICAL AND THERMAL PRETREATMENTS

The scientific literature contains many examples of wrong evaluations of the efficiency of pre- and post-treatment processes. The most frequent is performing an 'energy and mass balance', i.e., applying the first law of thermodynamics. Such an analysis is incomplete because it is necessary to consider the second law too, which is at the base of the concept of exergy. The academic definition of exergy is as follows:

> The exergy of a system is the maximum useful work possible during a process that brings the system into equilibrium with a heat reservoir, reaching maximum entropy.

In simple words. exergy is the useful fraction of a given quantity of energy, where 'useful fraction' means mechanical work or its equivalent, electricity. For simplicity, this section will not include the theory of exergy analysis, information that the reader can find in any book on technical thermodynamics. Because many biogas plant managers and investors are not necessarily familiar with thermodynamics and energy auditing, we will employ an accounting analogy to explain why many papers in the literature contain wrong or incomplete efficiency analyses. Let us consider the different forms of energy (heat, mechanical work, electricity, etc.) as currencies. The first law of thermodynamics states the possibility of converting energy from one form to the other by employing machines that perform thermodynamical transformations. Likewise, currencies are freely convertible at a bank or change agency. The second law of thermodynamics states the maximum (theoretical) efficiency in converting one form of energy into another. The most famous application of the second law is Carnot's theorem, which defines the maximum mechanical work that one could get from heat between two given temperatures. In our analogy, Carnot's theorem defines the exchange rate between the currencies 'thermal energy' and 'mechanical work'. On converting one currency to another, the bank or change agency applies a commission, so each transaction has a cost. In the physical world, each conversion of energy from one form to another always has irreversibilities (e.g., friction, electric resistance, and viscosity) that reduce the theoretical conversion efficiency. Every transformation will always lose a fraction of the initial energy. Likewise, part of the initial capital in a given currency will be lost as a change commission when converting it to another currency.

Imagine a process that employs high-temperature heat, electricity, mechanical work, and low-temperature heat. Analysing the overall efficiency requires giving each form of energy its actual value. In our accounting analogy, such a process is like a company that buys raw materials in a country, produces goods, and sells them in other ones, so it has costs and incomes in several currencies. Making the profit and loss balance requires converting all the cash flows into a reference currency. In our accounting analogy, electricity is the equivalent of euros; mechanical work is like dollars; high-temperature heat is Swiss francs; and low-temperature heat is rupees. The said comparisons are neither arbitrary nor casual: the exchange rates at the moment of writing this chapter (April 2022) reflect the thermodynamic equivalences between energy forms. How does methane appear in our analogy? Regardless of being fossil or 'bio', methane is an energy vector, i.e., a given quantity of chemical energy 'packed' in a unit of mass or volume. In real life, we can store or transport methane and employ it to produce high-temperature heat when and where we need it. In our accounting analogy, 1 m^3 of methane is the equivalent of a pre-paid credit card in Swiss francs.

The following case studies show how to employ the analogy of energy accounting for a more complete analysis of potential process improvements.

2.4.1 Redigestion of Digestate

A possible strategy for improving the efficiency of feedstock degradation is treating the digestate and redigesting it. A team of Swedish researchers [12] sampled digestate from ten biogas plants treating different kinds of waste. Then they divided each sample into four parts, treated three, measured the BMP of each treated sample, and compared it with the BMP of the untreated one. The thermal treatments were incubation at 55°C for 24 hours, pasteurisation at 70°C for 1 hour, and thermal hydrolysis process (THP), i.e., 20 minutes at 165°C and 8 bar. One would expect the THP to give the best results because its severe conditions break complex organic matter into smaller molecules that are easier to digest. Counterintuitively, the researchers found that milder pasteurisation gave similar or slightly better results. Following this monumental and extremely accurate study (10 digestate samples × 4 tests/sample × 3 replicates/test = 120 reactors = 8 months of work with an AMPTS II at full capacity!), the researchers compared the additional energy derived from the increased BMP to the energy consumed in each thermal treatment. For simplicity, let us concentrate on the most promising case of pasteurisation.

a. The thermal energy required to heat 1 m^3 of digestate from 38°C (the digester's temperature) to 70°C is 38.5 kWh (true).
b. The additional quantity of methane that further digestion of the treated digestate will yield is 0.6 Nm^3/ton w.w. (true, resulting from the tests). The lower heating value (LHV) of methane is 9.94 kWh/Nm^3, so the additional (thermal) energy that the digestate pasteurisation can yield is 5.96 kWh (true).
c. Since the energy balance is 5.96 kWh–38.5 kWh = –32.54 kWh, the researchers conclude that the thermal treatment and redigestion of digestate are not energetically convenient.

The last statement is an example of how one can come to false conclusions from true hypotheses or facts. The basic principle of logic – the syllogism – states that basing reasoning on facts is a necessary but insufficient condition for deriving a true conclusion. The logical flaw in the researchers' reasoning is that they did not consider the thermodynamic equivalences, as explained in the next paragraphs.

Let us reformulate the energy balance using our accounting analogy. Is spending 38.5 Indian rupees for a 6.1 Swiss francs cheque a good deal? In the financial world, it would be the best of the deals (when writing these lines, 38.5 rupees is equivalent to 0.47 francs). In the physical world, the quantity of electrical energy that is (theoretically!) possible to generate from 1 m^3 of water at 70°C is about 0.07 kWh. In practice, we all know that recovering electricity from hot water is almost impossible. On the other hand, burning 0.6 Nm^3 of CH_4 in a conventional gas generator will produce 2.4 kWh of electricity and 3.6 kWh of heat (at 80°C).

Are pasteurising and redigesting the digestate an improvement we can consider or not? It depends on the kind of biogas plant and feedstock. For instance, in a plant running on sewage sludge, each m^3 of feedstock with 3% VS w.w. will produce roughly 5 Nm^3 of CH_4. The thermal energy to treat the digestate is already there because all wastewater treatment plants (WWTPs) have a CHP group. Hence, it is possible to recover the heat required for pasteurisation to heat the digester or the post-digester. So, getting an additional 0.5 Nm^3 is an improvement worth considering. In a plant treating OFMSW, pasteurisation and redigestion of digestate would represent a marginal improvement from the energy point of view but may bring other economic benefits, such as better quality of the compost or shorter composting times. In cases where pasteurisation is mandatory for sanitary reasons (e.g., slaughterhouse waste), pasteurising the digestate before a mesophilic post-digestion step may save energy, compared to operating thermophilic digestion for 20 days (see Section 2.2.1).

2.4.2 Energy-intensive Pretreatments: Cavitation and Sonication

Cavitation is a phenomenon observed when the static pressure on a liquid suddenly falls below its vapour pressure, forming microscopic vapour bubbles (cavities). When the pressure increases, these cavities collapse (implode), generating shock waves. Such shock waves are intense enough to damage solid materials like metals in the immediate neighbourhood of the imploded bubble but rapidly weaken as they propagate away from the implosion. The rationale for treating slurry with cavitation is to break up the solids suspended in it. Suspended solids consist of two broad categories: living cells and dead biomass. Breaking apart the cellular walls of living cells will diffuse their enzymes in the aqueous medium, favouring the hydrolysis of dead biomass and turning it into soluble compounds. Disintegrating the macroscopic solids in the slurry facilitates the attack of enzymes, speeds up the anaerobic digestion, and makes it more complete. The main scope of cavitation treatments is increasing the BMP of the solid feedstock suspended in the slurry. A secondary scope is reducing the apparent viscosity of the slurry, which allegedly translates into less energy consumed for pumping and stirring. There are basically two commercial methods for inducing cavitation within a liquid: hydrodynamic cavitation and sonication. The first one employs a rotor turning at high speed within a confined space. Each manufacturer proposes a different rotor shape, but the final goal is the same for all: causing highly turbulent flows and big speed gradients that will lead to cavitation. The second method employs ultrasonic devices called sonicators. Ultrasonic waves travel along the sonication chamber and reflect against its walls, creating an interference pattern. At some points, the dynamic pressure fronts subtract or add up, causing sudden expansion and compression of the liquid. Both hydrodynamic and ultrasonic cavitators are energy-intensive machines. For instance, a cavitator employed for homogenising corn mash for pigs (PigFEED PFC-1000 https://pigfeed-cavitator.com/caratteristiche/) requires a motor with 100–300 kW rated power for treating 10 tons/hour.

There are many brands specialising in cavitators and sonicators for biogas plants, but it is difficult to evaluate the effectiveness of such technologies. All of them claim results 'validated by the university XY' (sponsored by the same manufacturer) during 'real-scale test campaigns'. In some cases, sonication or cavitation is not a pretreatment of the whole feedstock ration because the device treats a small flow of digestate and recirculates it. The rationale for this configuration is that the treated digestate will provide a kind of on-site enzymatic treatment. One of the marketing arguments is that 'the lower viscosity of the slurry saves stirring energy, offsetting the cavitator's energy consumption'. It is possible to argue that reducing the viscosity of the slurry will also facilitate the sedimentation of inert solids (a direct consequence of

Stokes law), so it will be necessary to increase the agitation power to avoid the accumulation of sediments.

The following case studies provide insights for checking the claims of equipment manufacturers and the scientific literature alike.

CASE STUDY 2.3: CAVITATION

Cavitation of silage of Virginia mallow (*Sida hermaphrodita*) for 20 minutes increased its BMP up to 439 – 45 Nl CH_4/kg total solids (TS). Nevertheless, the energy balance was negative (more energy was consumed for the cavitation than the increase in CH_4 obtained). A modest energy gain of 0.17 kWh/kg TS was obtained with just 5 minutes of cavitation [13], so the researchers concluded that the latter treatment is a viable option for commercial biogas plants. As usual, the researchers made the energy balance comparing the increase in methane production multiplied by its calorific value minus the electric energy employed for the treatment, obtaining a positive value of 0.17 kWh/kg TS. In other words, they spent euros for Swiss francs, without applying the conversion rate in their calculation. Let us do the correct balance, assuming that electrical generation from methane has 42% efficiency.

BMP of untreated sample = 340 Nml/g VS
BMP after 5 minutes of treatment = 410 Nml/g VS (the researchers declared they measured the BMP with an AMPTS but did not provide the exact BMP value, and the figure was retrieved from a bar graph).

Gain in methane production = 70 Nml/g VS
Potential gain in electricity production = 0.42×9.94 kWh/Nm^3 $\times$ 0.070 Nm^3/kg VS = 0.292 kWh/kg VS

VS = 93.2% TS $\rightarrow$ Electricity gain = $0.932 \times 0.292 = 0.272$ kWh/kg TS

Electricity consumed in the 5-minute treatment = 0.28 kWh/kg TS (the researchers do not provide information on how they calculated this figure).

So, the exergy balance of the 5-minute treatment of *Sida hermaphrodita* with the cavitator model employed in the test is almost null (slightly negative or slightly positive if we consider the measurement uncertainties).

Of course, the case study discussed above does not prove that other combinations of substrates and cavitator models will give better or worse results, but it shows how the reader should proceed before adopting the cavitation technology in his plant.

CASE STUDY 2.4: SONICATION

Sonication is mostly employed for wastewater sludge digestion, especially with viscous sludge. The efficacy of the process depends on the characteristics of both the sonicator and the sludge. In our case study, the researchers found a 25% increase in BMP with an energy consumption of 300 kJ/kg TS [14]. Sonication reduced the digestion time to 5 days and the viscosity of the sludge by 60%.

Is such an increase in BMP worth the energy consumed for the sonication?
1 kg TS = 0.77 kg VS(N.B.: the author's data on sludge. This figure is not defined in the article)

300 kJ of electricity consumed by the sonicator = 0.083 kWh
BMP of untreated sludge= 35 Nl/kg VS = 27 Nl/kg TS
BMP of treated sludge= 44 Nl/kg VS = 34 Nl/kg TS
Difference = 7 Nl of CH_4 /kg TS
Potential gain in electricity generation = 0.007 Nm^3 / kg TS × 9.94 kWh/ Nm^3 × 0.42 = 0.029 kWh/kg TS
Exergy balance = 0.029 – 0.083 = –0.054 kWh/kg TS

Sonication consumes more electricity than the one it produces. In the specific case of a sewage treatment plant, the potential advantage of sonication could be operational rather than energetic. Reducing the digestion time by 5 days allows for increasing the daily treatment volume of sludge without increasing the digester's size. Dewatering digestate requires energy for a centrifuge. Allegedly, sonicated digestate requires less energy for centrifugation (this should be checked experimentally). Since the primary income of a WWTP is the tipping fee per cubic metre of sludge treated and the electricity production is marginal, installing a sonicator to increase the plant's capacity could be economically convenient depending on the combination of capital expenditure (CAPEX), tipping fees, and energy cost.

2.4.3 Thermal Pretreatments: Pressure Cooking and Steam Explosion

A useful parameter to compare different thermal treatments is the severity factor (SF), defined as [15]:

$$SF = log\left[t\, e^{\left(\frac{T-100}{14.75}\right)}\right],$$

where

t = time of the treatment in minutes,

T = temperature in °C,

14.75 = activation energy (minimum energy to start a chemical reaction).

In general, BMP grows with the SF, up to about 4. Exceeding 4, the BMP drops because of the formation of inhibitors.

Steam explosion (aka pressure swing conditioning; 16) is a thermal process that has two effects on biomass: mechanical (pulverisation of the fibres) and chemical (hydrolysis). Pressure cooking causes mostly hydrolysis and has little effect on the biomass fibres. Allegedly, steam explosion provides the best results in terms of digestibility because of its double effect. In the case of the paper cited above, the highest BMP from fresh vegetal biomass (water hyacinth, *Eichhornia crassipes*) resulted from the thermal treatment at 170 °C for 60 minutes (SF = 3.8). For the said SF, steam explosion yielded 347 Nm^3/ton VS and pressure cooking 295 Nm^3/ton VS. The BMP of untreated biomass was 252 Nm^3/ton VS.

Is it worth pretreating the feedstock with any of the two methods? The answer depends on the border conditions:

1. The minimum quantity of water to heat, or heat and vaporise, required to treat a unit mass of substrate.
2. The thermal losses of the pressure or steam cooker during the time required to complete the treatment.
3. The fraction of heat eventually recovered at the end of the process.

The following example shows the calculation procedure for the conditions reported in the case study.

Arbitrary (optimistic) assumptions:

1. Treating 1 kg VS of water hyacinth requires adding 0.5 kg of water.
2. The initial temperature of the process water is 70°C.
3. It is possible to recover heat at the end of the process, so the final temperature of the effluent will be 90°C.
4. The treatment equipment is thermally insulated, and its thermal loss during the processing time is just 10% of the total energy.
5. The specific heat of the VS is neglectable compared to that of water.
6. The moisture of fresh water hyacinth is 90%.

From thermodynamic tables, the enthalpies of pressurised water and saturated vapour at different temperatures are as follows:

$$h_{w\ 170^\circ C} = 718.1\ \text{kJ/kg}$$

$$h_{V\ 170^\circ C} = 2.768.1\ \text{kJ/kg}$$

$$h_{w\ 90^\circ C} = 376.9\ \text{kJ/kg}$$

$$h_{w\ 70^\circ C} = 284.7 \text{ kJ/kg}$$

The total mass of water to heat, or heat and vaporise, is 1.4 kg (0.5 kg of process water plus the moisture of biomass, 0.9 kg).
LHV_{CH4} = 9.94 kWh/Nm^3 = 35,784 kJ/Nm^3
First case: pressure cooking

$$Q_{input} = 1.1 \text{ x } 1.4 \text{ kg } (718.1 - 284.7) = 667.4 \text{ kJ/kg VS}$$
$$Q_{recovered} = 1.4 \text{ kg} \times (718.1 - 376.9) = 477.7 \text{ kJ/kg VS}$$
$$Q_{loss} = 667.4 - 477.7 = 189.7 \text{ kJ/kg VS}$$

The increase in BMP is
I_{BMP} = 0.292 – 0.255 = 0.037 Nm^3/kg VS
Supposing to cover the heat losses with part of the production and that the heat generator is a boiler having 90% efficiency, the self-consumption of methane is
SC = 189.7 kJ/kg VS/(0.9 x 35,784 kJ/Nm^3) = 0.006 Nm^3/kg VS.
The net balance of methane production is then

$$B = 0.037 - 0.006 = 0.031 \text{ Nm}^3\text{/kg VS.}$$

Second case: steam explosion

$$Q_{input} = 1.1 \text{ x } 1.4 \text{ kg } (2{,}768.1 - 284.7) = 3{,}824.4 \text{ kJ/kg VS}$$
$$Q_{recovered} = 1.4 \text{ kg} \times (2{,}768.1 - 376.9) = 3{,}347.7 \text{ kJ/kg VS}$$
$$Q_{loss} = 3{,}824.4 - 3{,}347.7 = 476.7 \text{ kJ/kg VS}$$

The increase in BMP is

$$I_{BMP} = 0.347 - 0.255 = 0.092 \text{ Nm}^3\text{/kg VS}$$

Supposing to cover the heat losses with part of the production and that the steam generator is a boiler having 90% efficiency, the self-consumption of methane is

$$\text{SC} = 476.7 \text{ kJ/kg VS/}(0.9 \times 35{,}784 \text{ kJ/Nm}^3) = 0.015 \text{ Nm}^3\text{/kg VS.}$$

The net balance of methane production is then

$$B = 0.092 - 0.015 = 0.077 \text{ Nm}^3\text{/kg VS.}$$

2.5 CONCLUSIONS

High-temperature thermal treatments are potentially interesting for increasing a biogas plant's productivity because they involve only heat transfer. In our financial analogy, they are like investing a certain amount of Swiss francs and having a return on investment in the same currency. On the other hand, the balance can be positive or negative depending on many variables: the feedstock, the SF, the initial temperature of the feedstock + water mixture, the quantity of water required per mass unit of feedstock, the moisture content of the feedstock, the thermal loss during the process, the percentage of heat recovered after the process, and how the necessary heat is produced.

The success of a project involving pressure cooking or steam explosion requires thorough research in the laboratory and depends on the accuracy of the experimental data and the validity of the assumptions. It is highly advisable to perform a sensitivity analysis to estimate the influence each variable has on the result.

2.5.1 Pasteurisation and Cooking of the Feedstock

Investing in rupees and cashing revenues in Swiss francs or dollars is the dream of every magnate. Is this possible in the AD analogy? We saw in Section 2.4.1 that a mild thermal treatment of the digestate kills part of the living unicellular biomass, allowing its redigestion. We all know that cooked food is easier to digest than raw one. Does the same principle apply to AD and feedstock? The answer is not univocal because low-temperature thermal treatments have different effects depending on the substrate. The EU legislation mandates that some potentially hazardous feedstocks (e.g., slaughterhouse waste) must undergo pasteurisation at 70°C for 1 hour (i.e., SF = 0.89) for sanitary reasons. Pasteurisation slightly increases the BMP of slaughterhouse waste, WWTP sludge, and OFMSW but has little or no effect on the BMP of vegetal matter, cattle slurry, and other kinds of waste [17].

Treating high-solid swine manure at 70°C for 5 hours increased the methane production rate by 40% but had little influence on the BMP [18].

CHP groups are present in most biogas plants, but an important fraction of the available heat finds no use and is simply dissipated. The main advantage of pretreating the feedstock at 70°C–80°C is then recovering heat for some useful purpose instead of wasting it. Many CHP manufacturers offer the possibility of recovering low-pressure steam or overheated water from the exhaust gases. In general, pretreatment temperatures up to 120°C (SF > 2) produce higher BMP than untreated or pasteurised feedstock (SF < 1). Before deciding to invest in a heat exchanger for recovering low-pressure steam or overheated water from the exhaust gas, it is mandatory to conduct a campaign of BMP tests at different SFs. Once the most suitable pretreatment condition and the corresponding increase in BMP are known, the analysis method is the same as already described in the former section for the case of pressure cooking.

2.6 INCREASING THE DIGESTIBILITY OF THE FEEDSTOCK WITH ENZYMATIC TREATMENTS

Treating feedstocks with enzymes is a widely diffused technique, but, in the author's experience, the cost seldom justifies the benefits. The reasons are as follows:

1. Adding cellulase (the cheapest and most widely diffused product) only speeds up the process but does not increase the BMP. The reason is

evident to anybody who has ever performed a BMP test of cellulose: anaerobic inocula always contain hydrolytic bacteria capable of hydrolysing complex polysaccharides. Inocula derived from ruminant manure (bovines, sheep, goats, and camels) are generally more efficient in digesting cellulose than those derived from monogastric animals (swine, poultry, and humans). Hence, sourcing cow manure to inoculate a digester presenting low hydrolytic activity will always be cheaper than purchasing cellulase.

Massive additions of cellulase can be beneficial under emergency conditions such as the formation of a floating fibre mat in the digester's surface (for instance, when the stirrers are broken or the digester's diet is too rich in fibres with the ability to float, e.g., sorghum silage and straw) and/or the presence of a thick sediment layer that reduces the HRT.

2. In theory, adding lignin-modifying enzymes (LMEs) will increase the BMP because the indigestible component of vegetal organic matter (usually between 10% and 40%) will become digestible. LMEs are various types of enzymes produced by fungi and bacteria that catalyse the breakdown of lignin. Ligninases and lignases are older names for the same class, with 'lignin-modifying enzymes' currently preferred because the mechanism of these enzymes is not hydrolytic but rather oxidative (electron withdrawing). LMEs include peroxidases, such as lignin peroxidase, manganese peroxidase, versatile peroxidase, and many phenoloxidases of the laccase type. It can be argued that some of the products of lignin degradation by LMEs are humic substances which are potential inhibitors [19]. Research on this matter is contradictory and still subject to debate. In any case, pretreating the feedstock with LMEs surely represents a cost, but there is some probability that the digester's performance will worsen instead of improving. Laboratory tests with inoculum and feedstock samples from the plant are mandatory before purchasing such products.
3. Adding lipolytic enzymes may prove beneficial in AD plants treating waste streams rich in soap, fat, or oil, especially because the two latter are insoluble in water, making it difficult for the bacteria to hydrolyse them. For instance, stearin – a by-product of palm oil refining – is solid below 80°C, so it will float on the digester's surface, and its digestion may take more than 60 days. When dealing with fatty substrates, it may prove convenient to measure the BMP of the raw feedstock, the soap resulting from a pretreatment with NaOH solution, and the feedstock pretreated with commercial lipase. Raw bile and pancreatic juice from slaughterhouse waste are natural sources of lipases, but (at least in Europe, see ref. 11 already cited) their use could cause some legal issues, not to mention that slaughterhouses will probably not take any responsibility for separating pancreases and gall bladders. Commercial lipases, on the other hand, are expensive.

CASE STUDY 2.5

Within a research study that the author carried out together with subjects from academia and the biogas industry [20], it was found that the average BMP values of enzyme-treated samples were systematically higher than those of untreated samples. Nevertheless, this does not mean that such differences are significant. The BMP of the enzyme-treated corn silage was 34% higher than that of the untreated one, but the dispersion margin of the triplicate test was of the same order of magnitude. Hence, the difference is not statistically significant. In other words, it is not possible to conclude that the higher (average) BMP of the treated samples is the result of the enzyme's effect: it could be just the result of random errors adding (property related to a particular sample). On the other hand, the comparison of the average measured BMP with the maximum value calculated with the theoretical formula is well below the theoretical limit, which indicates that there were no errors in the tests. Hence, treatment with the enzyme indeed increases the anaerobic degradability of the silage. The calculated 34% increase is then an acceptable result. In the case of the digestate, the difference between treated and untreated samples was 48%, while the standard deviation was 12%. Hence, in the worst case, it is possible to assume that the enzymatic treatment increased the BMP of the digestate, by at least 36%. Whether the proposed treatment makes sense from the economic point of view is still to be proved. Apart from the enzyme's cost, one should invest in building an additional tank in the biogas plant, with enough capacity for at least 48 hours of treatment, plus the equipment for shredding, stirring, and pumping the treated slurry, which will consume additional energy.

2.7 INCREASING THE DIGESTIBILITY OF THE FEEDSTOCK WITH BIOLOGICAL TREATMENTS

A soft approach to improve the digestibility of fibre-rich vegetal feedstock could be the cultivation of oyster mushrooms (*Pleurotus ostreatus*) or any other edible mushroom of the genus *Pleurotus*, which possesses the ability to grow on lignin-rich substrates. The process would exploit the eventual excess residual heat available from the CHP for the sterilisation and growth chambers, and the solid fraction of digestate, which is naturally rich in lignin, as the cultivation substrate. Other mushrooms belonging to the genus *Fusarium* and *Trichoderma* can produce even higher quantities of LMEs than *Pleurotus* but present some drawbacks: they are potential plant pathogens, and their cultivation could be subject to strict restrictions.

The rationale for growing fungi on fibrous feedstock is that fungal biomass is made up of water, chitin – a complex but fully digestible polysaccharide, and LMEs. So, adding the shredded fungal biomass and its juice to fibrous feedstock will increase the overall methane production. In practice, such reasoning is too simplistic, and the production of methane may show no increase, or even decrease, compared to untreated feedstock. The two possible cases are

1. Mushrooms grown on the feedstock (pretreatment). Fungi are aerobic microorganisms and not 100% selective, so they will consume not only part of the lignin but also part of the cellulose, lowering the overall BMP of the feedstock.
2. Mushrooms grown on solid digestate fraction (posttreatment). In this case, the cellulose fraction has already been converted into methane, so any fungal mass obtained from an indigestible lignin-rich substrate will produce additional methane when added to the usual feedstock.

In both cases, there is a potential undesired side effect: as already stated above, degradation of lignin leaves humic substances as residue that could eventually inhibit the digestion process.

Rather than cultivating fungi to increase methane production, focusing only on edible species and harvesting them as food may be more profitable (see Chapter 8, Section 8.4).

BIBLIOGRAPHY

[1] Veronica Moset, Morten Poulsen, Radziah Wahid, Ole Højberg & Henrik Bjarne Møller, Mesophilic versus thermophilic anaerobic digestion of cattle manure: Methane productivity and microbial ecology, *Microbial Biotechnology*, 8:5, 787–800 (2015). https://doi.org/10.1111/1751–7915.12271

[2] Italian Standard UNI CEI 70098–3:2016, *Incertezza di misura – Parte 3: Guida all'espressione dell'incertezza di misura.* http://store.uni.com/catalogo/index.php/uni-cei-70098-3-2016

[3] ISO/IEC GUIDE 98–1:2009, *Uncertainty of measurement – Part 1: Introduction to the expression of uncertainty in measurement.* https://www.iso.org/standard/46383.html

[4] Piero Mattirolo, *Biometano da rifiuti*, Editoriale Delfino (2021). https://editorialedelfino.it/prodotto/il-biometano-da-rifiuti/

[5] Regulation (EU) 2019/1009 of the European Parliament and of the Council of 5 June 2019 laying down rules on the making available on the market of EU fertilising products and amending Regulations (EC) No 1069/2009 and (EC) No 1107/2009 and repealing Regulation (EC) No 2003/2003. Full text here: https://eur-lex.europa.eu/legal-content/en/TXT/PDF/?uri=CELEX:32019R1009

[6] Commission Regulation (EU) No 142/2011 of 25 February 2011 implementing Regulation (EC) No 1069/2009 of the European Parliament and of the Council laying down health rules as regards animal by-products and derived products not intended for human consumption and implementing Council Directive 97/78/EC as regards certain samples and items exempt from veterinary checks at the border under that Directive (OJ L 54, 26.2.2011, p. 1). Multi-language text: https://eur-lex.europa.eu/eli/reg/2011/142/oj

[7] IEC 60034–30–1:2014, *Rotating electrical machines – Part 30–1: Efficiency classes of line operated AC motors (IE code).* https://webstore.iec.ch/publication/136

[8] Guidelines accompanying Commission (EC) Regulations n. 640/2009 implementing Directive 2005/32/EC with regard to ecodesign requirements of electric motors and No 4/2014 amending regulation EC No 640/2009. https://ec.europa.eu/energy/sites/default/files/documents/20141211_GuidelinesElectricMotors%20cover.pdf

[9] Commission Regulation (EU) 2019/1781 of 1 October 2019 laying down ecodesign requirements for electric motors and variable speed drives pursuant to Directive 2009/125/EC of the European Parliament and of the Council, amending Regulation (EC)

No 641/2009 with regard to ecodesign requirements for glandless standalone circulators and glandless circulators integrated in products and repealing Commission Regulation (EC) No 640/2009 (Text with EEA relevance). https://eur-lex.europa.eu/legal-content/EN/TXT/?uri=CELEX%3A02019R1781-20230124

[10] Pubchem page of anacardic acid, CAS 11034–77–8 https://pubchem.ncbi.nlm.nih.gov/compound/Anacardic-acid

[11] Regulation (EC) No 1774/2002 of the European Parliament and of the Council of 3 October 2002 laying down health rules concerning animal by-products not intended for human consumption, official texts in all EU languages here: https://eur-lex.europa.eu/legal-content/EN/TXT/?uri=CELEX%3A32002R1774

[12] E Nordell, A Björn, S Waern, S Shakeri Yekta, I Sundgren & J Moestedt, Thermal post-treatment of digestate in order to increase biogas production with simultaneous pasteurisation, *Journal of Biotechnology*, 344, 32–39 (2022). https://doi.org/10.1016/j.jbiotec.2021.12.007. ISSN 0168–1656.

[13] M Zieliński, P Rusanowska, A Krzywik, M Dudek, A Nowicka & M Dębowski, Application of hydrodynamic cavitation for improving methane fermentation of *Sida hermaphrodita* silage. *Energies*, 12, 526 (2019). https://doi.org/10.3390/en12030526

[14] Jochen Bandelin, Thomas Lippert, Jörg E Drewes & Konrad Koch, Assessment of sonotrode and tube reactors for ultrasonic pre-treatment of two different sewage sludge types, *Ultrasonics Sonochemistry*, 64 (2020). https://doi.org/10.1016/j.ultsonch.2020.105001

[15] Enrique Regidor Alfageme, Masaaki Fujiwara & Tatsuki Toda, Steam explosion and thermal hydrolysis for pretreating *Eichhornia crassipes* to enhance anaerobic digestion, *The 30th Annual Conference of JSMCWM* (2019). https://www.jstage.jst.go.jp/article/jsmcwm/30/0/30_551/_pdf/-char/en

[16] B Schumacher, T Zerback, H Wedwitschka, S Weinrich, J Hofmann & M Nelles, The influence of pressure-swing conditioning pre-treatment of cattle manure on methane production, *Bioengineering (Basel, Switzerland)*, 7:1, 6 (2019). https://doi.org/10.3390/bioengineering7010006

[17] Xiaojun Liu, Souli Ikbel, Mohamad Amr Chamaa, Thomas Lendormi, Claire Sabourin, Yves Lemée, Virginie Boy, Nizar Chaira, Ali Ferchichi, Pascal Morançais & Jean-Louis Lanoisellé, Effect of thermal pretreatment at 70°C for one hour (EU hygienization conditions) of various organic wastes on methane production under mesophilic anaerobic digestion. *AIMS Environmental Science*, 5, 117–129 (2018). https://doi.org/10.3934/environsci.2018.2.117

[18] Jing Wu, Yu-ying Hu, Shi-feng Wang, Zhi-ping Cao, Huai-zhi Li, Xin-Mei Fu, Kai-jun Wang & Jian-e Zuo, Effects of thermal treatment on high solid anaerobic digestion of swine manure: Enhancement assessment and kinetic analysis, *Waste Management*, 62, 69–75 (2017). https://doi.org/10.1016/j.wasman.2017.02.022. ISSN 0956–053X.

[19] Cheng Zhao, Hongbo Liu, Xuedong Zhang, Yajie Li, Mingze Shi, Fang Huang, Lu Dong, Jiaxin Wen & He Liu, Humic acids promotion or inhibition of sludge anaerobic digestion depends on their redox potentials, *Chemical Engineering Journal*, 464, 142653 (2023). https://doi.org/10.1016/j.cej.2023.142653. ISSN 1385–8947.

[20] Giuliano Degrassi, Rodrigo Hoshino, Milan Koijc, Mariela Catone, Vibha Bhardwaj & Mario A Rosato, Increased biomethane production from endoglucanase-pretreated feedstock, *Journal of Research in Environmental Science and Toxicology*, 9:1, 1–8 (July 2020). Free download here: https://www.interesjournals.org/articles/increased-biomethane-production-from-endoglucanasepretreated-feedstock.pdf

3 Unexpected Results from Batch Tests

How to Interpret Them

3.1 PROBLEMS THAT ARISE WHEN PERFORMING BMP TESTS

Murphy's law states: 'Anything that can go wrong will go wrong', and anaerobic digestion laboratories are not an exception. Reliable laboratory tests are necessary for making the right decisions and increasing profit. But how can we assess the reliability of the tests when the results are apparently 'abnormal'? Was there something wrong with the instrument or with the test preparation, or, even worse, was the inoculum inhibited or did the feedstock inhibit it? After launching a test, one or more reactors may produce no gas at all, suddenly stop their gas production a few days after the start, or show erratic gas production. These types of troubles can have four different causes:

1. Problem with the instrument used to measure gas production. This one is seldom the cause of unexpected results, but people tend to always blame the instrument first.
2. Gas leaks from a loose or damaged seal, or a loose joint between tubes. This is the most frequent cause of fault, but at the same time, it is probably the most difficult to track.
3. Wrong experimental setup, mixing standards, adopting nonstandard test protocols, and incorrect preparation of the inoculum. These are the most frequent causes when users are either plant managers who 'urgently need the results', or researchers adopting 'creative protocols' instead of sticking to established ones.
4. A biological problem. This situation is the most relevant for a biogas plant manager since making the right decisions depends on its correct interpretation. Does the substrate inhibit the AD process? Does the inoculum possess sufficient methanogenic or fermentative activity? One can conclude that the problem has a biological origin only after being able to exclude with absolute certainty all the other causes listed above.

The present chapter shows how to check the reactors and gas flow instrument when one or more channels show no activity, or when one of the channels stops recording while its replicates continue to measure a gas flow, or when the cumulative volume curves have atypical shapes. Only when we can exclude cases 1–3 can we assume that there was a problem with the inoculum and figure out how to translate the laboratory

DOI: 10.1201/9781003609278-3

results into management decisions. In the unlikely case that the cause is a failure of the instrument, this chapter shows how to solve the problems by oneself, calling the manufacturer's customer service only when it is necessary.

3.2 FINDING AND FIXING FAULTS IN AUTOMATIC BMP MEASUREMENT INSTRUMENTS

The guidelines in Sections 3.2.2 and 3.2.3 apply to the AMPTS II and III, and, in some cases, also to Gas Endeavour II and III and Bio Reactor Simulator II and III (see disclaimer in the Preface). All the other troubleshooting procedures are based on logical deduction by trial and error, so it is possible to extend them to other brands of instruments. Electronic gas measurement systems are very reliable, so the probability that they can cause some trouble is low. On the other hand, checking their correct functioning can be done quickly and is straightforward, so we will start with it.

3.2.1 Symptom: No Activity after Some Days from the Test's Start

Proceed in the following order:

a. Open the 'Control' tab and check that the suspected faulty channel is indeed in the 'running' state (central button green). If it is in 'Pause' (right green button) or 'Stopped' (left blue button), it means that you forgot to start the test, or you or somebody else stopped it accidentally after the start.
b. Open the 'Graphs' screen. Uncheck the visualisation of all channels, except the one being tested. Lift the cover of AMPTS II Unit C.
c. Observe that the central black element has engraved numbers, corresponding to each measurement channel. Gently lift and drop the green flap corresponding to the channel in question. Wait 10 seconds and repeat this operation two or three times. If you have a third-generation instrument, like the Gas Endeavor, AMPTS III, BRS III, or BPC Go, unlock, lift, and relock the flow cell unit (FCU) two or three times, waiting at least 10 seconds between each unlocking.
d. Observe if a new line appears on the screen (or if the existing curve shows a new segment). If a new segment appears on the Graphs page shortly after manually lifting of the suspect flap, then there is nothing wrong with your unit. Please refer to the next sections.

If you observe no change in the curves after manually lifting the flap, then refresh the browser's page (click on the circular arrow at the left of the URL bar, or close the page and reopen it). Try again by manually lifting the green flap (or removing and repositioning the FCU if you have a third-generation instrument like AMPTS III, BRS III, GE III, or BPC Blue).

Should the problem persist even after performing the aforementioned procedure, then it is worth calling the manufacturer for support.

3.2.2 Symptom: 'Whimsical' Cumulated Volume Curves

This section applies only to AMPTS II and AMPTS II. In some cases, the user may find that the 'Graph' page displays curves with unexpected shapes, like the ones depicted in Figure 3.1. In a similar case, there is nothing to worry about: the instrument is working correctly. The cause of such curves is either a wrong date and time setting during the test (human error of the operator) or some random interference. In one case, it was found that the instrument was placed near an old Uninterruptible Power Supply (UPS), and each time there was a strong storm, the date and time of the system were upset. Since the UPS was not CE-compliant, most probably the reason was some electromagnetic interference. Moving the UPS 1 metre away from Unit C solved the issue. Connecting a surge suppressor between the electric socket and the instrument's power supply may be a good idea; a modest expense may save your instrument from voltage surges.

To reshape the curves correctly, go to the 'System' screen and correct the system's time and date. Note that the date format is YYYY-MM-DD. If you see a wrong date and time in the instrument's setup, then click on the button '*use this time for the instrument*' to synchronise the instrument with your browser. Uncheck the option 'synchronise time through the Internet' because this function

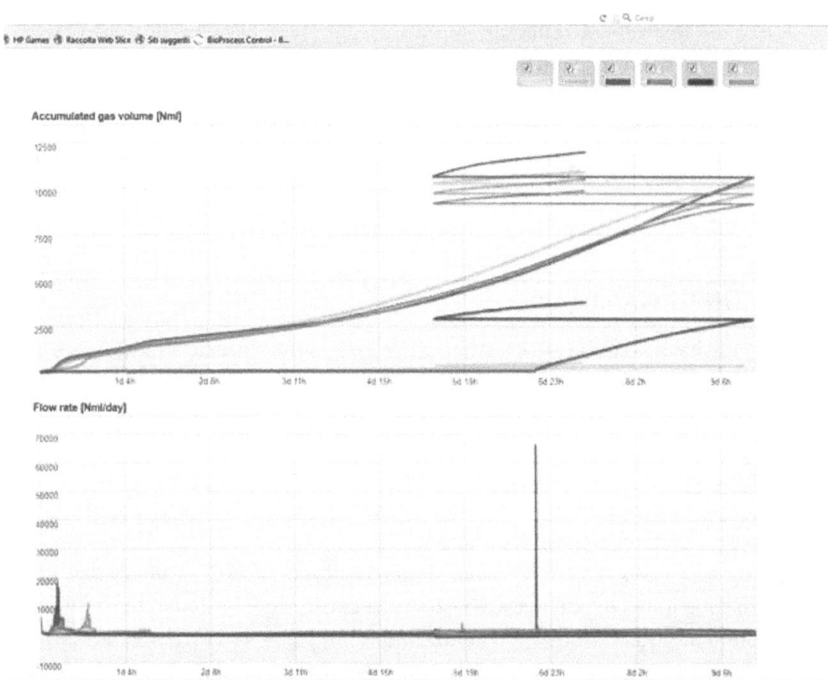

FIGURE 3.1 Examples of Whimsical Curves in an AMPTS II, Caused by Changes in the Date/Time Setting During the Test.

sometimes may lead to the problem described here, for instance, when the legal time changes.

3.2.3 Symptom: Access to the Instrument via the Usual IP Address Is Impossible; Sometimes the Stirrers Run Uninterruptedly Too

Check in the order described below:

a. Check that the instrument is on. The LEDs in the back panel should be green. If the LEDs are off, check the cables and power supply. Try connecting the power supply to another socket.
b. Unplug the power supply and wait for 10 seconds, then plug it in again and wait a minute for the system to restart. Try accessing the instrument again.
c. Check if the LAN cable is correctly inserted or try connecting the instrument with another cable.
d. From the Windows command line, type PING followed by the instrument's IP address and check if you get a pingback message. If you get a message of the type 'Packets: sent = 4, lost = 4', then the problem is either the network settings (contact the administrator for help) or the instrument's internal router (see below).
d. Reset the instrument. This operation will not erase the data of your experiment. The reset procedure only changes the password and IP address to the factory by default. Follow the procedure explained in the user's manual (download the last version from https://bpcinstruments.com/product-manuals/) and try accessing the instrument by connecting its LAN cable directly to the PC.
e. If all the above fails, contact the manufacturer's technical support.

3.3 FINDING GAS LEAKS

Gas leaks may have several causes. According to the International Water Association (IWA) guidelines on BMP assay [1], checking and identifying eventual gas leaks in all reactors before starting any test is advisable. The procedure described in this guide will help the reader with both routine checks according to IWA's guidelines and solving suspected faults when a channel is not showing gas production while its replicates do. It is assumed that the user already checked that there are no faults in the instrument, as described in Section 3.2.

Very tiny gas leaks, resulting from damaged Tygon tubes, reactors not closed correctly, or tube joints attached too loosely, can cause the instrument to show erratic gas production or no gas production at all. Pinpointing gas leaks from volumetric gas measurement instruments is difficult because they work at near-ambient pressure. Soap bubbles or similar solutions are not useful.

3.3.1 Standard Procedure for 500–2000 ml Reactors

a. Get a 100-ml syringe (without its needle). Pull the piston up to its course end to fill the syringe with air, open the flush gas valve in the reactor (red clamp in the photo), and connect the syringe to it (Figure 3.2).
b. Choke the tube that connects the reactor to the CO_2 filter flask, either by hand or with the aid of another laboratory tube clamp. Push the syringe's piston to the end – or as strongly as you can – and hold it for at least 60 seconds (Figure 3.3).
c. Release the piston and wait for some seconds. It should slowly return to its initial position as the air in the headspace expands to its original pressure (Figure 3.4). If the piston does not return to its initial position, then there is some gas leak in the reactor.

N.B. It is unlikely that the syringe's piston returns exactly to its initial position. This is perfectly normal because of the friction between the syringe's walls and the piston's rubber head. Note the difference in the piston's positions in Figures 3.2 and 3.4.

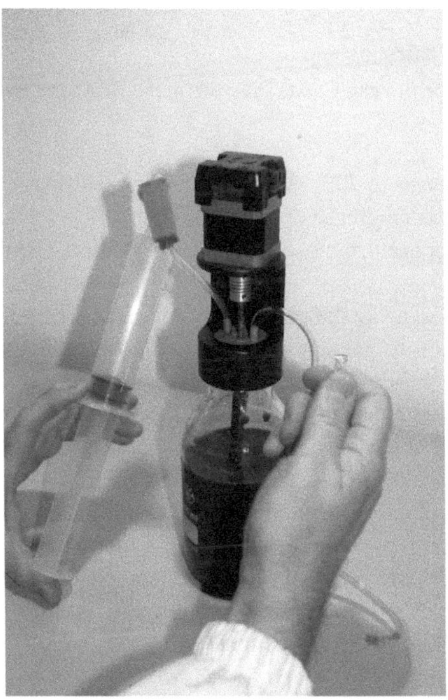

FIGURE 3.2 Using a Syringe to Create Pressure in the Reactor's Headspace: First Step.

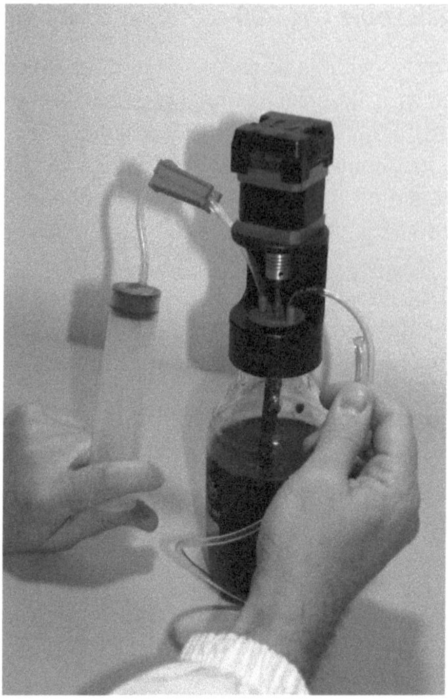

FIGURE 3.3 Using a Syringe to Create Pressure in the Reactor's Headspace: Second Step.

The piston usually returns to 70%–80% of its initial position. Lubricate the syringe's rubber piston with silicone grease or oil for best results.

If the piston does not return close to its initial position, then

1. Make sure to take away the pour ring (blue plastic ring in Figure 3.5) from the bottle's mouth because it interferes with the rubber seal of the stirring cap, preventing its hermetic sealing.
2. The most frequent cause of leaks is old tubes being deformed and not perfectly fitting with the cap's connectors. Disconnect both tubes from the reactor, cut 10 mm off from each end, reconnect them to the reactor, and repeat the test. Eventually, replace them with new tubes.
3. If the problem persists, then the stirrer cap's rubber seal or the stirring rod's rubber sleeve (Figure 3.6) is probably damaged. Replace the stirring cap with a new one. Also, small cracks in the glass bottle's mouth can lead to micro-leaks: in this case, replace the bottle.
4. Repeat steps 1–3 with CO_2 traps.
5. Repeat steps 1–3 with Unit C. In this latter case, connect the syringe to the suspected faulty measure channel, gently lift the corresponding green flap, cover the gas nozzle with your fingertip, hold the syringe with the other

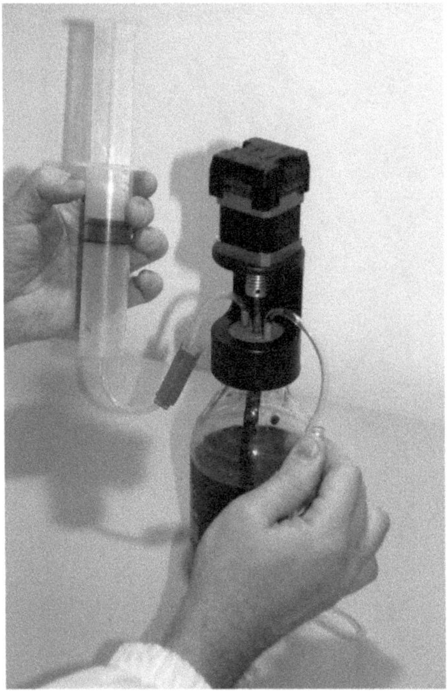

FIGURE 3.4 The Syringe's Piston Must Return to Its Initial Position When Released. If It Doesn't, Then There Is a Gas Leak Somewhere in the Reactor.

FIGURE 3.5 Pouring Ring on a Laboratory Bottle.

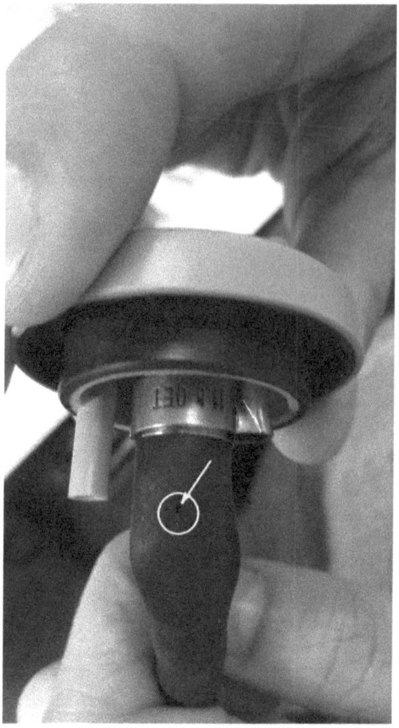

FIGURE 3.6 Almost Imperceptible Hole in the Stirrer's Sleeve.

hand, and apply pressure for at least 30 seconds (Figure 3.7). If the piston does not return to nearly its initial position when released, it means that the internal check valve or the connection tube is damaged. Do not attempt to open the unit, call customer service to arrange the reparation. Before calling, write down the unit's serial number or include it in your email to save time.

3.3.2 Modified Procedure for Bigger Reactors

One piston stroke of a 100-ml syringe will generate just a small pressure inside the headspace of a big reactor. The friction will then prevent the piston from returning when released, giving the false impression of a gas leak. It will be necessary to pressurise the headspace in steps: aspire air, then connect the syringe to the valve, open the valve, push the piston to the end, close the valve, retire the syringe, aspire air, and repeat the cycle for as many times as necessary to create enough the pressure in the headspace. Wait for 5 or more minutes (depending on the size of your reactor). Now open the valve and check if the syringe piston returns to its initial position. If it doesn't, then there is a leak. If you are testing the standard 2-litre batch reactors, the same considerations already explained for the 500-ml model will apply. If you are looking for leaks in 5-litre and 10-litre reactors, check the rubber O-ring under the cover, the reactor's level (it must be filled up to at least 80% of its nominal capacity), and the gas exit tube.

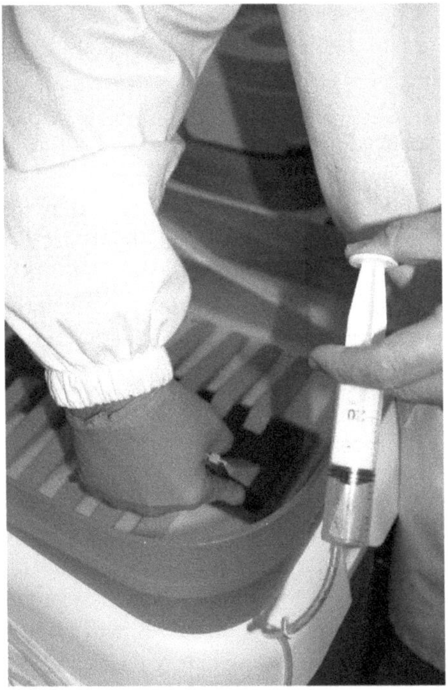

FIGURE 3.7 Searching for Gas Leaks in an AMPTS II.

A final consideration, valid for all reactor types: always use Tygon® or semi-rigid PU tubes because they are impervious to CH_4 and H_2. Other tubes usually employed for laboratory purposes (PVC, silicone, latex, etc.) may leak gas by molecular diffusion or are too rigid to perfectly fit the reactor's gas connector. The method described in this guide is unsuitable for detecting leaks caused by the molecular diffusion of CH_4 and H_2 through plastic tubes. In any case, leaks by molecular diffusion are so small that they fall below the detection limits of any instrument.

3.4 WRONG TEST PROTOCOLS

Sometimes, laboratories adopting the AMPTS technology after many years of working with other instruments try to maintain the same test protocols and setup procedures they have always employed. In other cases, people want to repeat experiments published in the literature and define an experimental setup trying to maintain 'equal conditions' for instruments designed with very different operational criteria. It is important to note that there are four types of respirometric methods for BMP tests: volumetric, barometric, gas-chromatographic, and most recently gravimetric methods. While the second and third typologies have a common feature – both require a large headspace in the reactor to minimise the errors, AMPTS belongs to the first type and requires the opposite setting: a headspace smaller than 20% of the reactor's volume and the same headspace volume and inoculum content in all reactors.

On the other side, laboratories with no prior experience in BMP tests tend to repeat experiences found in 'the literature', in an attempt to gain experience. Such an approach is questionable for several reasons:

a. Many papers present 'alternative' test procedures that may work in some cases, but these are not universally valid and are questionable.
b. There is no guarantee that the inoculum employed by the paper's author is comparable to the one available in your laboratory.
c. Substrate descriptions such as 'corn silage' or 'OFMSW' are vague. The composition and physical properties of complex or heterogeneous substrates can be very different from one place to another, or vary with time, making it difficult to compare their BMP.

If you are new to anaerobic digestion tests, start employing an established protocol, for instance, the IWA's recommendation (ref. 1 already cited), VDI 4630 (available in German and English), or UNI/TS 11703 (available only in Italian, see ref. 2 for a comparison with the other norms). The operation manual of the AMPTS II (Appendices A and B, freely available for download from the manufacturer's website https://bpcinstruments.com/product-manuals/) includes a summarised description of the method. Following one of the said procedures will often give good results but may require some adaptation when testing hardly degradable substrates or assessing the efficacy of commercial boosters and additives.

If you already followed the procedures described in Sections 3.2 and 3.3 but you still get unexpected results from your experiment, then the most probable cause is either a wrong protocol or a biological problem (inhibition), or both.

3.4.1 Setting Your Tests Correctly

The following guidelines provide some additional information for a correct experimental setup with an AMPTS instrument:

a. Always connect the instrument to the reactor through the standard CO_2 trap. This latter can be filled with a 3M NaOH solution (120 g/l) if you want to measure the net methane production, or just with acidulated distilled water if you intend to measure the total biogas production (which is not advisable, but many do). The reason not to connect the reactor directly to the instrument is to avoid foam or other impurities clogging the check valve (placed inside the AMPTS II or before the FCU in AMPTS III).
b. Do not set the stirring too high. Too much stirring stresses the bacteria in the reactor and may lead to reduced biogas production. No stirring at all may underestimate the BMP by 20% and make the classical BMP sigmoids look 'flat', requiring long digestion times. As a rule, start by setting the stirring at 50% power, 10 minutes on and 10 minutes off, and check if it is enough to avoid the formation of floating layers and/or sediments. Increase or decrease the agitation intensity or the on/off times as necessary, by 10%–20% at a time.

c. Like all volumetric methods, AMPTS requires that the headspace in the reactor is as small as possible to minimise measurement errors. Do not waste your time trying to compare your new AMPTS with your old barometric method, defining the same headspace volume in both instruments: it will not work. As a rule, always fill your AMPTS reactors to at least 80% of their nominal capacity but keep the level at least 5 cm below the bottle's mouth to avoid eventual clogging of the biogas tubes with foam.
d. The accuracy of 9-ml cells is warranted for gas flows from 1–6,000 Nml/hour. Do not attempt to connect your AMPTS unit to reactors with more than 10-litre nominal capacity.

3.4.2 Measuring Total Biogas with Batch Methods

Note that whatever 'the literature' and the 'experts' may say, the composition of biogas is seldom 60% CH_4 and 40% CO_2. A particular feature of batch tests is that the inoculum's alkalinity will capture most of the CO_2 produced during the digestion of the biomass, so this CO_2 will never leave the reactor and, consequently, the instrument cannot measure it. This is particularly true for batch tests with sodium acetate as a reference substrate, which usually yields biogas containing at least 90% CH_4. Such a phenomenon is common to all batch methods. Should you need to measure total biogas for any research reason, consider performing continuous tests instead of batch ones. An alternative to continuous tests is to saturate the inoculum and the water in the flow measurement cell with CO_2 by bubbling it in the reactor for an hour or so before starting the test (see Chapter 8). The stirrer cap has an inner tube connector; connect a piece of flexible tube to it and the outer flushing tube and clamp it to a CO_2 bottle (Figure 3.8). See Chapter 9, Section 9.2, for a detailed error propagation analysis in this particular kind of tests.

3.4.3 Memory Limit When Performing Long-duration Tests (for Instance, Continuous Tests)

Note that AMPTS II and GE II are instruments conceived for batch operation, which means that all tests are usually run for less than 60 days, and the datalogger memory is enough for that. If for any reason you need to perform long-duration tests, then the options are as follows:

1. Download a report monthly, stop the test, and restart it immediately. Then copy and paste the monthly reports into a single file. When stopping and starting the experiment, the instrument resets its memory, preventing any risk of data overflow.
2. Consider the purchase of an AMPTS III, a Bio Reactor Simulator III, or a GE III, all instruments having enough memory capacity for long-term continuous tests.

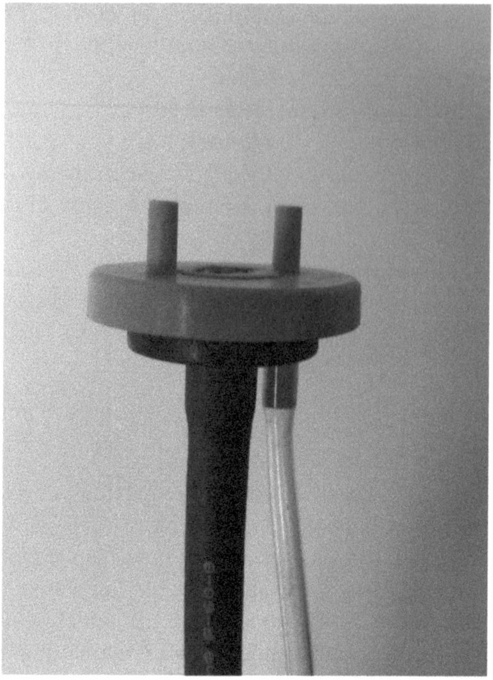

FIGURE 3.8 How to Connect a CO_2 Sparger Tube to the Special Stirrer Cap for GL45 Bottles.

3.4.4 Employ Positive Controls Whenever Possible

Including a reactor loaded with sodium acetate and another reactor loaded with cellulose is useful to assess the biological activity of the inoculum. Such reference substrates have well-defined BMP values: not reaching the reference BMP during a test is an important clue for diagnosing eventual problems with the inoculum (see Chapter 5 for details).

3.4.5 Assessing the Quality of a BMP Test from the Shape of the Curve

An excellent open-access paper by Koch et al. [3] proposes a simple method to assess the reliability of BMP tests and identify eventual procedural problems by simply checking the shape of the SMP curve. In extreme synthesis,

a. keeping the inoculum at room temperature for a long time or pre-incubating it too long will result in lower BMP values and flat SMP curves;
b. diluting the inoculum will result in SMP curves that show a slightly negative slope instead of a flat or slightly growing plateau;
c. if the inoculum to substrate ratio (ISR) < 2, the shape of the SMP curve may show a long initial lag phase and some 'humps'. This particular case is discussed in detail in Section 3.5.3.

The following are some further quality control criteria from ref. 1 already cited and from the Italian norm UNI/TS 11703:2018:

1. The VS concentration of the inoculum after sieving it should be in the range 1%–3% (10–30 g/l).
2. The average endogenous methane production of the blank reactors should be less than 20% of the average gas production of the sample reactors. A higher percentage indicates that the preincubation of the inoculum was too short or missing. Apart from functioning as an amplification factor of the metrological error (Chapter 2 of ref. 2), an excessive endogenous production may give SMP curves like the one depicted in Figure 3.9.
3. The dispersion between replicates must be less than 5% in the case of tests with sodium acetate, less than 10% for more complex substrates, and less than 20% for heterogeneous substrates (e.g., OFMSW).
4. The minimum quantity of inoculum is 400 g.
5. The headspace must be the same in all reactors to eliminate the error induced by moisture and flush gas. In this case, you may deactivate the function 'eliminate over/underestimation' in AMPTS and GE and leave blank the cells 'assumed CH_4 content' and 'CO_2 in flush gas' as they do not influence the BMP value. The said function is useful when your aim is assessing the shape of the curves during the first days, for instance, when you are performing an SMA test.

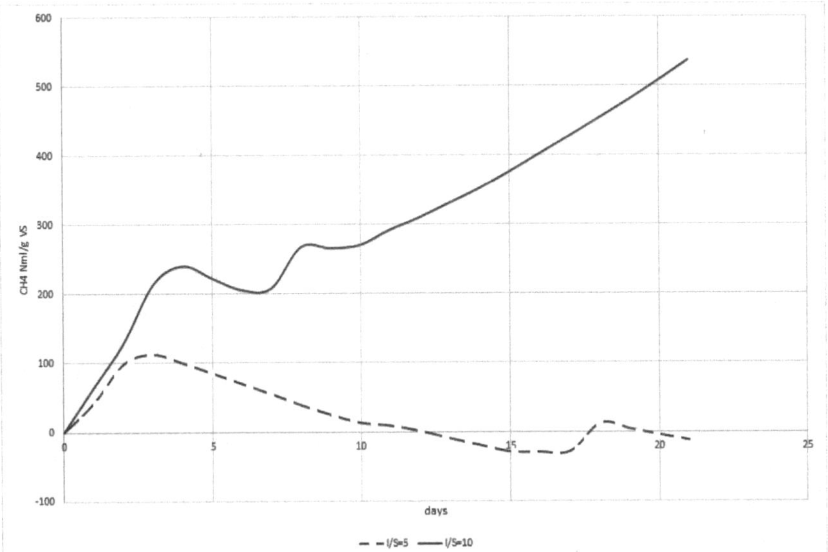

FIGURE 3.9 Anomalous Shapes of the SMP Curves During the Digestion of Waste Glycerol from a Biodiesel Production Plant. The high content of sodium chloride or sulphate inhibits the inoculum's methanogenic activity.

3.5 BIOLOGICAL PROBLEMS

If you are still getting unexpected or erratic results despite having checked all the details explained till now, then the cause is certainly biological. The most frequent causes of inoculum inhibition are either the substrate (e.g., contains toxic compounds) or the inoculum itself (lack or low bioavailability of trace elements). The present section provides hints on how to prepare the inoculum before starting any test and the criteria to adopt in programming the BMP assay when the substrate is potentially inhibiting.

3.5.1 Lack or too Low Methanogenic or Fermentative Activity

When the inoculum lacks methanogenic activity, the reactor produces only CO_2. Consequently, the NaOH solution in unit B will trap it, and hence, no gas will flow through Unit C. High CO_2 production results in depression within the headspace, leading to a higher rate of evaporation and water droplets condensing in the tube that connects the reactor to the CO_2 trap. The causes of low methanogenic activity may be the lack of trace elements, the presence of antibiotics, too low pH, too high concentration of NaCl, or a combination of factors. Or, more simply, the inoculum has no methanogenic flora. See Chapter 5 on how to select and test the inoculum.

If the inoculum has good methanogenic activity but low or no fermentative activity, the bacteria are unable to convert the substrate to acetic acid. Hence, the archaea will not produce any methane, or they will produce very little over a very long time. This happens when trying to digest a substrate that the inoculum's bacterial flora is not adapted to. For instance, inoculum from a plant running on corn silage and bovine manure will often be unable to digest soapy or fatty substrates, no matter how good its methanogenic activity may be (see Section 3.5.2).

If the inoculum comes from a source you have not tested before and its microbial activity is unknown, it is advisable to check it using reference substrates (sodium acetate, cellulose, protein, or vegetable oil. Refer to Chapter 5).

3.5.2 Inhibiting Substrate

Despite the inoculum coming from a 'healthy' biogas plant or having run positive controls that prove its biological activity, sometimes the SMP curve of the substrate is atypical. Its shape may be concave, flat, or even negative, which is a sign of inhibition. Then, the probable cause is substrate-induced inhibition. Examples of inhibitory substrates are those containing high concentrations of oils and fats, soap, phenolic compounds, ammonia, sulphur, antibiotics, chlorides, chemicals, and propionic acid, or organic matter having a low carbon-to-nitrogen ratio ($C/N < 8$). The following are some examples of such substrates: glycerol from the biodiesel industry, wastewater from palm or olive oil extraction, fatty slaughterhouse waste, tannery waste, chicken dung, and swine slurry treated with anti-odour products. The easiest way to check substrate-induced inhibition is to perform three or more tests with increasing ISRs, for instance, 3, 5, and 8. Figure 3.9 shows an example of a test on waste glycerol from the biodiesel industry. The inoculum was obtained from a biogas plant running on cow manure and corn silage. Positive controls showed it had good hydrolytic and methanogenic activities. Although the inoculum was 'healthy', note that even an ISR of 5 was insufficient to avoid inhibition.

Correct preparation of the inoculum before starting the test is a requirement to obtain reliable results. The standard procedure consists of sieving it through a 5- to 10-mm mesh to eliminate macroscopic solids. Then it is mandatory to degas (pre-incubate) it for at least 1 week or until the daily production becomes less than 10% of the cumulative production from the beginning of the preincubation. Figure 3.10 shows an example of a test performed without having degassed the inoculum in advance. Despite the substrate-induced inhibition not being strong enough to stop the anaerobic digestion, the decreasing portion of the SMP curve from the 16th day onwards indicates that endogenous methane production of the inoculum was higher than the substrate's production in the same period. The reason is that the inoculum in the blank reactor still contained half-digested organic matter at the start of the test, and the bacteria converted it into methane at a higher speed than they could convert the substrate. Degassing the inoculum in advance would have avoided such a phenomenon.

Figure 3.11 is an example of total inhibition caused by a fat-rich substrate during a test performed with an ISR of 2 and inoculum that had not been degassed in advance. BMP tests of fatty substrates require ISR > 5.

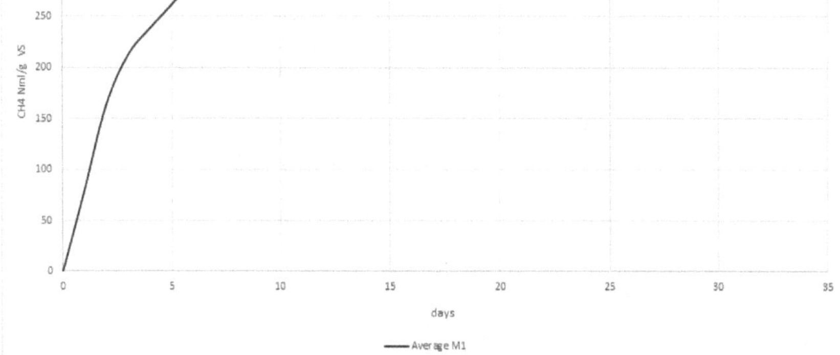

FIGURE 3.10 Omitting the Inoculum's Degassing Step Leads to an Anomalous Shape of the SMP Curve.

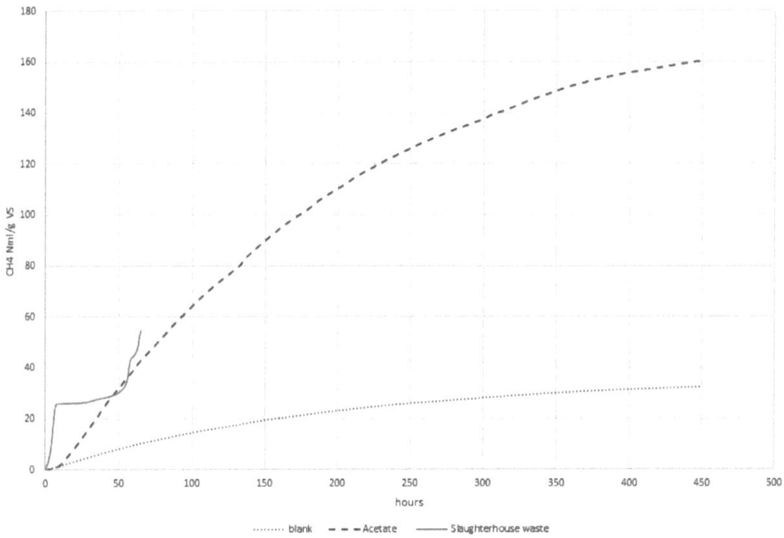

FIGURE 3.11 Example of Total Inhibition Caused by a Fat-Rich Substrate During a Test Performed with ISR = 2 and Without Having Degassed the Inoculum in Advance.

3.5.3 Diauxia and Polyauxia

SMP curves sometimes may show multiple plateaus – the phenomenon is called *diauxia* if there are two plateaus and *polyauxia* if there are three or more plateaus. According to the VDI 4630 standard and ref. 3 already cited, di- or polyauxia indicates an unreliable test. In the author's opinion, the said criterion is valid when the ISR employed for the test is too low, causing some degree of inhibition, but it does not invalidate the test in all cases. An SMP curve can present polyauxia and be however perfectly valid. The author discussed in the first volume the case of Substrate X, a 'secret formula' marketed in Italy as 'special feedstock for digesters' (Chapter 4, Figure 4.3 of ref. 2 already cited). Figure 3.12 shows the example of a substrate causing polyauxia, although the ISR was 2, the inoculum was undiluted, and its SMA was in the norm. The reason for the unexpected behaviour was probably the complexity of the substrate tested: industrial wastewater containing alcohols and organic acids as well as traces of chemicals employed in the process. The BMP reached 85% of the theoretical value, the digestion time was short, the metrological error was 3%, and the standard deviation between replicates was 2.5%. Everything indicates that the test is valid, although the shape of the SMP is 'undulated'.

The reader can experience him- or herself by preparing a mixture of sugar and chopped cotton flocks or filter paper, measuring the resulting VS and performing a BMP test with ISR = 2. It is easy to forecast the result: bacteria will degrade sugar in a few days, while they will need more time to degrade cellulose. The resulting SMP curve will exhibit two distinct plateaus corresponding to the percentages of sugar and cellulose in the mixture.

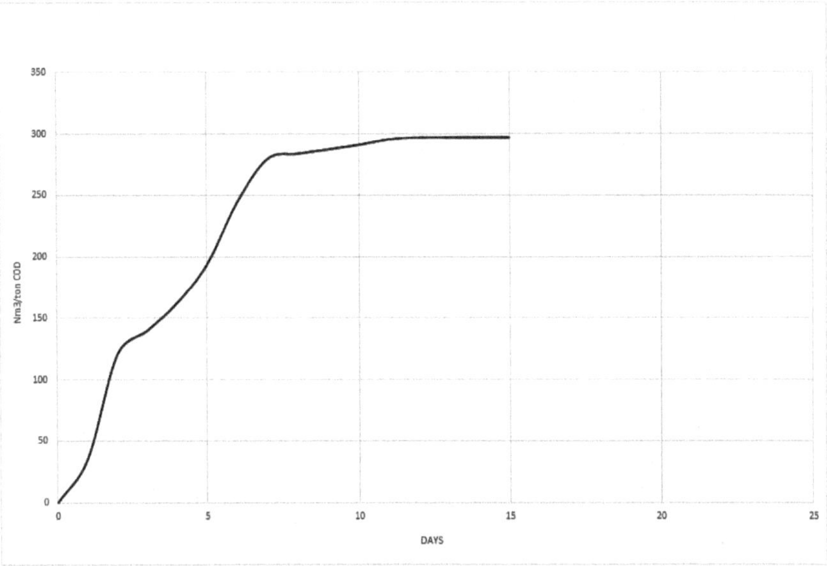

FIGURE 3.12 Example of an Undulated SMP Curve Showing Incipient Diauxia.

In the author's opinion, di- or polyauxia is not an absolute indicator of test bad quality or problems with the inoculum: it must be judged in the context of the assay, deciding case by case whether the result is reliable or not.

3.6 CONCLUSIONS

Dubium sapientiae initium (Doubt is the beginning of wisdom)

René Descartes

Cartesian scepticism is the best method to ensure highest profitability of a biogas plant. Making the correct decisions requires formulating process management hypotheses, checking them in the lab, and discarding substrates that can potentially upset the digester. The condition for a profitable decision-making process is that the results of the laboratory tests are reliable, which in turn requires absolute certainty about the inoculum's quality, the instrument's correct functioning, and the robustness of the experimental setup and the method employed. Calculation should then overcome Murphy's law, but you still have to overcome Rosato's laws:

1. Under controlled laboratory conditions, a biological system will do whatever it wants.
2. What the biological system decides to do is always the contrary of what one expects.

The apparent self-consciousness of the anaerobic inoculum playing pranks on the researcher is usually the result of our own cognitive biases: oversimplification, wrong assumptions, generalisation of particular cases, etc. The following hints may help the reader draw correct conclusions from apparently inconsistent or fuzzy test results:

1. If you periodically test a given substrate and obtain unexpected results at a given moment, the problem could be the batch itself, for instance, the presence of a contaminant that causes partial or total inhibition, or a batch of feedstock that contains a higher fraction of lignin or other undegradable matter. Including a positive control in each test will help easily determine whether the cause of the anomalous behaviour is related to the biological activity of the inoculum or the substrate composition.
2. If you take the inoculum from a plant that recirculates the liquid fraction of the digestate, be aware that the accumulation of salts and/or ammonia will inexorably hamper the biological activity, which translates into decreasing values of BMP over time. If this is your case, you should always include a positive control in each run and keep a historical record of its BMP. If you observe that the BMP of the control substrate decreases steadily over time, then remove some of the liquid digestate, eventually replacing it with fresh water.
3. Do not 'simplify' the established test protocol. The inoculum must always be sieved and preincubated to minimise the endogenous gas production but must not be exhausted. If you run many tests and cannot afford to delay the test because of the preincubation, then buying an additional thermostatic bath and a set of bottles and stirrers will help (see Chapter 4). You can start preincubating the inoculum in the auxiliary thermostatic bath 1 week before ending the current test, so it will be ready to start a new test without delay.
4. There are two schools of thought about dosing the inoculum to each bottle: measuring the volume with a graduated cylinder and weighing it. In general, weighing the inoculum instead of measuring its volume with a graduated cylinder reduces dispersion among replicates, i.e., increases the precision of the test. Weighing is also preferable when the inoculum contains a relevant fraction of solids tending to sediment. All technical balances have a tare button that allows you to place the bottle on the weighing pan, set the reading to zero, and then weigh exactly the quantity of inoculum poured into the bottle. If you employ reactors of at least 500 ml capacity, as requested by the UNI/TS 11703:2018, a kitchen balance with +/− 0.5 g precision is already enough. Stir the inoculum in the jar or jerrycan before pouring it into the bottle as this will help in distributing the bacterial biomass between the bottles more uniformly.
5. Always remember that the BMP formula is based on assuming that the quantity of gas produced by the blank reactor is the same endogenous gas production in the reactor with the sample. If the inoculum contained in the blank reactor has more or less bacterial biomass than the inoculum in the sample reactor, then the fundamental supposition on which all the methods are based will be (partially) false. Consequently, the BMP calculated with such biased measurements will be less accurate. Sometimes (e.g., when you

are testing very diluted wastewater or similar substrates), it is necessary to fill the sample reactors with less inoculum than the blank. In such cases, your BMP formula must include the correction factor *f*, as follows:

$$BMP = \frac{S - f\,B}{VS},$$

where *S* is the average methane production of the sample reactors, *B* is the average methane production of the blanks, *f* is the quotient between the quantity of inoculum in the sample and blank reactors, and *VS* is the quantity of VS of the substrate added to the sample reactors.

Another school of thought holds that when the substrate is so dilute that just a tiny amount of inoculum is necessary to obtain the desired ISR, then all reactors should be filled with the same amount of inoculum and then the blank reactors should be tipped up with water up to the same total volume of the sample reactors.

The author has checked both criteria and considers that adding water to the blank to the desired volume and correcting the BMP formula with the *f* factor are equivalent. Choosing one or the other is then just a question of convenience.

6. Feedstock is not equal to feedstock. General descriptions like 'effluent from brewery', 'municipal solid waste', or 'dairy manure' mean nothing. The composition and features of a given feedstock may vary largely from one source to the other, or over time for a given source. So, you cannot expect to obtain the same results always: differences in the order of 20% or higher are very common when comparing BMP tests of apparently similar substrates.
7. Batch tests provide a model of the AD process. But a model is not the reality. A given substrate could be fully digestible when tested in batch mode, but this does not grant its digestibility in the long-term operation of a plant, especially when the said substrate is the staple feedstock of the biogas plant. When in doubt, perform replicates with the addition of different proportions of trace elements and observe if there are significant differences between them.

BIBLIOGRAPHY

[1] Christof Holliger, Madalena Alves, Diana Andrade, Irini Angelidaki, Sergi Astals, Urs Baier, Claire Bougrier, Pierre Buffière, Marta Carballa, Vinnie de Wilde, Florian Ebertseder, Belén Fernández, Elena Ficara, Ioannis Fotidis, Jean-Claude Frigon, Hélène Fruteau de Laclos, Dara S. M. Ghasimi, Gabrielle Hack, Mathias Hartel, Joern Heerenklage, Ilona Sarvari Horvath, Pavel Jenicek, Konrad Koch, Judith Krautwald, Javier Lizasoain, Jing Liu, Lona Mosberger, Mihaela Nistor, Hans Oechsner, João Vítor Oliveira, Mark Paterson, André Pauss, Sébastien Pommier, Isabella Porqueddu, Francisco Raposo, Thierry Ribeiro, Florian Rüsch Pfund, Sten Strömberg, Michel Torrijos, Miriam van Eekert, Jules van Lier, Harald Wedwitschka & Isabella Wierinck, Towards a standardization of biomethane potential tests, *Water Science and Technology*, 74:11, 2515–2522 (December 2016). https://doi.org/10.2166/wst.2016.336

[2] Mario Alejandro Rosato, *Managing biogas plants – A practical guide*, Chapter 6, CRC Press (2018). https://www.routledge.com/Managing-Biogas-Plants-A-Practical-Guide/Rosato/p/book/9781138626614. ISBN 9781138626614.

[3] Koch Konrad, Sasha D Hafner, Sören Weinrich & Sergi Astals, Identification of critical problems in biochemical methane potential (BMP) tests from methane production curves, *Frontiers in Environmental Science*, 7, 178 (2019). https://www.frontiersin.org/article/10.3389/fenvs.2019.00178. https://doi.org/10.3389/fenvs.2019.00178

4 Tricks and Hacks in the Lab

4.1 CHAPTER OVERVIEW

This chapter is indirectly related to the rest of the book. It contains some simple do-it-yourself (DIY) solutions to small problems that one may encounter in the day-to-day operation of an anaerobic digestion laboratory. Although the tricks and hacks presented herein will not make a difference in the overall economy of a biogas plant, the author decided to include them anyway, hoping that they will save time for the readers in similar situations.

4.2 MAXIMISING THE INSTRUMENT'S PRODUCTIVITY: TIME IS MONEY

By definition, the BMP of biomass is the result of a calculation based on independently measured magnitudes. Each of such measured magnitudes – fresh weight, dry weight, and ash of the sample; total methane production of each sample reactor; and total methane production of each blank reactor – is subject to measurement uncertainty. The uncertainty of each measurement adds in a non-linear manner, potentially decreasing the overall accuracy of the BMP test. Preincubating the inoculum is fundamental for maximum accuracy because reducing its endogenous methane production reduces the overall uncertainty of the BMP test (ref. 1 and 2). Preincubation of the inoculum is hence mandatory for many standards, such as the German VDI 4630, the Italian UNI/TS 11703, or the IWA recommended BMP protocol. The economic consequence for the anaerobic digestion lab is having to choose between two possible operational strategies:

a. Accepting an unproductive time between BMP tests for preincubating the inoculum. This implies longer times to get results, usually 5–10 days for the inoculum's preincubation plus 25–45 days for the BMP test itself.
b. Using two thermostatic baths simultaneously: one for performing the BMP tests with the AMPTS and the second for preincubating the inoculum for the next batch.

The first case can be acceptable in a small laboratory performing BMP tests for didactic or research purposes, where the average time to conclude a test is not a priority. The second case is the solution of choice for industrial laboratories, where carrying out the maximum number of tests in the shortest time is necessary to ensure profit. A thermostatic bath is a cheap piece of equipment available for purchase everywhere. Any model of adequate capacity will be suitable for the preincubation of the inoculum

DOI: 10.1201/9781003609278-4

as ±2°C accuracy of temperature control is enough for this preparation step. A small investment will allow you to avoid occupying a piece of expensive equipment like the AMPTS for an unproductive routine operation like inoculum preincubation. In the following examples, suppose you are running tests with an AMPTS III Standard or AMPTS III Light, so you will need 15–16 litre of inoculum for each batch run. Check your current tests, and when you observe that the cumulated volume curve begins to become flat or the flow curve begins to become constant, it is time to start preparing the inoculum for your next batch. Sample 20 litre of fresh digestate from the plant. Sieve it through a mesh (5–10 mm) to eliminate macroscopic solids. At this point, there are different options for preincubating the inoculum with the auxiliary thermostatic bath:

1. **Cheapest solution**: Incubation in plastic jerrycans with daily supervision. Take two jerrycans of 10 litre capacity and pour 8 litre of the sieved inoculum in each. Then, place the jerrycans in the auxiliary thermostatic bath set at the desired incubation temperature. At least once a day, agitate the jerrycans, open the caps slightly to release the overpressure, and then close them immediately.
2. **Intermediate cost solution**: Three 5-litre ISO GL45 bottles with independent stirring. In this case, you can purchase three stirring caps for 5-litre ISO GL45 bottles and three BPC Move stirrers. The BPC Move stirrer has the advantage that each unit has an independent power supply, and the speed control is integrated into the motor's block. Since each stirring cap has two tube connectors, the three bottles can be interconnected with short tube pieces. Close one end of the bottle with a piece of tube choked with a plastic clamp. Then, connect the bottle on the opposite end to a longer piece of tube (50–80 cm), placing its free end in a jar full of water. The jar will work as a water seal, allowing the gas to escape and preventing the air from entering the bottles and damaging the inoculum. Always place the end of the tube in a plastic or glass jar or a container full of water; do not place it in the thermostatic bath because the eventual H_2S in the gas will dissolve in the water and may corrode metallic parts.
3. **Efficient but more expensive solution**: Use the same quantity of ISO GL45 bottles and stirrers as your AMPTS. In this case, you need to purchase a thermostatic bath with adequate capacity, a kit of brushless motors, a DC power supply (min. 3 V, max. 5 V) or a two-battery holder for standard 1.5 V batteries, and a kit of stirring caps. The two-battery holder or DC power supply is necessary to provide the signal input to the control cable for the motors to run. The speed of the motors is proportional to the voltage applied to the signal cable. Note that 0 V DC (control cable not connected or shunted) means the motors have stopped, and 5 VDC applied to the control cable means the motors are running at 250 RPM, their maximum speed. If you use two AA batteries, i.e., 3 V DC as the control signal, the motors will run at 150 RPM. Figure 4.1 shows the setup in the author's laboratory. In this case, the thermostatic bath holds six 2-litre bottles (for an AMPTS II Light), all stirred with an agitation kit of six brushless motors (not shown in

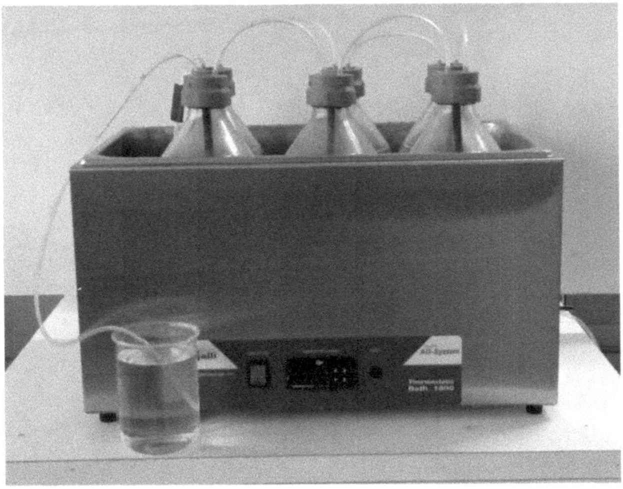

FIGURE 4.1 Setup of an Auxiliary Thermostatic Bath with Six Reactors. The Stirrers Were Removed for Clarity.

the photo for clarity). Figure 4.2 shows the detail of the battery holder connected to the signal cable of the motor speed controller. Pay attention to the polarity: the positive (red cable) must be inserted into the connector socket, as shown in the photo. Inverting the polarity poses no risk to the controller, but the motors will not run.

Like in the former case, the bottles must be interconnected with short tubes; one end of the bottle should be closed with a short tube choked with a clamp, and the opposite end should be connected to a longer tube submerged in a jar full of water. The advantage of this solution is saving substantial amounts of time when finishing a batch and starting a new one. Take the bottles with exhausted inoculum away from the thermostatic bath of the AMPTS, replace them with the bottles from the preincubation bath, introduce the required amount of sample, and start the new run. Once the new test is running, you can proceed at your convenience to discard the spent inoculum from the previous test and wash the bottles for the next batch. Minimising the exposure of the preincubated inoculum to air, which is unavoidable when transferring the inoculum from the preincubation container to the test bottles, also reduces the lag phase and improves the quality of the test results.

4.3 OPTIMISING THE POWER SUPPLY OF THE ANAEROBIC DIGESTION LABORATORY

4.3.1 Dealing with Power Outages

Murphy's law is valid in every aspect of human life, and BMP assays are no exception. We tend to believe that everything is under control during a test just because the

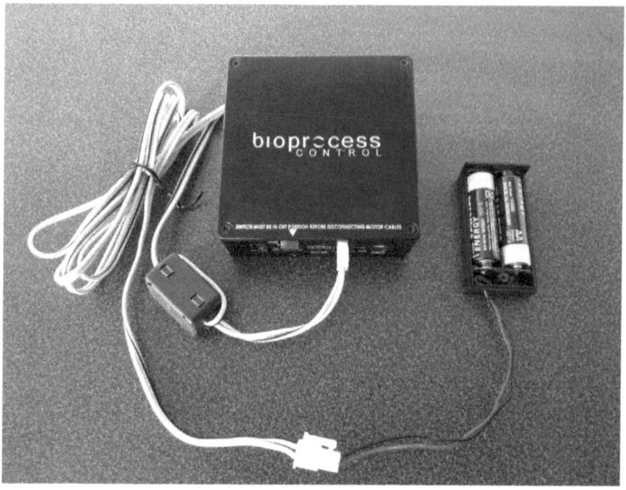

FIGURE 4.2 Connection of Two Batteries to the Signal Input Port of the Stirrer Control Unit of an AMPTS III Used for the Auxiliary Thermostatic Bath. Note the Polarity.

function of a lab is to provide controlled conditions to ensure the test's repeatability. Although both common sense and the manufacturers of the instruments strongly recommend feeding all of them through an Uninterruptible Power Supply (UPS), many small laboratories – especially those installed in anaerobic digestion plants for process monitoring purposes – usually do not have one. All BPC Instruments' products are equipped with non-volatile memories, so in case of a power outage, there is no risk of losing all the experiment's data. If the blackout lasts less than 2 hours, then the thermostatic bath will cool down just a few degrees, and the reactors will continue to produce biogas. The total volume of gas that the instrument will not be able to read is irrelevant, and the final BMP will be just a bit underestimated. If the power interruption is long (more than 2 hours), the final BMP will be inaccurate and eventually useless because it is underestimated. Fortunately, AMPTS II and Gas Endeavour II have a feature allowing data recovery from a test that otherwise would be lost because of a blackout: the */data* function (AMPTS II) and the *system log* (AMPTS III). Third-generation instruments AMPTS III and GE III have different software features, but the procedure to recover lost data by interpolation is the same.

In this section, we will deal with two possible situations: recovering information lost during general blackouts of long duration, out of the user's control, and preventing power interruptions caused by excessive power uptake during normal operation of the laboratory.

4.3.2 Long-duration Blackouts

The trick described in this section allows 'reconstructing' the unrecorded data during the outage, providing a final corrected volume that will be more accurate than the uncorrected value. The test's overall accuracy will, of course, be inversely proportional to the duration of the blackout but acceptable for most industrial purposes.

Long power interruptions during the night or the weekend may affect the test's quality because nobody will notice them. A keen instrument user will detect a long power loss from the flow curves because all will converge to zero at a given time and then rise from the instant when the power supply resumes. Reciprocally, all the cumulated volume curves show a flat step corresponding to the power interruption interval. Figure 4.3 is an eloquent example of such a situation: a 10-hour-long blackout.

In the event of power interruption during a test, the instrument will log the restart time as soon as the power supply is resumed. If you own a third-generation instrument with Aurora software (AMPTS III, GE III, BPC Blue, or BPC Go), finding out the blackout's duration is straightforward. Just open the System page and scroll down to the Log textbox. You will see a text like the one shown in Figure 4.4, so you can calculate the exact duration of the outage.

In the case of second-generation products (AMPTS II and GE II), it is necessary to download the log to search for the blackout information. To access the log, type the instrument's IP address followed by */data* in the URL bar, e.g., ***192.168.10.11/data***, and then press enter. The screen will look as shown in Figure 4.5.

Double-clicking on the *log.txt* file will open it. It is advisable to save the *log.txt* file on the PC's hard disk. If the date at which the test was started is known, you can find the start of the experiment by searching it first (the format is YYYY-MM-DD). If you are not sure about the date, another possibility is to use the 'search' function of

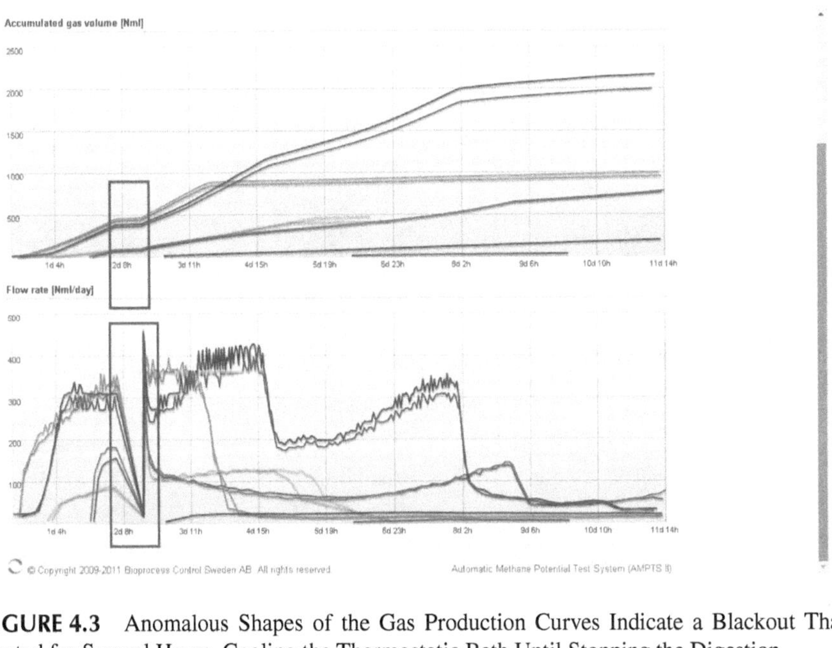

FIGURE 4.3 Anomalous Shapes of the Gas Production Curves Indicate a Blackout That Lasted for Several Hours, Cooling the Thermostatic Bath Until Stopping the Digestion.

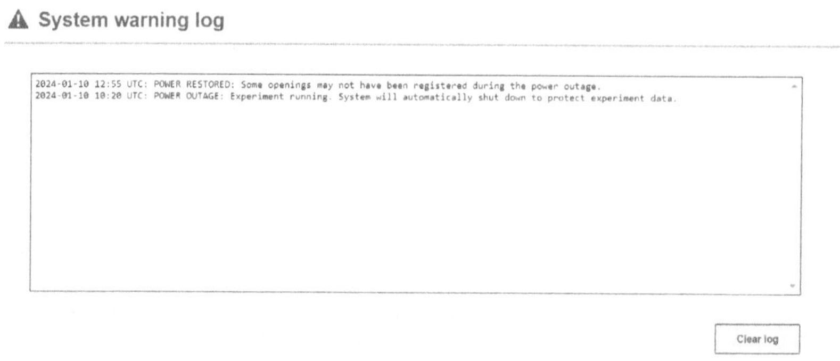

FIGURE 4.4 Example of System Log Warning.

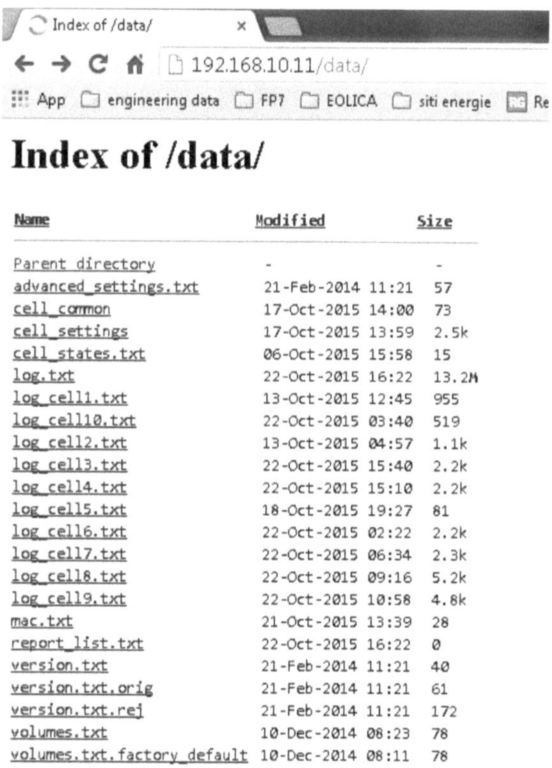

FIGURE 4.5 Screenshot of the Data Retrieved with the ***/data*** Instruction of an AMPTS II.

the text editor to look for the text '*Start registering openings on line #*', and the text lines will appear as follows:

```
2015-10-06 15:58:24 :: [ssamples] Start registering openings on
line #1

2015-10-06 15:58:27 :: [ssamples] Start registering openings on
line #2

2015-10-06 15:58:29 :: [ssamples] Start registering openings on
line #3

2015-10-06 15:58:31 :: [ssamples] Start registering openings on
line #4

2015-10-06 15:58:33 :: [ssamples] Start registering openings on
line #5
```

From the said lines, we deduce that the test was started on 6 October 2015 at 15:59. For our purposes, ignore the seconds and round to the next minute. It is now advisable to delete all the lines above the said start point to reduce the size of the *log.txt* file. Now we will search for the word 'exception', which in our example brings us to the following lines:

```
2015-10-08 20:53:31 :: [motor] Stopping motor for 300 sec.

2015-10-09 07:02:17 :: [exception] Exception application started
```

We deduce then that the outage started shortly after stopping the stirrers, on 8 October 2015, at 20:53 (i.e., 52 hours after starting the tests), and the power returned 10 hours later, on the following day at 07:02.

Once you know the exact duration of the power interruption, it is very easy to estimate the volume of gas lost during the power interruption: generate a report, choosing 'hours' as the time scale. For each of the instrument's channels (each column of the report), it is enough to calculate the average hourly flow of the 5 hours before the power interruption, F_1, and the average flow of the 5 hours after the power was resumed, F_2. The average of F_1 and F_2 represents a linear interpolation of an unknown transient state – flow decreasing in time as the reactors got colder and eventually stopped their biogas production – between two known steady states. By multiplying the average (F_1; F_2) by the blackout duration (in this example, 10 hours), we get the estimated volume of gas lost because the instrument was off, which we must add to the final volume to obtain a more approximate value of total cumulative volume. In our example, a routine test with sodium acetate was underway to check the activity of *inoculum A* when the power was interrupted. The average normalised flow and cumulative volume before the power outage were 3 Nml/h and 82 Nml, respectively. The corresponding values after the blackout were 5.5 Nml/h and 98 Nml, respectively. Hence, volume V_1 that was not counted during the blackout is approximately

$$V_1 = \frac{3+5.5}{2}\left[\frac{\text{Nml}}{\text{h}}\right] \times 10\ \text{h} \approx 42\ \text{Nml}\ \cdot$$

Figure 4.6 shows the difference between the corrected and uncorrected curves (after subtracting the volume produced by the inoculum).

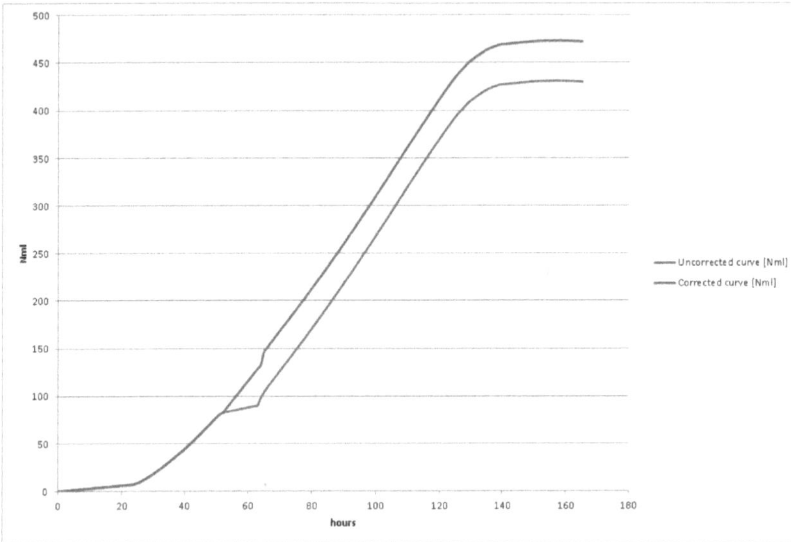

FIGURE 4.6 Comparison between the Uncorrected and Corrected Curves.

N.B.: This example demonstrates why one should always include a control reactor in any batch test. Since the BMP of acetate is known, the difference between its theoretical value and the BMP calculated after introducing the correction V_1 is the error resulting from the blackout-induced underestimation and the correction applied. We can assume with good approximation that the BMP of all the other samples under test corrected in the described manner will have a similar error margin. We can then decide if the error is acceptable or if the test must be repeated. In our example, we know that the BMP of sodium acetate is a reference value equal to 350 Nml/g COD, and the amount of acetate used was 1.42 g eq. of COD. Calculating the BMP with the uncorrected measured volumes of the sample and blank gas productions yielded $BMP = 430 / 1.42 = 303$ Nml/gCOD, which means a 13% underestimation error. By applying the correction algorithm described above, the corrected curve yields an estimated $BMP = 332$ Nml/gCOD (–5% error). Such a result shows that the accuracy of the proposed method is acceptable and that the log is a powerful and easy-to-use feature of AMPTS II and AMPTS III instruments. The author recommends connecting Unit C (core unit in the case of third-generation instruments) to a UPS with at least 8 hours of autonomy at 15 W power output (see Section 4.3.3 on how to select one). By installing a UPS, the results in case of blackout will be more accurate than those obtained with the interpolation method described earlier. Furthermore, the UPS will provide additional protection against voltage surges that could damage the microprocessor or the local area network (LAN) interface.

4.3.3 Blackouts Caused by Power Uptake Peaks during Normal Operation

The solutions proposed in Section 4.2 present a new problem: each thermostatic bath absorbs about 1.5 kW of power, and the heating elements switch on and off randomly.

It may happen that in a given moment, both thermostatic baths are switched on, and their combined power adds to other equipment eventually in operation, such as the muffle oven, convection dryer, moisture analyser, or air conditioner. The total absorbed power can then exceed the installation's rated power, leading either the protection circuit breaker in the central electrical board of the building or the power-limiting relay of the electric meter board to switch off. To overcome this problem, we can employ a simple electromechanical timer controlling the coil of a relay. Relays usually have two ports: normally open (NO) and normally closed (NC). If each thermostatic bath is connected to each port, when one is on, the other will be off. The high thermal inertia of water ensures that none of the thermostatic baths will show significant temperature variations if the programmed cycles are 30–60 minutes long. Figure 4.7 presents the wiring diagram, and Figure 4.8 shows the power toggling system assembled in a standard DIN rail plastic distribution board.

LIST OF MATERIALS

- Six-module DIN rail distribution board, IP 40 protection or higher (1 piece)
- Bipolar DIN rail circuit breaker, 220 V × 10 A or 110 V × 20 A, depending on your national standard (one pc.). Optional: for further protection, you may also employ a circuit breaker with thermal + magnetic protection
- DIN rail 24-hour timer (1 piece)
- DIN rail monopolar relay with 220 V AC (or 110 V AC) coil and NC + NO contacts (1 piece)
- DIN rail sockets for your national standard (2 pieces)
- Copper wire, at least 2.5 mm^2 section, blue (neutral), brown (phase), and yellow-green (ground), necessary quantity
- Chord of suitable length with plug according to your national standard.

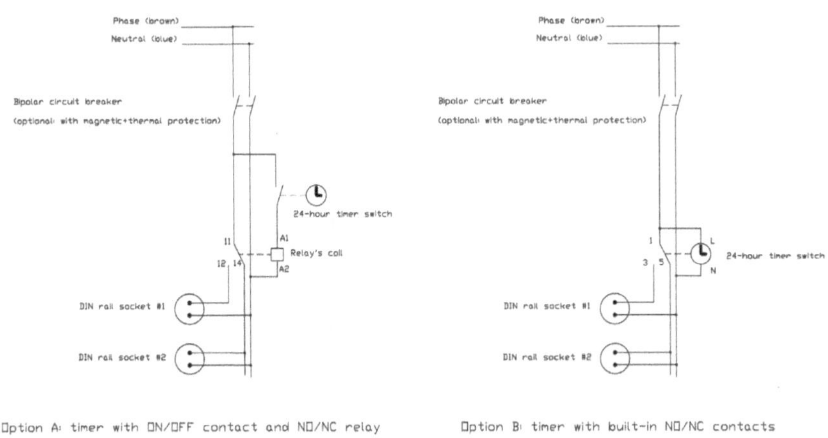

FIGURE 4.7 Wiring Diagram of the Power Interlock System, Which Allows Turning On Only One Thermostatic Bath at a Time.

FIGURE 4.8 Wiring of an Interlocking System Is Simpler When Using a Timer with Contacts Rated 16 A.

N.B.: Timers with built-in NO and NC switches with 10 A capacity or higher are also available in the market. If you choose this solution, then it is not necessary to include the relay. Figure 4.8 depicts a system assembled with an on/off timer with contacts rated 16 A, hence with no relay. In this case, the bipolar circuit breaker was not necessary because the board is connected to another board that already has it.

4.3.4 Tip: How to Choose a Suitable UPS for the Lab

Feeding your AMPTS through a UPS will protect delicate electronics from voltage peaks. There is no need to connect the thermostatic bath and the stirrers because a temperature drop of 2 or 3°C and a few hours without stirring will not affect the quality of the test. Both BPC Core of AMPTS III and Unit C of AMPTS II absorb roughly 15 W. Readers unfamiliar with electrotechnical units may find it difficult to choose a suitable UPS because these are usually rated in VA (volt-ampére) without stating the output power in W. To add complication, some manufacturers have invented a fancy unit called VAi (informatic volt-ampére) to give the impression that their products are more powerful. The following definitions will help clear the ideas.

Apparent power, measured in VA. It is just the product of the AC voltage applied to a load and the current it absorbs.

Active power, measured in W (watt). If the power source is DC (e.g., a battery) or the power source is AC, but the load is a pure resistance, then the active power coincides with the apparent power. If the power source is AC and the load is partly reactive – which is usually the case, then the active power is the product of the apparent power by the power factor:

$$P = U\,I\,cos\,\varphi,$$

where U is the root mean square voltage (220 V or 110 V, depending on your national grid), I is the current, and $cos\ \varphi$ is the 'power factor', with φ being the phase difference between the voltage and current waves.

Electronic devices are usually fed with AC through power adapters that convert AC (usually 110 V or 220 V) to DC (usually 12–24 V). Such AC power adapters usually behave as reactive loads; i.e., the current and voltage sinusoids are not in phase because of capacitors and inductors in the electronic circuitry.

Voltage spike: A sudden increase in voltage lasting for less than 3 nanoseconds.

Voltage surge: A sudden increase in voltage lasting for 3 nanoseconds or more.

Offline UPS: It is the cheapest technology. The PC or any other device is fed directly from the grid, and the battery stays in the floating mode (standby). In a blackout case, a relay will switch the grid connection off and feed the device from the battery. If possible, avoid this technology because it will provide little or no protection against surges or spikes of the grid voltage (e.g., in case of storm) that may damage the instrument's motherboard.

Online UPS: It is the best choice for lab instruments because it provides both maximum protection against voltage spikes or surges and operation autonomy. The devices are always fed from an inverter that gets power directly from a battery, regardless of the grid voltage.

Like in any other market, there are different qualities and prices of UPS models. Low-cost UPS usually use large capacitors and inductors to filter the output voltage wave, so even if you connect them to a purely resistive load (e.g., an incandescence lamp), their power factor will anyway be about 0.5–0.6. Upper-range products have more sophisticated circuitry, and their power factor is very close to 1. Upper-quality manufacturers usually state both the apparent power and the active power rating.

Another important technical datum is the backup time. This is the time gap in which the integrated battery of the UPS will be able to maintain the nominal voltage at the nominal maximum active load (or at another specified load level).

The following example shows how to check a UPS model's suitability to your case. The technical data are those published on the web page of a popular online shop and correspond to a low-cost model (€80 in January 2024).

Nominal power 1000 VA (in another line, it states 1 kVA, which is the same)
Output voltage: $220-240$ V, $50-60$ Hz
Battery: 12 V, 9 Ah

The page does not state the power factor, but the photo of the back of the unit shows the label with the rated input $220-240$ V, $50-60$ Hz, 5.4 A and output $220-240$ V, $50-60$ Hz, 560 W. Note the following: the 'commercial' nominal (apparent) power stated by the manufacturer is the value resulting from the minimum acceptable input voltage, 185 V (stated in another site where the technical description is more detailed), multiplied by the maximum rated input current, 5.4 A. As usual, the 'commercial' technical sheet specifies the apparent power to give the impression of a powerful device. If the photo of the back had not been available, we could have assumed $\cos\varphi = 0.55$, and the result would not have changed significantly. In any case, 560 W >> 15 W, so the rated power is more than enough. What you need to calculate is the longest operation time that this UPS model could provide to the AMPTS during a long blackout to check if the thermostatic bath will cool down enough to stop any gas production or if the said time is longer than the usual duration of blackouts in your area.

We know that the battery has 12 V DC × 9 Ah, so the maximum energy that it can provide is E = 12 × 9 = 108 VAh = 108 Wh (remember, for DC loads, VA = W).

If we connect only the AMPTS, the operation time will be t = 108 Wh/15 W = 7.2 h.

Just for comparison purposes, let us analyse an upper-range model (cost: €1,720 in January 2024). From its technical specification sheet, we know that

nominal apparent power: 2,200 VA
nominal (active) output power: 1,980 W
backup time at 50% output power: 11.6 minutes.

The total energy this UPS can yield is then

$$E = 0.5 \times 1980 \times 11.6/60 = 191.4 \text{ Wh}.$$

If we connect only the AMPTS, the UPS will keep it running for $t = 191.4/15 = 12.76$ hours.

For most applications, a low-cost UPS of 1000 VA with 15 minutes of autonomy at full power will be enough. Some middle-range models have an integrated communication module that will send a warning message (email or SMS), allowing the intervention of the laboratory operator or the maintenance personnel before the battery is exhausted.

4.4 REPAIRING BRUSHLESS STIRRER MOTORS

Conventional DC motors have two graphite sliding contacts, called brushes, that feed the rotor's windings when their position is orthogonal to the stator's magnetic field, providing the maximum torque for a given current. The rotation speed and sense are proportional to the applied voltage and polarity. Laboratory equipment manufacturers usually employ conventional DC motors for stirring reactors smaller than 20 litres because they are cheap and the controller is just a variable voltage source. DC motors are a commodity widely available for applications where energy efficiency is not a key factor. Their biggest disadvantage is that the brushes will wear out with time, producing graphite dust that can cause short circuits and damage the DC power source. While changing the graphite brushes is possible for DC motors larger than 200 W, many small models are sealed and cannot be repaired. Since 2013, all AMPTS instruments come with brushless stirrer motors, a particular type of DC motor having a permanent magnet as the rotor and several windings in the stator that are sequentially fed with DC pulses from an electronic switching array, making their operation principle similar to that of a synchronous AC motor, but with the advantage of nearly constant torque as a DC motor. Brushless motors have several advantages:

- Higher efficiency
- Longer service life because the only moving part is the rotor
- Built-in protection against short circuits or blocked rotors
- No sparks
- Torque is constant throughout the whole speed range
- The speed is proportional to the frequency of the pulses and not to the voltage, so there is no speed difference between the first and the last stirrer of a row
- Can be easily repaired.

Sometimes, one of the AMPTS stirrer motors may cease to work. The causes can be mechanical (e.g., an accidental fall or shock that deforms the motor axle) or electrical (e.g., voltage spikes or surges, random damage to any electronic component). When one or more stirrers of the AMPTS do not turn, the following procedure will help in identifying which motor(s) is (are) broken and whether it is possible to repair it (them):

1. Disconnect all the short cables that interconnect the motors. Connect the long cable from the control box to each single stirrer motor and identify which one(s) is (are) causing the problem.
2. Remove the faulty motor from the bottle. Move the motor axle with your fingers and check if it turns freely. If you notice that it is blocked or makes some strange noise, as if the rotor was scratching, then most probably the motor suffered a shock or some other mechanical damage, and the axle is deformed and cannot be repaired (1% of cases). You will need to purchase a new stirrer motor. Nevertheless, you could still remove the printed circuit mounted on top of the motor and keep it as a spare part for the future.
3. In 99% of the cases, the rotor turns freely when moving it with the fingers. Connect the motor directly to the long cable from the control box again and turn it on. If it runs when being out of its seat, then the problem is inside the reactor. Most probably, the motor was not correctly inserted on its seat, and the stirring rod got stuck. Reposition the stirrer on its seat, check if the axle turns freely when moving it with your fingers, and reconnect it to the long cable from the control box. If it turns correctly, the problem is solved. You can now reconnect all the cables and resume normal operation.
4. If the motor does not turn when taking it out of the bottle and connecting it directly to the control box, then the causes of malfunctioning are a damaged cable (0.1% of cases) or a broken/burnt printed circuit (98.9% of cases; Figure 4.9). Remove the plastic cap that covers the top of the motor. Then, remove the white connector by inserting a nail or a small flat screwdriver and sliding it out slowly. *Never pull from the coloured cables*: they are thinner than human hair and can easily break. Remove the four screws that hold the printed circuit (motor driver), replace it with a new one, secure it with the four screws, reconnect the white connector with the coloured cables, and place the protective plastic cap. Check that the motor runs correctly and resume normal operation.

FIGURE 4.9 Built-in Stirrer Driver Board for GL45 Stirrer Motors.

Sometimes, all motors run correctly when connected individually to the control box, but one or more of them may not run when interconnected. This situation is not frequent, and its causes can be either some dirty contact in one of the sockets placed on the driver board or in the cable connectors. Clean all sockets and cable connectors with contact cleaner spray. In older AMPTS II and GE II units (before 2020), all 15 motors are connected in series, and this problem may arise more frequently. The easiest solution is to purchase an additional cable and a power splitter, as shown in Figure 4.10.

4.5 THE THERMOSTATIC BATH'S ALARM KEEPS BUZZING DESPITE THE WATER LEVEL BEING CORRECT

Thermostatic baths supplied with AMPTS III, GE III, and BPC Blue from 2023 to the end of 2024 had a safety alarm that would sound when the water level is too low. The operation principle of the said alarm is based on water conductivity: if the water level is low, the resistance between the probe's electrodes will be higher because the section crossed by the probe's current is smaller, triggering the alarm. Sometimes, the alarm may continue to sound despite a correct water level because of some undesired bias in the probe or the batch of distilled/demineralised water having a lower conductivity than usual. The solution is to increase the water's conductivity slightly. The options are as follows:

a. Adding 0.2 mg of table salt per litre of distilled/demineralised water in the bath (8 mg for the full bath).
b. Adding 20 ml of tap water per litre of distilled/demineralised water in the bath. Salt dissolved in the tap water will be enough to increase the conductivity and mute the alarm. Do not exceed 2 l of tap water for the whole bath.

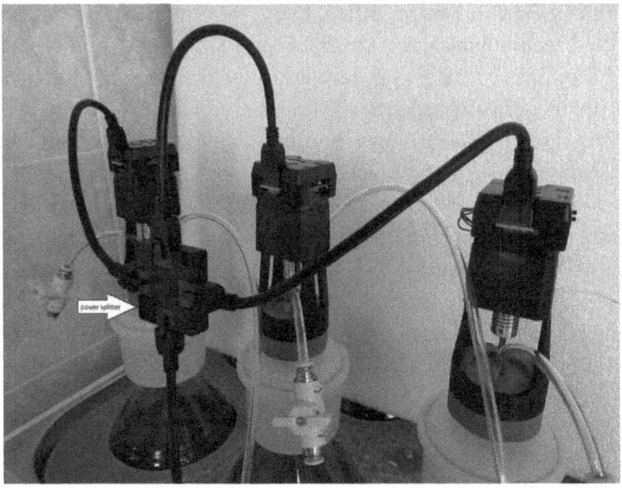

FIGURE 4.10 Power Splitter Connected to the First Three Stirrers of a Row.

In any case, when the water evaporates, and it becomes necessary to add more, do not repeat the addition of salt or tap water. Keep the water level by adding exclusively distilled/demineralised water. Never use tap water to fill the thermostatic bath because the alarm will not work until the bath dries off completely, and there is a high risk of burning the heating element.

The author employs ethylene glycol (EG) instead of water. This liquid is available in any petrol station or DIY supermarket as an anti-freeze additive for car radiators. EG does not evaporate, and the level sensor will not buzz when filling the bath with EG. In the author's opinion, filling the thermostatic bath with EG is the best option to avoid any risk of complete evaporation and damage to the incubation unit, especially during the summer holidays when the laboratory may remain unattended for several weeks.

4.6 HOW TO CONFIGURE THE ROUTER FOR ACCESSING YOUR AMPTS OR GE FROM THE INTERNET

Connecting your AMPTS, GE, BPC Go, or BPC Blue to the Internet is the best way to exploit its full potential. Bioprocess Control has pioneered the Internet of Things (IoT) technology applied to the biogas laboratory, so all of its products have built-in connectivity based on the IPV4 communications protocol. In a few words, there is no need to buy a dedicated PC for the instrument and install special communication software (e.g., TeamViewer® and AnyDesk®, or similar) to access it. All you need is to assign your router a public IP address and route it to your instrument's port, allowing any browser to find and access the instrument through the Internet. Such a feature allows the laboratory operator to monitor the tests from a remote location and to download the experiment's data in XLS, XML, and CSV formats, saving time and displacement costs. It is then possible to manage several biogas plants, each with its local laboratory, from a central office. Local laboratory operators do not need to be highly skilled technicians since the BMP protocol is easily implemented thanks to the AMPTS built-in software. A few hours of training is enough for a worker with basic skills to perform accurate BMP tests under the supervision of a skilled technician. This saves personnel costs because a single skilled engineer or biologist can supervise and optimise many plants without moving from his/her office or home. This section will show how to configure two AMPTS units installed in a central laboratory and a third unit placed in a biogas plant. All three units will be then accessible from anywhere in the world.

The first step is to assign a local IP address to your unit that is coherent with your router's private IP domain. Check your router's configuration manual to find out the LAN domain. For example, if the manual states to access your router's menu page with the address 192.168.0.1, then you need to configure your AMPTS units with IP addresses of the form 192.168.0.XXX, where XXX is any number different from the router's one or any other device connected to the LAN (for instance, printers or similar peripherals). Theoretically, the maximum value of XXX is 255, but some routers may be restricted to a smaller range. Hence, a good practice is to check your router's administration system before assigning the IP address to your AMPTS. In our example, the router's manual states that its default internal IP is 192.168.0.1. With your PC connected to the Wi-Fi or LAN of the lab, typing the said address in the browser's URL bar allows access to the router's configuration menu (Figure 4.11, callout A). In this case, the configuration menu corresponds to a router of the brand D-Link®. Although the graphics will vary depending on the router's brand and model, the configuration functions are universal.

FIGURE 4.11 Example of a Router's Homepage Showing the Configuration Menu. The graphic layout varies from one model to the other, but the functions are always the same.

In our example, we will arbitrarily assign 192.168.0.10 to the instrument called AMPTS A and 192.168.0.20 to the one called AMPTS B. The procedure for changing the default IP address of an AMPTS is described in detail in the user's manual [3]. When setting the IP addresses of each instrument, remember to set also the gateway. The default gateway is 0.0.0.0; you must enter the router's IP in each instrument. Figure 4.12 shows the configuration of AMPTS A. Once you have set the internal IP addresses and gateway in both instruments, it is time to connect them to the router's LAN connectors or the LAN wall socket if you have a larger network. Now you can access either AMPTS A or AMPTS B from any device connected to your LAN's Wi-Fi or wall socket just by typing its local IP address.

The second step is to check what kind of public IP address your router has. There are three possible cases:

1. The router has a static public IP. This is a feature that most Internet service providers (ISPs) offer for a small extra charge.
2. The router has a dynamic public IP. This is the most common default option offered by ISPs.
3. The router has a network address translation (NAT) IP. This case is often encountered when accessing the Internet through a 3G or 4G router because telephone companies often treat SIM cards as peripherals of their web servers. NAT IP addresses are usually of the forms: 10.XXX.XXX.XXX or 100.XXX.XXX.XXX or 172.XXX.XXX.XXX.

Before attempting the configuration of your AMPTS for remote access, you should check your case with your ISP. In this example, we will deal with the first two

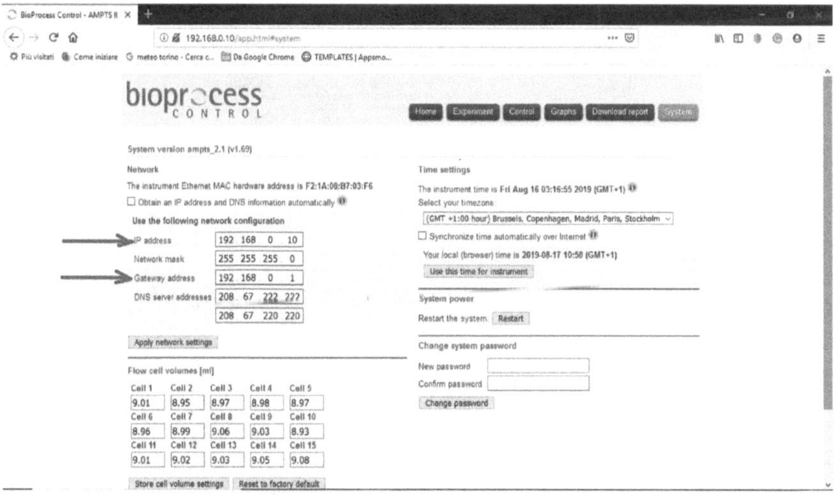

FIGURE 4.12 IP and Gateway Addresses in AMPTS II.

cases, which are the most frequent. Connecting your AMPTS through a NAT router requires a more complex procedure. If your router connection is via a NAT, then the easiest option is to connect the instrument(s) to the router as explained above, then connect a PC, and install a program like TeamViewer® or AnyDesk® to access the instruments via the dedicated PC. Note that the PC must remain on and connected to the same LAN or WiFi as the instrument.

In our example, the laboratory's router connects to the Internet through a static IP assigned by the ISP: 91.229.188.103 (Figure 4.11, callout B). Since many devices may be already connected to the router, it is necessary to configure it to send the traffic either to AMPTS A or AMPTS B as needed. In the jargon, this is called 'opening a port' of the router and 'forwarding' it to a port of your AMPTS instrument. Router manufacturers call this procedure 'port forwarding', 'port mapping', or 'virtual server' configuration. The following configuration procedure is the same for any router, although the screens will look different from one model to the other. In our example, the port forwarding menu is found by clicking on the 'advanced' icon and then 'virtual server' (Figure 4.13).

By default, AMPTS instruments have their internal port number 80 open. In our example, we arbitrarily chose to open router port number 81 (called 'service port') and route its traffic to port number 80 of AMPTS A, which was assigned the internal IP 192.168.0.10. In the same way, router port 82 is directed to AMPTS B port 80, having the internal IP 192.168.0.20.

FIGURE 4.13 Example of the Virtual Server Configuration Menu. The graphic layout may vary depending on the router's brand and model.

Now, you can easily access any of both instruments from a remote location. Just open your browser and type in the URL bar:

91.229.188.103:81 to access AMPTS A
91.229.188.103:82 to access AMPTS B

If you find it difficult to remember strings of numbers, no problem: there is a more human way. The simplest option is defining a subdomain of the type http://your-AMPTS.yourProvider.com, which will be easier to remember. Several companies, called DNS service providers, offer such services for free. In our example, we created a free account at www.noip.com. Then, we assigned the hostnames AMPTS-lab1.ddns.net and AMPTS-lab2.ddns.net. Figure 4.14 shows how to create two subdomains within a DNS service provider and assign them to the corresponding IP addresses. Note that we could have chosen any arbitrary name for the instruments and any available domain. For instance, MyAMPTS.pointto.us would have worked perfectly, too.

Now that both AMPTS II instruments in the lab are accessible from anywhere, it is time to configure the AMPTS Light installed in the biogas plant. In this case, the ISP is Vodafone, which provides standard Internet connections with dynamic IP addresses. The router's internal IP is 192.168.1.1. Figure 4.15 shows the configuration menu. We must assign our AMPTS Light with an IP address, for instance, 192.168.1.10, and type the router's IP address in the box 'gateway', and then click on 'save settings'.

Once the instrument's local IP is configured and connected to the router through the LAN cable, we must open a port in the router – in our example,

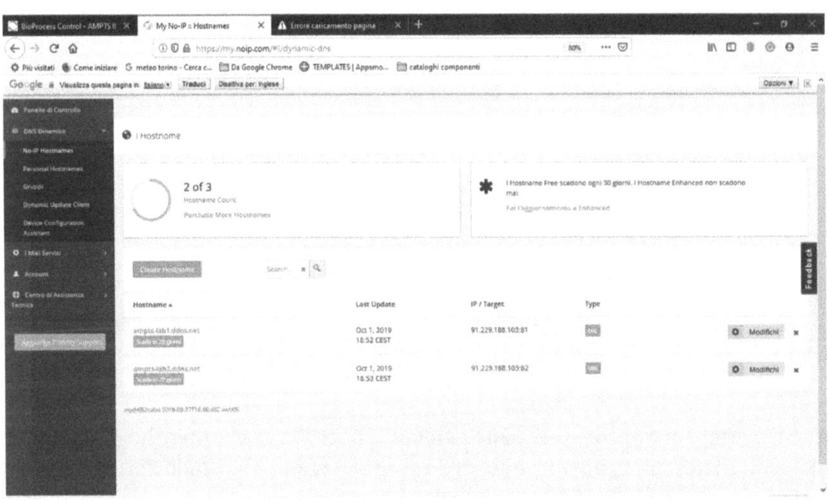

FIGURE 4.14 Example of the DNS Service Provider's Configuration Page.

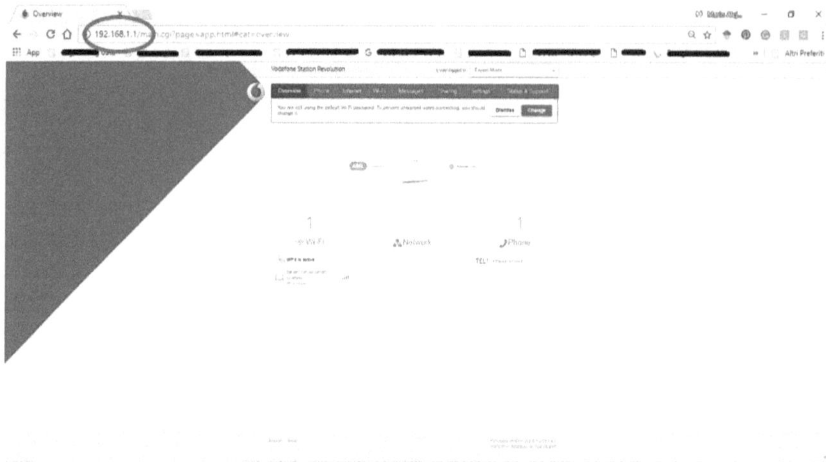

FIGURE 4.15 Router's Configuration Page Showing the Private IP Address.

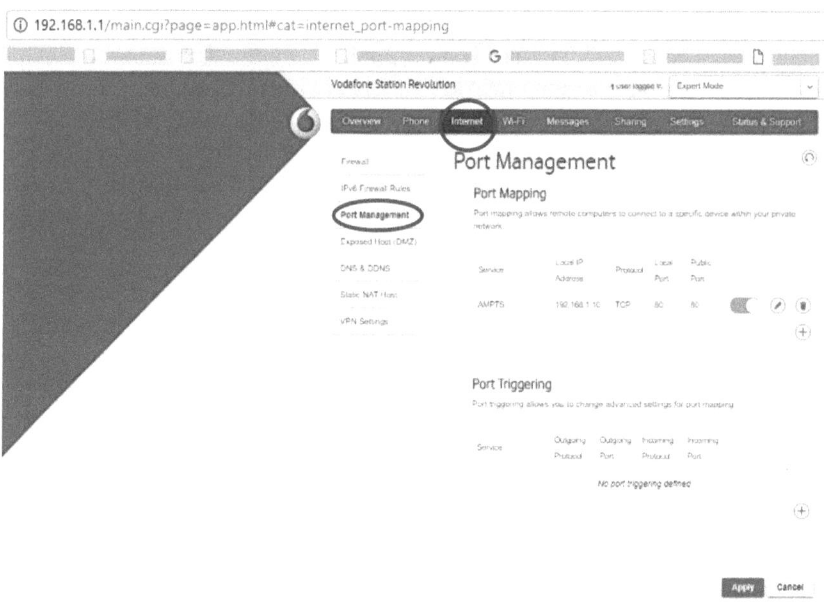

FIGURE 4.16 Port Management Configuration Menu.

port 80 – and forward it to the instrument's port 80. Note that the graphic may look very different from the one in Figure 4.11, but the logic is the same. In the Vodafone router, the function is called 'port mapping' and is found under the menu 'Internet' and submenu 'port management' (Figure 4.16). In this case, the router menu asks for the protocol of connection, which for AMPTS should be defined as 'TCP' or 'WEB' or 'Web Service' (again, the name depends on the router's brand and model).

Till now, the procedure is the same as in the case with a static IP address. Connections with dynamic IP require an additional setting, which is called 'DDNS' or 'DNS Settings' or just 'DNS'. In our example, the configuration page is found under the top bar menu 'Advanced' and the submenu 'DNS & DDNS' in the left bar (Figure 4.17).

Since we already created a free account at www.noip.com for the two instruments in the lab, we will choose NOIP from the list of DNS providers. Then, we will define the hostname AMPTS.hopto.org and also introduce our NOIP account name and password (Figure 4.18). Then click on 'apply' to save your settings. This will allow your router to synchronise with NOIP's server. That's why your AMPTS port routing is called a 'virtual server' in many router models.

Now, we must open a separate window in the browser, type www.no-ip.com, and enter the working area of our free account with the user and password chosen and the moment of signing up. Click on the 'create hostname' button and then choose whatever name and domain you prefer among the offered options. In our example, we chose ampts.hopto.org. Linking it to your router requires introducing your current public IP address. The router's public IP address can be found in the router's main menu (Figure 4.19). In our example, the public IP at the time of configuring the DNS service was 37.182.58.219. Save your settings. From now onwards, each time the router's public IP changes, the subdomain ampts.hopto.org will remain synchronised with it, and the router will redirect all the traffic arriving through the said subdomain to the AMPTS Light instrument.

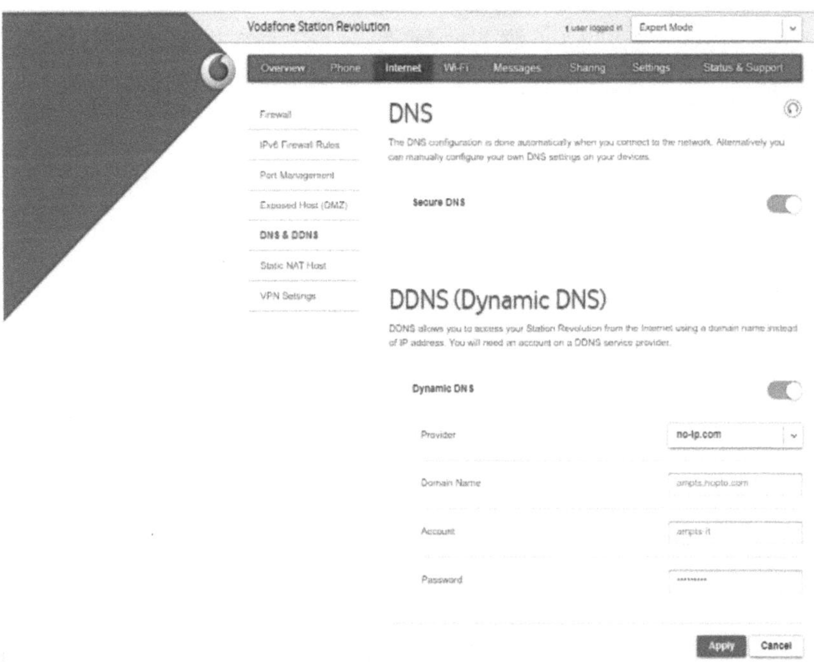

FIGURE 4.17 Example of the Dynamic DNS Configuration Page.

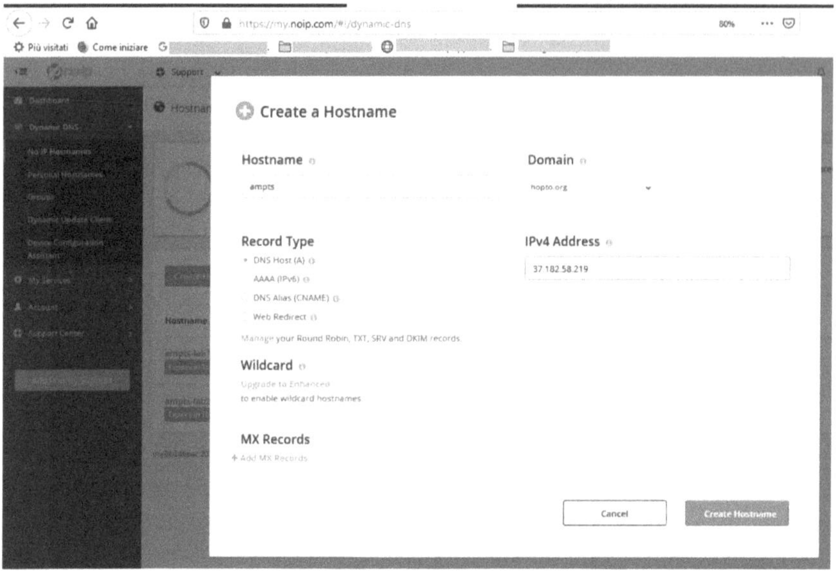

FIGURE 4.18 Example of Hostname Creation and Synchronisation.

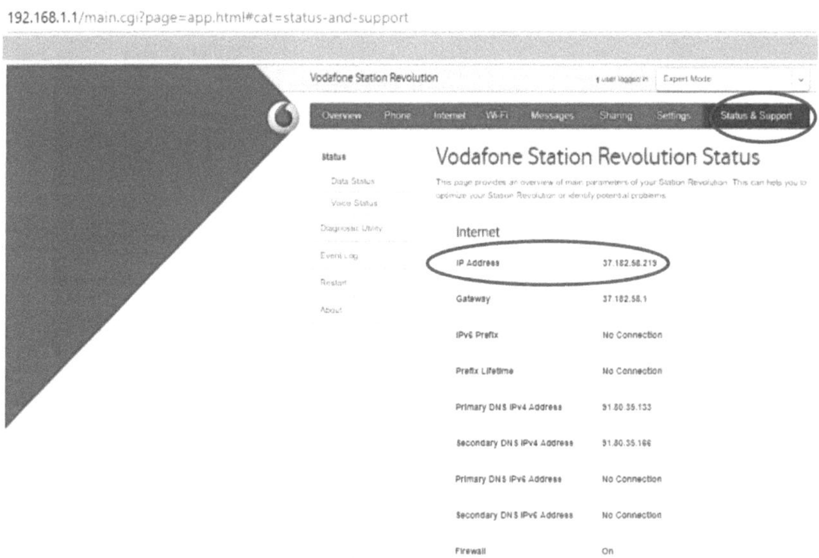

FIGURE 4.19 Router's Dynamic Public IP Address.

Remark: Note that DNS addresses take some time to propagate on the Internet. Hence, wait some minutes, or up to an hour, before attempting access to your instrument from the subdomain. The time for the activation may depend on many factors, including the local network speed and traffic volume.

Connecting an AMPTS to the LAN and configuring it for operator remote access require less than half an hour. Several DNS service providers offer free plans for dynamic DNS configuration and subdomains. Allowing remote access to the experiments saves travel time and money, especially when the biogas plant is far from the home or office of the person managing the anaerobic digestion process. It is enough to train an unskilled worker of the biogas plant to prepare and launch the tests with the AMPTS instrument, and then the engineer or biologist can supervise the same in real time from his or her headquarters.

4.7 FLUSHING THE REACTOR'S HEADSPACE WITH A LIGHTER REFILL CAN

The standard procedure to create an anaerobic environment in the reactor before starting the BMP test consists in flushing its headspace with N_2 or a mixture of 60% N_2 + 40% CO_2. Using a mixture is more complex and expensive than employing pure N_2 and does not warrant better results in terms of CO_2 underestimation (see Chapter 9, Section 9.1). Using pure CO_2 is unacceptable when employing a CO_2 trap to measure the CH_4 production only because the caustic soda solution will absorb almost all of the gas in the headspace, creating a partial vacuum volume and causing the inoculum to boil during the first hours of the test. The scope of flushing the headspace is just to eliminate the O_2 in the initial air for the anaerobic microbes to start their job immediately. So, it is not necessary to flush the headspace with an inert gas, as many researchers believe. For instance, one could easily employ biogas taken from another reactor. The author employs propane from a lighter refill can for several reasons:

a. Low cost;
b. It fits in the palm of your hand and can be kept in a desk drawer;
c. No need for an external deposit of gas bottles, pressure reducer, and gas distribution line, with all the associated safety compliance procedures;
d. Propane is as inert as N_2 for anaerobic microbes and equally insoluble in water.

Someone may argue that propane is explosive and its use may compromise safety in the laboratory. Let's check if such a statement is true:

1. The lower explosion limit (LEL) of propane in air is 2.1% (21,000 ppm).
2. The sum of the headspaces of 18 reactors of an AMPTS III is nearly 5.4 l. Supposing that the operator releases three times the flushed volume, the total volume of propane vented to the room is 16.2 l.
3. The worst laboratory environment possible to install an AMPTS III is a 6" container office, having a volume of roughly 20 m^3.

4. The resulting concentration of propane is then: $X = 0.0162\ m^3/20\ m^3 \times 100 = 0.08\ \% << 2.1\%$.

Hence, it is impossible to create an explosive atmosphere in the laboratory as a result of flushing the reactors' headspaces with propane.

4.8 USING THE *HOSTNAME.LOCAL* FUNCTION FOR AUTOMATIC NETWORK CONFIGURATION

One of the most useful features of AMPTS, GE, BPC Blue, and BRS is their ability to operate in the stand-alone mode or connected to an Ethernet network (LAN). In the first case, the standard way to access the instrument is by a peer-to-peer connection: you must set your PC with an IP address (usually 192.168.10.10) and type the instrument's default IP (192.168.10.11) in your browser. In the second case, the most common procedure is that the network administrator assigns a fixed IP address to the instrument. Then, it is necessary to access it with the default IP address as described above, then change the default address (along with the subnet mask and gateway), save the settings, exit, revert the PC's IP definition (either to the original IP address or to the 'assign IP automatically' option), and then connect the instrument and the PC to the LAN. Now, it is possible to access the instrument from any device connected to the LAN just by typing the assigned IP address in the browser's URL bar.

The advantage of a fixed IP address is the high stability of the data connection and saving it among the favourite URLs for immediate access. Nevertheless, people working in the stand-alone mode but needing to use the PC for other scopes must change the PC's IP address each time. Users who do not have a network manager in their company or institution may find it challenging or be wary of modifying the network configuration. If you recognise yourself in any of such cases, then the following procedure will solve all your troubles.

AMPTS III and its relatives come with three reset buttons. They are placed inside three pinholes in the back panel of the instrument. The central one is the Dynamic Host Configuration Protocol (DHCP) reset. When in the DHCP mode, the instrument's IP address will be assigned automatically. The procedure is as follows:

1. Connect the Ethernet cable to the PC (or to the LAN socket if you intend to use it in a network).
2. With the help of a pencil or pin, press the DHCP reset button for a few seconds (until you see the LEDs flashing).
3. Wait for 2–5 minutes. Then press the OK button or the Menu button in the instrument's front panel one or more times until you read in the display the word 'DHCP' and a set of numbers separated by dots below it. This is the IP that was assigned automatically to the instrument. Note that this IP address is dynamic, so it may vary over time.
4. Open the PC's browser (or any browser of any device connected to the network), type the IP address, and then press 'enter'.
5. Enter your password (the default is bpc) in the login and then go to the *Settings* page.

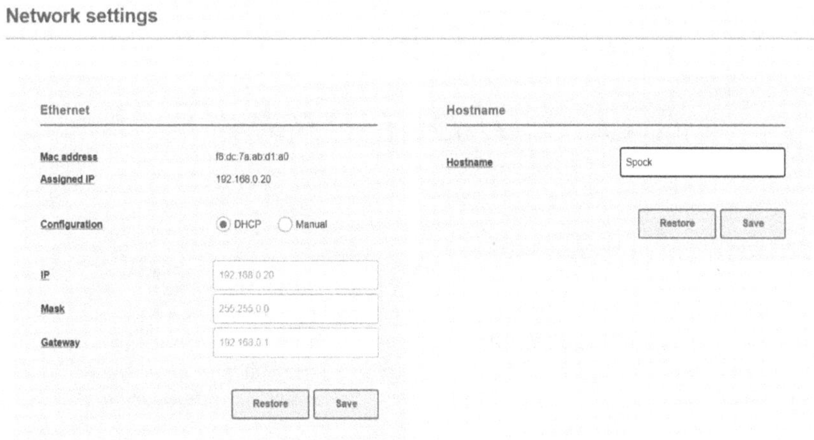

FIGURE 4.20 Defining the Hostname.

6. Scroll to the end of the *Settings* page until the section titled *Network Settings* (see Figure 4.20).
7. The default hostname is the MAC address written with hyphens instead of dots. Of course, it is unique but a bit difficult to remember. Fortunately, you can replace it with any name, on the condition that such name is unique to the network, i.e., no other device connected to the LAN has that name. In our example, we will call it Spock. Click on the Save button and close the browser.
8. From now onwards, any time you need to access your AMPTS, just open the browser of the PC (or the browser of any device connected to the LAN if you are working in a network) and type the chosen name followed by *.local* (in our example Spock.local). You can eventually save it in your browser as if it were a favourite webpage.
9. If you have two or more instruments, you can assign a different name to each, for instance, Spock, Kirk, and Uhura. There is no rule for the names; the only condition is that each one is unique to the network. Do not use names like 'My device', '#1', or your name because there is a high probability that a smartphone or any other device is already named like that, and they would cause a network conflict.

The author thanks Dr Sten Stromberg of BPC Instruments AB for the instructions on how to use the *hostname.local* function.

BIBLIOGRAPHY

[1] MA Rosato, *Managing Biogas Plants: A Practical Guide*, CRC Press (2017).
[2] Italian Standard UNI/TS 11703:2018. https://store.uni.com/uni-ts-11703-2018
[3] The manuals of all BPC products are freely available for download from https://bpc instruments.com/product-manuals/

5 The SMA
A Useful but Tricky Test

5.1 WHY MEASURING THE SMA

When you can measure what you are speaking about, and express it in numbers, you know something about it. Still, when you cannot measure it when you cannot express it in numbers, your knowledge is of a meagre and unsatisfactory kind: it may be the beginning of knowledge, but you have scarcely, in your thoughts, advanced to the stage of science, whatever the matter may be.

William Thomson, Lord of Kelvin, Popular Lectures and Addresses vol. 1 (1889) 'Electrical Units of Measurement', delivered 3 May 1883

Selecting the most suitable inoculum for the startup (or restart) of a biogas plant is often an overlooked problem. A biogas plant is a bacterial farm, so its management principles are the same as those of any other animal farm: selecting the most productive breed and thoroughly monitoring the health of each head. The difference between dairy and meat breeds is obvious to any farmer, and a veterinary visit is enough to check the health of the animals. But how can a biogas plant manager apply these concepts to 'farming bacteria'? The first volume of *Managing Biogas Plants* was devoted to checking the bacterial flora's health through

a. measuring the BMP of feedstock – or mixtures of different substrates – allowing one to find the optimum diet;
b. measuring the hydrolytic and proteolytic activities, allowing one to discover any signal of process disruption in progress;
c. knowing the most suitable instruments and laboratory techniques, and their error margins, allowing one to perform and interpret small-scale tests and translate them into management strategies; and
d. developing a Cartesian approach to critically interpret those methods that 'the literature' and 'the experts' recommend, saving time and money.

Choosing the best inoculum to start (or restart) a biogas plant is equivalent to choosing the best animal breed for a given farming context. The only difference between both activities is that a farm usually breeds a single species, while breeding microbes in the anaerobic inoculum is similar to managing a wildlife park with high biodiversity. For industrial scopes, assessing the biological activity of the acetoclastic

 DOI: 10.1201/9781003609278-5

methanogens is the simplest way of determining the quality of an inoculum. The following facts justify such an approach:

1. The acetoclastic methanogens constitute the last link of the anaerobic digestion chain.
2. They are responsible for roughly 75% of the methane produced in a digester.
3. They are the microbial group that can more easily suffer disruptions. If their biological activity is normal, then all the other bacterial groups are likely healthy, and the anaerobic digestion process will be stable.
4. The acetoclastic stage is usually the bottleneck of the anaerobic digestion process. The organic loading rate (OLR) that a digester can admit without suffering disruption is directly related to the abundance and activity of acetoclastic archaea.

The simplest way to quantify a bacterial group's vitality consists in measuring its respiration rate. Performing such a test periodically is very recommendable because of the following:

a. It is very selective. Depending on the reference substrate employed (carbohydrates, proteins, lipids, acetic acid, and a mixture of CO_2 and H_2), it is possible to measure the consistency and vitality of each bacterial group composing the inoculum: hydrolytic, proteolytic, and lipolytic bacteria, and acetoclastic and hydrogenotrophic methanogens.
b. It is easy to perform. There is no need to engage a biochemist or a microbiologist: anybody can perform a batch digestion test on the condition of having suitable instruments and following a precise method.

The specific methanogenic activity (SMA) assay consists of adding a given quantity of acetic acid (or sodium acetate) to a digestate sample and observing the total volume and flow of methane production. Conceptually speaking, this seems to be a simple solution, but, unfortunately, we have to deal with the 13th law of Murphy: 'The easy solution of a problem always generates a bigger problem to solve'. In the present case, the most difficult problem is how to reach a universal consensus on which substrate to employ, how much of it, how long the test must last, and how to define the threshold of acceptable methanogenic activity.

5.2 COD, VS, BMP, AND SMA: A SHORT CONCEPTUAL REVIEW

The first step to measuring something is to define its measurement unit. In the specific case of the SMA assay, every researcher seems to have invented a different way of carrying out the test and expressing its results. Although the most cited reference substrate in the literature is sodium acetate, followed by acetic acid, many studies about the methanogenic activity report having used other substrates: mixtures of volatile fatty acids (VFA), their salts, methanol, ethanol, glucose, saccharose, phenol, etc. A quick survey of the literature found attempts to measure the SMA with substrates that cannot be considered 'reference substrates' because of their highly variable composition: molasses, brewery malt, and unrefined cane sugar. Such a broad disparity of criteria represents a challenge for anybody trying to manage a biogas plant on objective data instead of tabulated values and self-referenced claims. It is then mandatory to standardise measurement units and methods to make the comparison of different tests possible. Since there is much confusion, even among professionals with many years of experience, it is useful to spend a few lines reviewing some basic notions and definitions. Then we can study the SMA test procedure in detail, finally defining how to employ its results in practical cases.

a. The following stoichiometric formula summarises the anaerobic degradation of acetic acid by the acetoclastic archaea:

$$C_2H_4O_2 \underset{AD}{\rightarrow} CH_4 + CO_2$$

Here, 1 mole of acetic acid (60 g) yields 1 mole of methane $\left(16\ g \equiv 22.4\ Ndm^3\right)$ plus 1 mole of carbon dioxide $\left(44\ g \equiv 22.4\ Ndm^3\right)$. In unit values, the methane conversion ratio of pure acetic acid is373 Ncm^3/g.

b. Considering sodium acetate instead of acetic acid:

$$C_2H_3NaO_2 + H_2O \underset{AD}{\rightarrow} CH_4 + NaHCO_3$$

Similarly to the former case, 1 mol (82 g) of pure sodium acetate yields 22.4 Ndm^3 of methane, i.e., 273 Ncm^3/g. The only difference between the digestion of acetic acid and sodium acetate is that the sodium ion in aqueous solution 'captures' carbon dioxide, forming sodium hydrogen carbonate (aka sodium bicarbonate). Hence, in theory, the biogas that leaves the reactor during the digestion of sodium acetate will be pure methane.

c. A more general stoichiometric calculation can demonstrate that the anaerobic digestion of organic matter will always produce CH_4 and CO_2 in quantities and proportions that depend on the content of C, H, and O in the biomass. There are three measurement units of the C concentration in the organic matter: chemical oxygen demand (COD, i.e., the minimum quantity of O_2 necessary for the total oxidation of the substrate), volatile solids (VS, i.e., the fraction of the dry matter except ashes), and total organic carbon (TOC). The first two are the most employed in the biogas industry because measuring them is simpler. Let's analyse the differences between them:

1. **COD**: Considering sodium acetate as the 'model molecule' of anaerobic digestion, we can represent its total oxidation with the following equations:

$$C_2H_3NaO_2 + 2\ O_2 \rightarrow 2\ CO_2 + H_2O + NaOH$$

$$2C_2H_3NaO_2 + 4\,O_2 \rightarrow 4CO_2 + 3H_2O + Na_2O$$

In the first case, 2 mols (64 g) of oxygen are needed to completely oxidise 1 mol (82 g) of sodium acetate. In the second case, oxidising a double quantity of acetate requires double the oxygen. Hence, in both cases, the COD of sodium acetate is the same, i.e., 64/82 = 128/164 = 0.78 g of O_2/g acetate despite the reaction products being different.

If we assume acetic acid as our 'model molecule',

$$C_2H_4O_2 + 2\,O_2 \rightarrow 2\,H_2O + 2\,CO_2$$

So, the COD of acetic acid is 1.066 g O_2/g, or if we consider its density, 1.12 g O_2/cm^3.

2. **VS**: If we want to characterise sodium acetate as a function of its VS, we will encounter a problem because the equations explained in the above paragraph represent two solutions, depending on the oxidation reaction, which, in turn, depends on the experimental conditions of the VS measurement. Hence, 82 g of anhydrous acetate will leave 40 g of ashes in one case, or 164 g of acetate will produce 62 g of ashes in the other. By definition, the VS are equal to

$$VS_1\,[\%] = 100 \times (TS - A)\,/\,TS = 100\,(82\text{ g} - 40\text{ g})/82\text{ g} = 51.2\,\%$$

$$VS_2\,[\%] = 100 \times (TS - A)\,/\,TS = 100\,(164\text{ g} - 62\text{ g})/164\text{ g} = 62.2\,\%$$

In practice, we will obtain VS percentages comprised within both theoretical limits, depending on the calcination temperature of the acetate sample.

If we assume acetic acid as our 'model molecule', it is impossible to characterise it in terms of VS: by definition, it is 100% volatile because it is an organic liquid containing no mineral fraction.

d. From the paragraphs above, we can deduce the following practical equivalence factors:
   1. BMP of sodium acetate = 273 [Ncm^3/g acetate]/0.78 [g COD/g acetate] = 350 Ncm^3/g COD. Since the VS of acetate can vary depending on the calcination temperature, VS is not a reliable parameter to define the BMP of acetate.
   2. BMP of acetic acid = 373 Ncm^3/g / 1.066 g COD/g = 350 Ncm^3/g COD.

e. It is possible to mathematically demonstrate that the BMP of any organic compound is always 350 Ncm^3/g COD, on condition that it is composed of only C, O, and H, and is 100% digestible too (see Table 5.1). The author assumed the total oxidation of each substrate to calculate the COD by stoichiometry, and the theoretical BMP deriving from the Buswell–Symons formula (from the Italian norm UNI/TS 11723, appendix B.2). It is impossible to apply similar reasoning to the VS because the result is not univocal: as shown in c. 2 above, the resulting VS depend on the calcination temperature of the acetate sample. The unambiguous determination of the SMA should then require loading the reactor with 'n g of COD' (regardless of the COD representing sodium acetate, vinegar, alcohol, palm oil, or any 100% digestible carbohydrate) and then observing if the methane production is $n \times 350$ Ncm^3.

5.3 THE SMA ASSAY IN PRACTICE

The scientific literature about the SMA test mentions many substrates (ref. 1). The first step in normalising the SMA is then defining which one to employ as the reference. From Table 5.1, it may seem that any pure substance composed of only C, O, and H will be suitable because all of them will produce 350 Ncm^3/g COD. From a formal point of view, a test done with any substrate different from sodium acetate or acetic acid is not properly an SMA test. For instance, using sugar as the reference substrate for a BMP or SMA test will show the global biological activity of three microbial groups: fermentative and acetogenic bacteria, and acetoclastic archaea. In the case such a test gives an unacceptable result, how can we know which group is inhibited or missing?

Rule: Only tests performed with sodium acetate or acetic acid can be properly called 'SMA tests'.

The second aspect to consider is the quantity of acetic acid or sodium acetate to employ. For instance, the Italian norm UNI/TS 11703:2018 [2] and the Anaerobic

TABLE 5.1
BMP of Some Pure Substances Calculated with the Buswell–Symons Formula

Compound	Formula	Molecular Weight	COD (g O_2/g Product)	BMP (Ncm^3/g Product)	BMP (Ncm^3/g COD)
Ethanol (anhydrous)	C_2H_6O	46	2.087	730	350
Acetone	C_3H_6O	58	2.207	772	350
Glycerine (glycerol)	$C_3H_8O_3$	92	1.217	426	350
White sugar (saccharose)	$C_{12}H_{22}O_{11}$	342	1.123	393	350
White vinegar (alcohol vinegar)	6% $C_2H_4O_2$ + 94% H_2O	19.56	0.064	22	350
Isopropanol	C_3H_8O	60	2.400	840	350
anhydrous acetic acid	$C_2H_4O_2$	60	1	373	350
Methanol (anhydrous)	CH_3OH	32	2	525	350
Palmitic acid	$C_{16}H_{32}O_2$	256	3	1006	350
Maleic anhydride	$C_4H_2O_3$	98	0.980	343	350
Lysine (amino acid)	$C_6H_{14}N_2O_2$	146	1.86	537	288

Digestion Task Group of IWA [3] both recommend including a positive control with sodium acetate in every BMP test to assess the quality of the inoculum. If the BMP is less than 332 Ncm^3/g COD, then either the inoculum may have an insufficient methanogenic activity or the test procedure is wrong. The cited norm does not define the inoculum/substrate ratio (ISR) to choose, implying that it should be 'equal to or higher than 2' as in any BMP test. How to choose the most suitable ISR then? Sodium acetate will not alter the pH of the inoculum, but an ISR close to 2 may produce 'false negatives' because sodium is an inhibitor of the methanogenic archaea, and we usually do not know the initial concentration of sodium in the inoculum. Figure 5.1 shows the results of testing a sample of inoculum with sodium acetate at different ISRs. Note that the ISR influences the added sodium concentration, but its initial content in the inoculum was unknown. This experiment demonstrates that the BMP of sodium acetate depends on the ISR. From the point of view of a biogas plant manager, this uncertainty may have negative economic consequences. For instance, testing an inoculum and failing to reach the theoretical 350 Ncm^3/g COD (or at least 95% of the said value) may lead to the wrong conclusion that the methanogenic activity is weak, but the cause may be just having chosen the wrong ISR. A wrong interpretation of the test may lead to wrong operational decisions and, in the last instance, to a loss of profit. On the other hand, performing several tests at different ISRs to avoid the uncertainty of eventual negative results is not practical. As an alternative, it is possible to employ acetic acid instead of sodium acetate, but the ISR should exceed 10. Otherwise, there is a risk of lowering the pH of the inoculum too much. Furthermore, the acid will react with the bicarbonates already dissolved in the inoculum, overflooding the reactor with foam (the so-called beer effect).

Regardless of choosing sodium acetate or acetic acid as the reference substrate, which ISR is safe enough to avoid producing false-negative results because of substrate-induced inhibition? The scientific literature does not provide a single value. Some researchers recommend keeping the concentration of sodium acetate below 7 g/l of inoculum [4]. Others state the range 4–8 g/l as valid only if pH > 7.5 [5], which, in general, is true for most inocula from agricultural biogas plants. Finally, one paper recommends a minimum concentration of 2 g acetate/litre of inoculum [6]. It is worth pointing out that the said dosage ranges result from experiments with anaerobic sewage sludge. The author obtained good results with acetate concentrations as high as 12 g/l of inoculum (granular sludge). The concentration limit for checking the methanogenic activity of digestate from agricultural biogas plants seems to be 10 g acetate/l digestate. Since the author has an AMPTS II with 500 ml reactors, he has adopted 3 g anhydrous sodium acetate in 400 ml of inoculum (i.e., a concentration of 7.5 g/l) as an internal standard for his laboratory. When employing pure acetic acid (aka glacial acetic acid), the total concentration should not exceed 2.4 g/l of inoculum; otherwise, the pH could drop too much [7]. For routine tests, cheap alcohol vinegar (aka white vinegar, usually employed as a household anti-limescale) instead of pure acetic acid gives good results. Since vinegar usually contains 6% acetic acid, reaching the recommended concentration requires the addition of 40 ml of vinegar/l of inoculum.

5.4 THE MEASUREMENT UNITS OF THE SMA

The respiration rate of any living being is usually expressed in Nm^3 of air (or oxygen) per live weight unit (usually kg) per unit of time. In the case of an anaerobic process, the specific respiration rate of the methanogenic archaea population is then the methane flow per unit of microbial mass in the inoculum (which includes archaea, bacteria, and any eventual microorganism group present). The usual measurement unit of the mass of live microorganisms is their VS concentration (g/dm^3 or kg/m^3). The literature usually reports the gas flow in Nml/hr or Nml/day, although the correct SI unit should be Nm^3/s (or its multiples or submultiples, e.g., $Ncm^3/hour$). The methanogenic activity is usually a curve plotted with the daily (or hourly) methane flow per unit of VS on the *y*-axis as a function of time on the *x*-axis. The SMA is the numerical value of the curve's peak, usually expressed in Ncm^3 of $CH_4/g\ VS_{inoculum}.day$ or the equivalent $g\ COD/g\ VS_{inoculum}.day$ $\left(\text{remember } 1\ g\ COD = 350 Ncm^3 \text{ of } CH_4\right)$. Figure 5.2 shows the SMA curves corresponding to the same tests reported in Figure 5.1.

Comparing Figures 5.1 and 5.2, we can draw some interesting conclusions:

a. The highest ISR yielded the theoretical BMP of sodium acetate but did not produce the highest SMA.
b. The highest values of SMA correspond to two ISRs that yielded acceptable BMP results (5% smaller than the theoretical one, as per UNI/TS 11703:2018 and IWA recommendation).

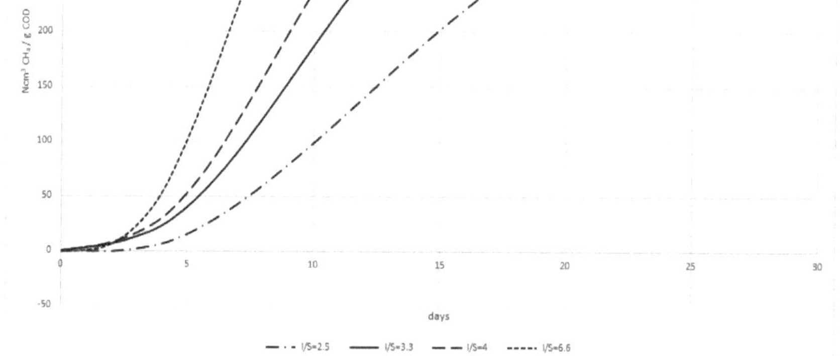

FIGURE 5.1 SMP Curves of Sodium Acetate Measured with Different ISRs.

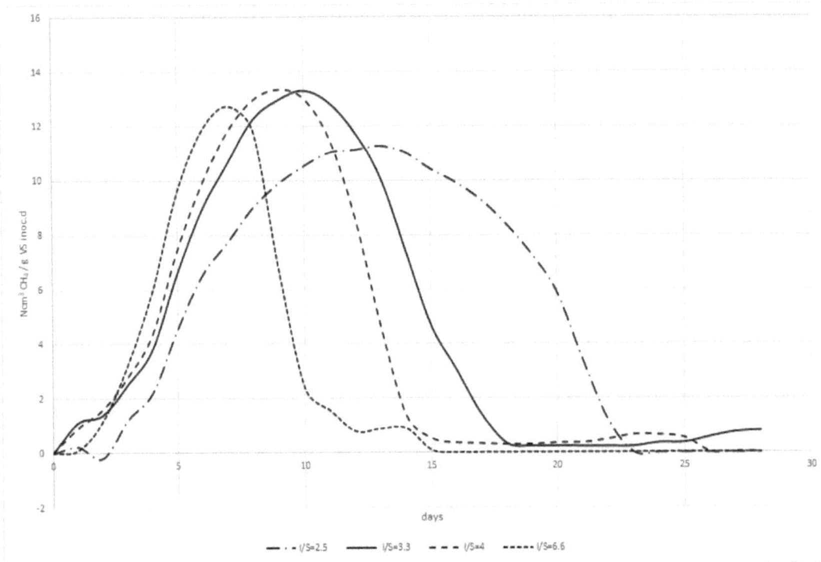

FIGURE 5.2 SMA Curves Resulting from Tests Performed at Different ISRs.

c. The lowest SMA corresponds to the lowest (unacceptable) BMP result, i.e., the lowest ISR.
d. All SMAs are higher than 10 $Ncm^3/g\ VS_{inoculum}/day$.
e. The time necessary for reaching the peak of each curve is inversely proportional to the corresponding ISR.

We can then define a rule of thumb to eliminate the uncertainty when the test with acetate fails reaching a BMP of at least 330 Ncm^3/g COD. If SMA > 10 $Ncm^3/g\ VS_{inoculum}$.day, then most probably the inoculum's methanogenic activity is acceptable. The reasons for the low BMP could be either procedural problems (weighing errors, gas leaks, or the inoculum was not preincubated) or partial inhibition resulting from a too low ISR. A BMP < 330 Ncm^3/g COD together with SMA > 10 $Ncm^3/g\ VS_{inoculum}$.day may eventually indicate a high initial concentration of sodium in the inoculum. Repeating the test with acetic acid instead of sodium acetate may be useful to assess the said eventuality.

It is important to clarify how to measure the methane flow. Some authors log the maximum hourly flow of CH_4 and multiply its peak value by 24 to obtain the maximum (theoretical) daily flow. Values in the literature exceeding 30 Ncm^3 / g $VS_{inoculum}$.day derive from this convention (see for instance [8]). There is no rule to decide which convention is better to adopt. If you need to select the inoculum for a UASB reactor that will operate at constant flow and COD concentration, then the maximum hourly methane flow from the batch test is the most useful selection parameter. In this particular case, the maximum SMA under continuous flow operation will be 24 times the maximum hourly flow measured under batch conditions. If, on the contrary, you need to select the inoculum for a digester fed with agricultural or municipal waste, then an average SMA calculated on the maximum daily methane flow is an adequate criterion.

CASE STUDY 5.1: A DUBIOUS POSITIVE CONTROL WITH SODIUM ACETATE AND HOW TO INTERPRET IT WITH THE SMA CURVE

A recurrent problem in the laboratory is the plant owner's hurry. In this case, the customer wanted the results of several BMP tests 'urgently' because he needed to decide whether the inoculum of his plant would be capable of digesting each of the candidate feedstock. Preincubating (aka degassing) the inoculum is an important step of the BMP test procedure. It usually requires at least 1 week, especially when the inoculum is 'fresh' digestate, and minimises the endogenous background methane production, which is a source of uncertainty. The customer asked not to preincubate the inoculum to get the BMP results 'as soon as possible'. The customer also chose to perform the tests with the 'minimum protocol' of the UNI/TS 11703 (two replicate samples instead of the customary three) and a control reactor with acetate (a single reactor instead of two replicates) to 'contain costs'. In absolute compliance with Murphy's law, some tests on the candidate substrates showed either signs of inhibition or too high dispersion between replicates. The problem in such cases is to find out the cause of the dubious results. Was there a problem with the test preparation or with the instrument? Are the feedstocks intrinsically inhibiting for the anaerobic bacteria? Is the methanogenic activity of the inoculum too weak?

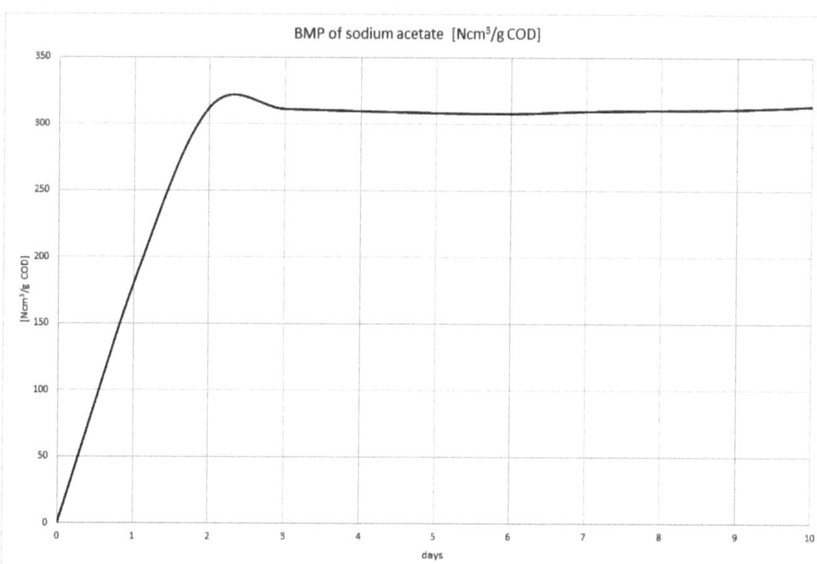

FIGURE 5.3 SMP Curve Has a 'Hump', Leading to an Uncertain BMP Value.

The positive control allows answering the last question when the test results are uncertain. In our case, the control test gave an uncertain BMP result, as shown in Figure 5.3.

The curve's shape is atypical: in a BMP test, the curve must reach a flat asymptote (aka plateau). A segment with a negative slope usually indicates inhibition, so was an acetate concentration of 8 g/l high enough to cause inhibition by sodium excess? Or is the atypical shape of the curve the result of bypassing the preincubation step? Even ignoring the shape of the curve, the maximum BMP is lower than the quality threshold defined in the UNI/TS 11703 norm. The result of the acetate BMP test seems to indicate that the inoculum suffered partial inhibition. Having data from a single reactor makes it impossible to discard the possibility of human or instrumental errors, although the probability, in this case, was extremely low. Fortunately, plotting the SMA curve (Figure 5.4) explains the mystery.

The SMA curve has a sharp peak, indicating good methanogenic activity. If there had been some degree of inhibition, the curve would be flatter and wider (like the one in Figure 5.2 corresponding to ISR = 2.5). The peak value of the SMA is higher than 10 Ncm^3 / g $VS_{inoculum}$.day. We can then conclude with reasonable certainty that the anomalous shape and low value of the BMP curve derive from having bypassed the preincubation step. Another quality criterion, not mentioned in the UNI/TS 11703 standard but defined in the literature (ref. 3 already cited), corroborates the last supposition. On checking the data from the AMPTS II instrument, it was possible to assess that the blank reactors produced more than 30% of the total methane volume produced by the reactor containing

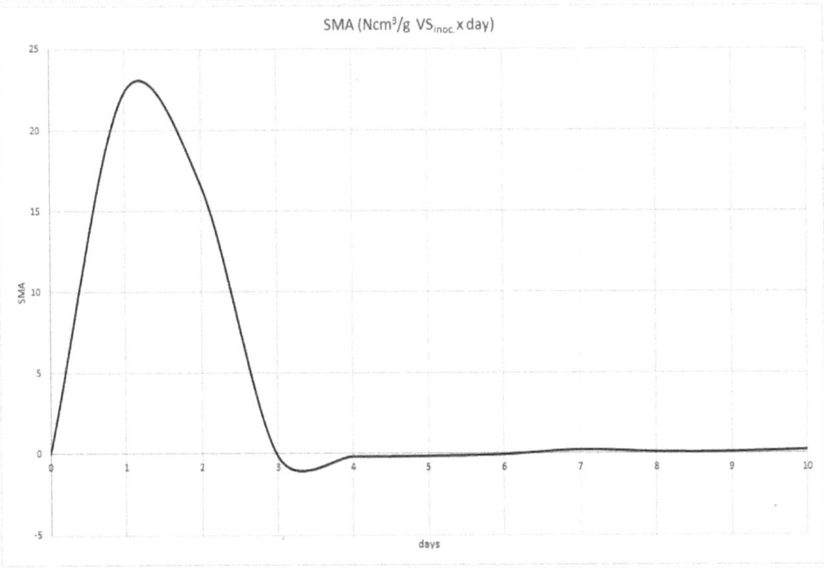

FIGURE 5.4 A Curve Having a Sharp Peak Indicates a Correct SMA.

sodium acetate, which is too much. The IWA proposes that the average methane production of the blank should be smaller than 20% of the average of the sample reactors. Since this test does not reach the quality threshold of BMP tests, it is possible to exclude the hypothesis of insufficient methanogenic activity of the inoculum. We can then conclude that some of the tested substrates are potential inhibitors for the inoculum in question, so their use in the customer's biogas plant must be avoided, or at least limited to small quantities.

5.5 COST OPTIMISATION USING THE SMA

Knowing the SMA of an inoculum allows one to optimise the operational costs of the biogas plant. This section explains how.

CASE STUDY 5.2: SELECTION OF THE INOCULUM FOR THE STARTUP OF A BIOGAS PLANT

Starting a new biogas plant or restarting one after a stop for maintenance requires sourcing an adequate quantity of inoculum of good quality. In this real-life case, there were two biogas plants at a reasonable distance from the one that needed the inoculum. The transport cost of the digestate from plant A is €3.65/m^3 and that from plant B is €4/m^3. Samples of both digestates were tested in the lab, obtaining the SMA curves shown in Figure 5.5. Which inoculum should the customer employ to start his/her digester?

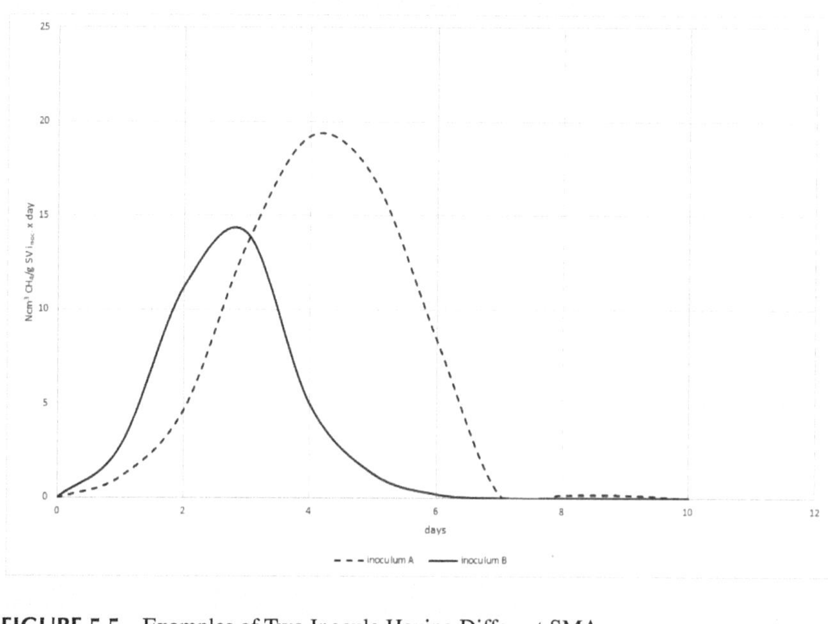

FIGURE 5.5 Examples of Two Inocula Having Different SMAs.

From the point of view of the SMA, inoculum B is better than inoculum A, but the absolute value of the SMA is not enough to decide: after sieving the macroscopic solids, it turns out that inoculum A has 5% VS, while inoculum B has only 2%. Comparing the absolute SMA of 1 m^3 of inoculum, we obtain

$$SMA_A = 0.05 \times 1 \text{ ton/m}^3 \times 14 \text{ Nm}^3\text{/ton VS.day} = 0.7 \text{ Nm}^3\text{/m}^3 \text{ inoculum.day}$$

$$SMA_B = 0.02 \times 1 \text{ ton/m}^3 \times 19 \text{ Nm}^3 \text{ / ton VS.day} = 0.38 \text{ Nm}^3\text{/m}^3 \text{ inoculum.day}$$

The cost of each SMA unit is then

$$C_A = €3.65\text{/m}^3\text{inoculum /0.7 Nm}^3\text{/m}^3 \text{ inoculum.day} = €5.21\text{/Nm}^3\text{.day}$$

$$C_B = €4\text{/m}^3\text{inoculum / 0.38 Nm}^3\text{/m}^3 \text{ inoculum.day} = €10.5\text{/Nm}^3\text{.day}$$

Despite inoculum A having the lowest SMA, its unit cost per Nm^3 of methane produced is half that of inoculum B. Hence, inoculum A is the best choice in this case.

5.5.1 Extension of the Concept of SMA

The SMA test provides useful information about the biological activity of the acetoclastic archaea population in the digester. Nevertheless, nobody feeds digesters with acetic acid or sodium acetate. By its nature, anaerobic digestion is the most sustainable process for treating organic waste. Waste streams usually consist of polysaccharide mixtures, VFA, proteins, lipids, partially saponified fats, phenols, etc. Checking the inoculum's capacity to degrade such complex substrates is crucial. Figure 5.6

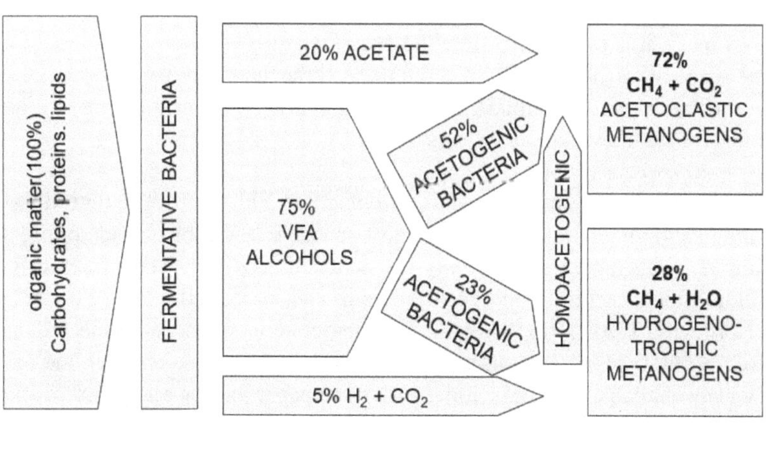

FIGURE 5.6 Interaction between Different Bacterial Groups in the Anaerobic Digestion Process.

shows the degradation paths of organic matter along the whole anaerobic digestion process and the microbial communities involved in each step. The reader can realise that the SMA alone only provides a measure of the robustness of the final step of the whole process, but as the old proverb says: *A chain cannot be stronger than the weakest of its links.* As we will see in the following paragraphs, there are cases in which the SMA is good, but the inoculum cannot digest a given substrate because the other microbial groups are not active enough.

Like in any ecosystem, the microbial communities in a digester's inoculum evolve and adapt to obtain the maximum nourishment from the feedstock they receive. In Italy and other European countries, over 90% of the biogas plants operate with manure in variable proportions and agricultural waste. Selecting the most suitable inoculum for a given feedstock or mixture is a question of good sense: if the feedstock consists mostly of vegetal matter, sourcing digestate from a plant running on the same or similar feedstock is the most logical option. A good alternative inoculum is manure from ruminants (i.e., cows, buffaloes, sheep, goats, or camels) because their intestinal flora is adapted to digesting lignocellulosic matter. If, on the contrary, the biogas plant must digest animal or industrial waste (slaughterhouse residues, glycerol from biodiesel or soap production, etc.), then manure from swine or poultry farms is a good option because the gut flora of monogastric animals is adapted to digest proteins and lipids. Sometimes, the quality of the digestate or manure available locally is not the best. This does not automatically mean that inoculum from a digester fed with vegetal matter will not be able to degrade slaughterhouse waste or vice versa. In such cases, it is necessary to check if the available inoculum has the initial capability of digesting the biomass the new plant is meant to treat. If it doesn't, start the new plant at low OLR: time and Darwinian selection will do the rest. Performing several continuous tests at different OLRs will help shorten the adaptation period.

Organic matter consists of variable proportions of carbohydrates, proteins (or amino acids), and lipids. The bacterial groups that degrade the said molecules are called hydrolytic, proteolytic, and lipolytic, respectively. The rational way of selectively checking the specific biological activity of each bacterial group composing the inoculum consists of launching BMP tests of reference substrates. In the same way that the SMA is conventionally defined as the daily flow of methane per unit of inoculum VS mass, it is possible to state the specific hydrolytic, proteolytic, or lipolytic activity. The author has not found references to the said approach in the literature, but he employs it as a benchmark of inoculum quality. The most common reference substrates for such tests are as follows:

1. **Cellulose**: It is the most complex carbohydrate, although 100% digestible. If the inoculum degrades cellulose, most probably it will be capable of digesting any vegetable biomass (only the hemicellulose and pectin fractions) or chitin (the structural component of fungi and insect cell walls). The Italian Norm UNI 11703:2018 [9] recommends employing monocrystalline cellulose (CAS 9004–34–6). The author has successfully employed cheaper and more widely available materials: toilet and tissue paper (only white paper, with no printed patterns), white cotton (unwaxed, finely chopped), and laboratory filter paper. The theoretical BMP of cellulose is $415\ \text{Ncm}^3/\text{g VS}$; the inoculum

is acceptable if it can yield at least 80% of the said value, i.e., 332 Ncm^3/g VS. The test requires 2 < ISR < 3, lasting usually 15–20 days. N.B.: when employing toilet or tissue paper, the BMP can be as low as 320 Ncm^3/g VS. Such a low value does not necessarily mean anything wrong with the inoculum: the cause is just that the sample may not be pure cellulose.

2. Animal gelatine (aka flavourless jelly, usually sold in foils or powder form): It is pure protein from fish or bovine skin and cartilage, practically anhydrous, ashless, and easy to dose. Do not employ vegetable jelly (being it either pectin or alginate, i.e., a polysaccharide). Alternatively, egg white, tofu (soy protein), and casein are widely available in supermarkets. The theoretical BMP of protein is 496 Ncm^3/g VS. Like in the former case, the inoculum is acceptable if it yields at least 80% of the theoretical value, i.e., 397 Ncm^3/g VS. The test requires 2 < ISR < 3, taking usually 15–20 days.
3. Vegetal oil (sunflower, olive, corn, peanut, or soy): It is impossible to measure the VS of oils. Pre-dosed vials for wastewater COD tests are unsuitable for testing oils and fats because oil is immiscible in water. We must assume that oil has 2.9 g COD/g of dry oil. The theoretical BMP of oils and fats is 1,014 Ncm^3/g dry product. The test requires ISR > 5 to avoid the inoculum's inhibition, usually requiring 30–60 days for completion.

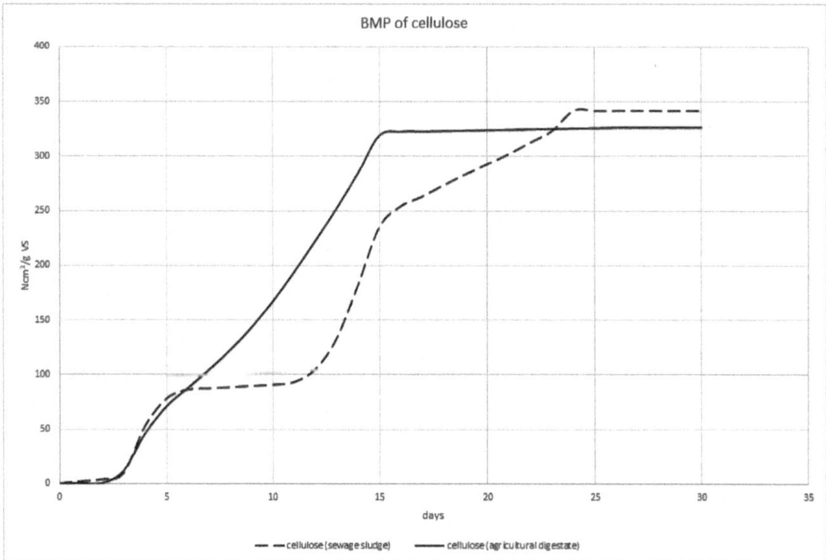

FIGURE 5.7 Using Sludge and Agricultural Digestate as Inocula Results in Different SMP Curves of Cellulose.

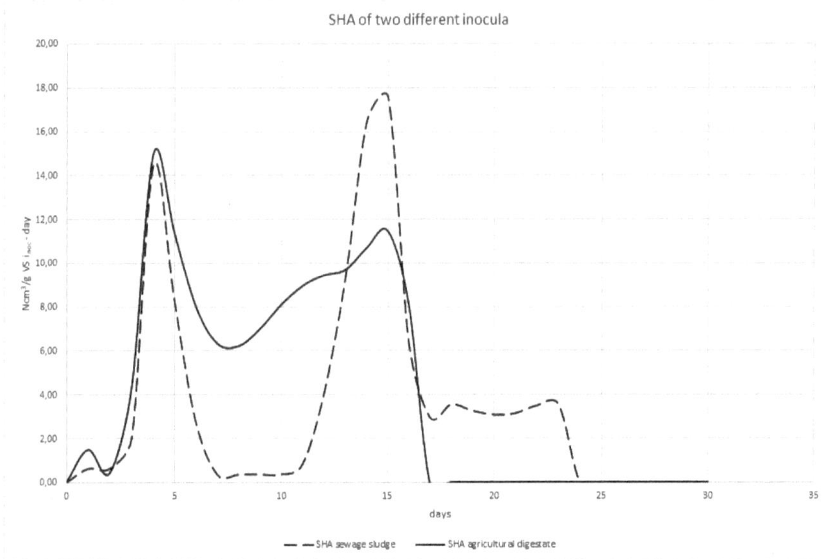

FIGURE 5.8 SHA of Cellulose Retrieved from Figure 5.7, Showing Two Peaks Each.

Figures 5.7 and 5.8 show how to employ the concept of specific hydrolytic activity (SHA) to evaluate the quality of any anaerobic inoculum meant to digest vegetal biomass.

The inoculum from agricultural digestate yielded only 326 Ncm^3/g VS of cellulose, apparently an unacceptable value. Nevertheless, the said value is just 2% less than the conventional minimum, a difference within the error margin of the test. Unexpectedly, measuring the BMP of the same cellulose sample with inoculum from sewage sludge yielded 342 Ncm^3/g VS, which is just 3% over the conventional minimum and, again, within the error margin of this test. The BMP differences may be then irrelevant and both inocula equivalent, but the shape of the curve resulting from sewage sludge is atypical: it reaches the final value 10 days later than the one of agricultural digestate.

Plotting the quotient between the daily methane flow and the VS of each inoculum, i.e., the SHA, provides some further information. Both curves show peaks on the 4th and 15th days, but the sewage inoculum presents an almost-null SHA between these days. On the other hand, the agricultural inoculum shows positive SHA throughout the test. Both inocula are barely acceptable, although the agricultural one seems slightly better. A 'healthy' inoculum should present a wide SHA curve with only one peak, like the one shown in Figure 5.9. The absolute value of the peak should be higher than $12\,Ncm^3 / g\ VS_{inoc}.day$.

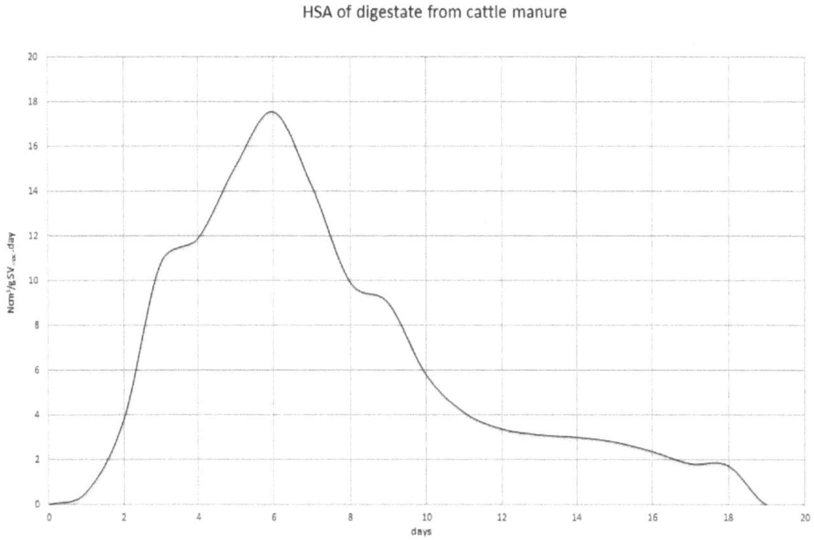

FIGURE 5.9 SHA Curve of a Healthy Inoculum Must Have Only One Peak.

CASE STUDY 5.3: FINDING OUT THE CAUSES OF PROCESS DISRUPTION

A biogas plant fed with a mix of cereal silage and agricultural by-products had been running stably for a year after its commissioning. Suddenly, the plant's productivity reduced to 50%–60% of its nominal power, despite the quantity and quality of the mixture remaining the same. After measuring the VFA/total alkalinity (TA) ratio and without conducting any digestion test, the 'biologist' in charge of the aftersales service proclaimed that the cause of the problem was 'too a heterogeneous feedstock mix'. He upheld feeding the digester with corn silage only. The diet change brought a further productivity reduction. After several months of low productivity, the plant's owner decided to test the inoculum's biological activity. Figures 5.10–5.12 will allow us to understand what happened.

Despite the test with cellulose reaching the minimum acceptable BMP, the curve's shape is irregular and does not fit into the classical sigmoid. The test with gelatine started normally, but the shape of the curve indicates some inhibition from the 10th day. The SMA test undoubtedly shows the inhibition of the inoculum's methanogenic flora. The said combination explains why the

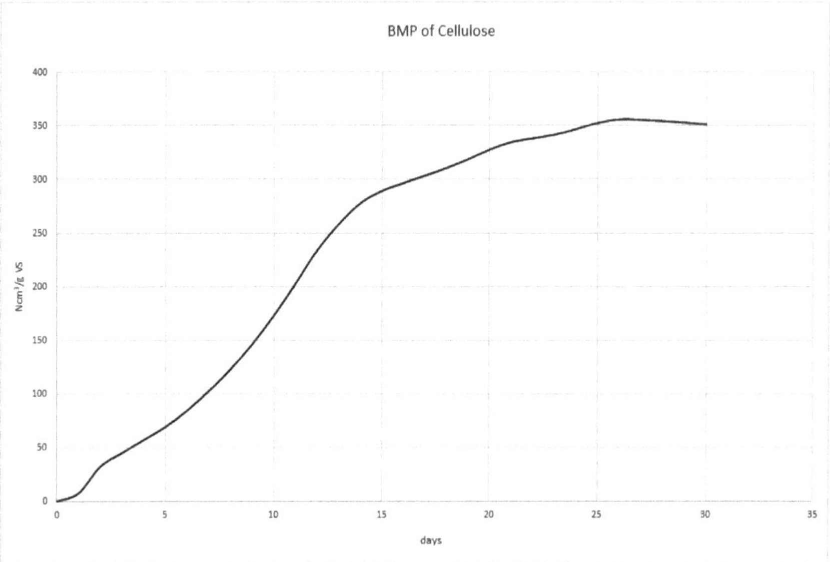

FIGURE 5.10 SMP Curve of Cellulose Obtained with the Dubious Inoculum. The hydrolytic activity is still acceptable.

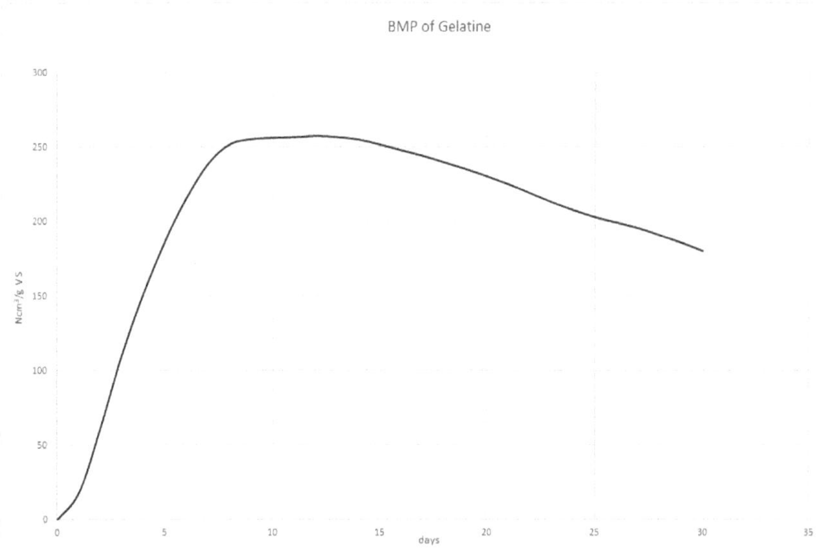

FIGURE 5.11 SMP Curve of Gelatine Showing a Negative Slope. This indicates a low proteolytic activity.

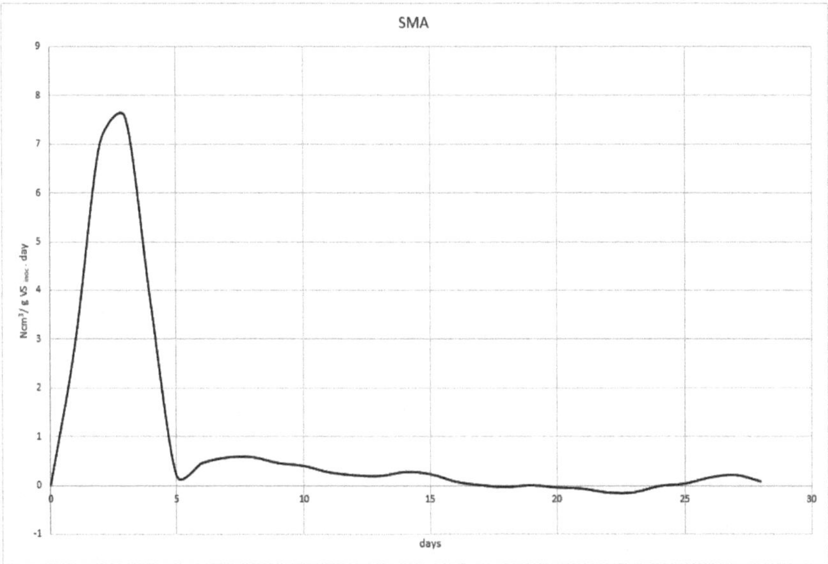

FIGURE 5.12 SMA Curve of the Inoculum. Note That the Peak Value Is Lower Than the Minimum Conventionally Assumed as 'Acceptable'.

plant could maintain 50% of its productivity when fed with a mix of silage and other lignocellulosic agricultural residues. The acetic acid production rate of hydrolytic bacteria was approximately equal to the conversion rate of methanogenic archaea when the mixture included some slowly degradable feedstock. Switching the diet to corn silage, an easily degradable substrate, caused a sudden rise in the acetic acid production rate. The archaea were unable to keep pace, so acetic acid began to accumulate. This lowered the pH and finally disrupted the process. Further investigation [10] showed that the cause of the low acetoclastic activity was a deficit of trace elements, especially nickel and cobalt. Adding a few kilogram of soluble salts of the necessary trace elements was enough to resume the plant's productivity within a day.

5.6 CONCLUSIONS

The SMA is a useful routine test because the acetoclastic stage is usually the weakest link in the chain of anaerobic digestion. This assumption is not a universal truth, so there are cases in which the SMA test alone does not provide enough information about what is going wrong with the digestion process. Screening the specific biological activity of each bacterial group composing the inoculum's ecosystem gives a panoramic view of the process's robustness. The reference substrates necessary for carrying out such tests are widely available at any supermarket. The test procedure is easy, and getting reliable results requires only a modest investment in laboratory instruments.

BIBLIOGRAPHY

[1] Athar Hussain & Shashi Kant Dubey, Specific methanogenic activity test for anaerobic degradation of influents, *Applied Water Science*, 7, 535–542 (2017). https://doi.org/10.1007/s13201–015–0305-z

[2] Italian standard UNI/TS 11703:2018, *Metodo per la misura della produzione potenziale di metano da digestione anaerobica ad umido – Matrici in alimentazione*, Technical Commission: CTI – Biogas da fermentazione anaerobica e syngas biogenico (in force from February 22, 2018). http://store.uni.com/catalogo/index.php/uni-ts-11703-2018

[3] Christof Holliger, Madalena Alves, Diana Andrade, Irini Angelidaki, Sergi Astals, Urs Baier, Claire Bougrier, Pierre Buffière, Marta Carballa, Vinnie de Wilde Florian Ebertseder, Belén Fernández, Elena Ficara, Ioannis Fotidis, Jean-Claude Frigon, Hélène Fruteau de Laclos, Dara S. M. Ghasimi, Gabrielle Hack, Mathias Hartel, Joern Heerenklage, Ilona Sarvari Horvath, Pavel Jenicek, Konrad Koch, Judith Krautwald, Javier Lizasoain, Jing Liu, Lona Mosberger, Mihaela Nistor, Hans Oechsner, João Vítor Oliveira, Mark Paterson, André Pauss, Sébastien Pommier, Isabella Porqueddu, Francisco Raposo, Thierry Ribeiro, Florian Rüsch Pfund, Sten Strömberg, Michel Torrijos, Miriam van Eekert, Jules van Lier, Harald Wedwitschka & Isabella Wierinck, Towards a standardisation of biomethane potential tests, *Water Science and Technology*, 74:11, 2515–2522 (2016). https://doi.org/10.2166/wst.2016.336

[4] Arjen Rinzema, Jules van Lier & Gatze Lettinga, Sodium inhibition of acetoclastic methanogens in granular sludge from a UASB reactor, *Enzyme and Microbial Technology*, 10:1, 24–32 (1988). https://doi.org/10.1016/0141-0229(88)90094-4. ISSN 0141–0229.

[5] Satoshi Fukuzaki, Naomichi Nishio & Shiro Nagai, Kinetics of the methanogenic fermentation of acetate, *Applied and Environmental Microbiology*, 56:10, 3158–3163 (October 1990).

[6] TF Souto et al., Influence of incubation conditions on the specific methanogenic activity test, *Biodegradation*, 21, 411–424 (2010). https://doi.org/10.1007/s10532-009-9311-x

[7] IH Franke-Whittle, A Walter, C Ebne & H Insam, Investigation into the effect of high concentrations of volatile fatty acids in anaerobic digestion on methanogenic communities, *Waste Management*, 34:11, 2080–2089 (2014). https://doi:10.1016/j.wasman.2014.07.020

[8] Jules B Van Lier, Jose Luis Sanz Martin & Gatze Lettinga, Effect of temperature on the anaerobic thermophilic conversion of volatile fatty acids by dispersed and granular sludge, *Water Research*, 30:1 (1996). https://doi.org/10.1016/0043-1354(95)00107-V

[9] Italian Norm UNI/TS 11703:2018, *Metodo per la misura della produzione potenziale di metano da digestione anaerobica ad umido – Matrici in alimentazione*. http://store.uni.com/catalogo/index.php/uni-ts-11703-2018

[10] MA Rosato & S Stromberg, Quick restart of a biogas plant and micronutrients cost minimisation with AMPTS II Light, Application note, *Bioprocess Control*. https://www.researchgate.net/publication/318787289_Quick_restart_of_a_biogas_plant_and_micronutrients_cost_minimisation_with_AMPTS_II_Light

6 Quick Startup and Thermal Stability of the Digester

6.1 OVERVIEW: TIME IS MONEY!

The startup of an industrial digester is usually an expensive operation. Increasing the profit of the plant requires minimising three cost items during the startup phase:

1. **Heating and energy costs**: Heating the digester during its startup requires burning fossil fuel in an auxiliary boiler. The pumps, stirrers, and other ancillary equipment consume electricity from the grid.
2. **External staff cost**: In some cases, commissioning contract costs include the salary and tuition of an external supervisor.
3. **Loss of productivity**: Tipping fees are the principal income of waste treatment plants. Feeding the digester at a fraction of its capacity, or not feeding it for a long time, means collecting only a fraction of the budgeted yearly revenue.

Figure 6.1 shows the essence of this chapter. Starting an industrial biogas plant may cost as much as €29,000/month (cost of heating oil €1,69/l in December 2022; boiler with 300 kW thermal power, digester in Northern Italy, startup in winter). Many plant constructors and some researchers consider the startup a delicate and fragile process that requires a very long time to avoid upsetting the bacterial ecosystem. This chapter will prove why they are wrong and present a robust method, already tested on several occasions, that allows calculating with fairly good accuracy how long the startup of a digester will take, depending on the border conditions of each case.

6.2 THE TRADITIONAL APPROACH TO BIOGAS PLANT STARTUP

The startup of biogas plants is an aspect usually overlooked. For instance, the Italian norm UNI 10457:2010 only provides some general directions, useful for redacting a contract between the plant constructor and the buyer but of little use to the engineer or the plant manager.

DOI: 10.1201/9781003609278-6

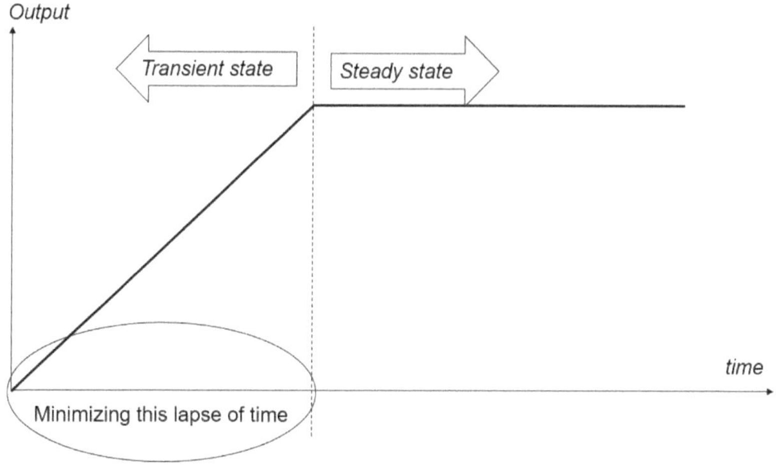

FIGURE 6.1 Energy Production Transient and Steady States During the Digester Startup.

The following text is an excerpt of the said norm.

Definition 3.4: the startup is the set of procedures and tests on the plant before assessing its performance, which is necessary to allow the development of the microorganisms and to reach the design operational conditions.

Article 6.1.7: The startup phase is concluded when the plant has reached at least 70% of the (nominal) gross biogas production during at least 15 days. The conclusion of the startup phase coincides with the beginning of the plant's steady operation.

N.B. The duration of the startup phase may vary, depending on the feedstock and the digester technology.

Article 7.5:. . . the following must be declared in the description of the plant:

. . . maximum startup time;

Article 8.1: The final commissioning tests must be:

. . . performed at the end of the startup phase

. . . After 6 months from the end of the startup (and having reached steady operation), it is the right of the Parties (customer and supplier) to request the performance trial.

Article 8.4: The performance trial must begin at the end of the startup phase, and its duration must be at least equal to the HRT (hydraulic retention time) and not less than 30 days. At least the following parameters must be checked:

Feedstock: Quantity and input flow along the time, pH, total solids (TS), volatile solids (VS), daily input flow (for continuous plants), total nitrogen (TKN, aka Kjeldahl nitrogen).

Digestate: pH, total solids (TS), volatile solids (VS), total nitrogen (TKN, aka Kjeldahl nitrogen), microbiological analysis (where the sanitary norms require them after waste treatment).

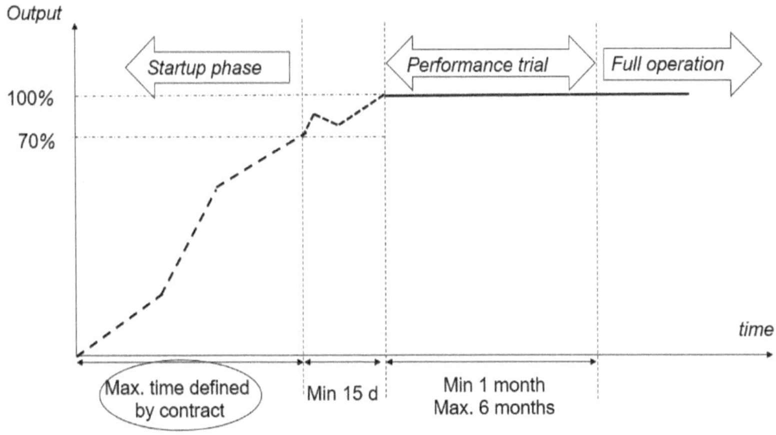

FIGURE 6.2 Power Production During the Startup Phase according to the UNI 10457 Norm.

Biogas: Gross daily production, average of the gross daily production over at least 15 days, composition, any other parameter agreed by the Parties
Energy section: Gross electricity generation, gross and recovered thermal power and process T, self-consumption . . .

Figure 6.2 summarises the startup conception contained in the Italian norm UNI 10457:2010. As the reader can appreciate, the constructor must define the startup duration in the contract, but the standard states neither a method to calculate the same nor a time limit.

6.3 AN OVERVIEW OF THE TRADITIONAL METHODS

The following list does not pretend to be exhaustive. It simply reflects the most common methods of digester startup that the author has observed during his professional career or those he has found in the literature. The startup methods differ slightly, depending on the type of anaerobic plant – sludge, wastewater, or solid waste treatment – and also the digester technology.

6.3.1 Sludge from Wastewater Treatment Plants (CSRT Digester)

A widely diffused method consists in the following:

1. Filling the digester with sludge.
2. Heating it to 35–40°C.
3. Awaiting the methane production to peak and afterwards to decrease. Allegedly the flow decrease indicates that the anaerobic flora has consumed most of the available 'dead' organic matter and the sludge is now teeming with 'live methanogens'.
4. Starting operation at the nominal organic loading rate (OLR).

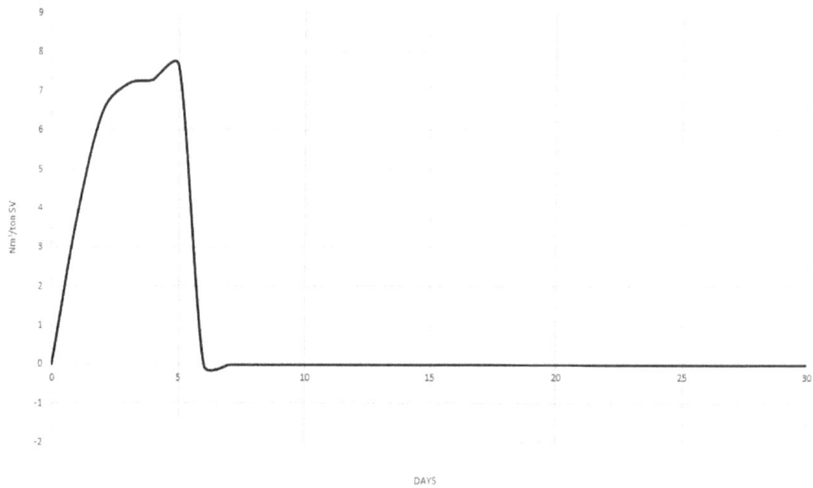

FIGURE 6.3 SMP Curve of an Inoculum Having a Neglectable Specific Methanogenic Activity (SMA).

This method has a logical flaw: plant operators assume that sewage sludge is a suitable anaerobic inoculum, which is merely a conjecture *not* a fact. Such an inoculum often contains mainly aerobic and facultative bacteria and few or no methanogens. Consequently, it will take several months to reach steady-state operation, and the final CH_4 flow will be low, or even null. Figure 6.3 shows why assuming that sewage sludge is a suitable anaerobic inoculum, even after keeping it for a month under anaerobic conditions, is incorrect. In this example, the BMP test shows that digestion of sewage sludge using the said inoculum failed after a few days. The inoculum stopped producing methane on the 5th day and carried on producing only CO_2. Furthermore, methane production was unacceptably low.

6.3.1.1 An Improved Variant of the Method

Adding a sufficient quantity of verified anaerobic inoculum (i.e., sludge or digestate having adequate SMA and hydrolytic activity) will ensure a shorter startup and a more robust microbial ecosystem. After the peak of biogas flow, it is possible to start feeding the digester with the foreseen feedstock. Figure 6.4 shows the BMP curves of the same sewage sludge as the former example. The inocula employed for this comparative test were three samples of digestate taken from three similar wastewater treatment plants that were 'running stably' and from one agricultural digester. In this special case, the inoculum/substrate ratio (ISR) was not in the customary 2–3 range. For logistic reasons that we will discuss later, the customer wanted to compare how the digestion would proceed if the inoculum were just 20% of the total volume of the digester. Since the VS concentration of the agricultural digestate was twice that

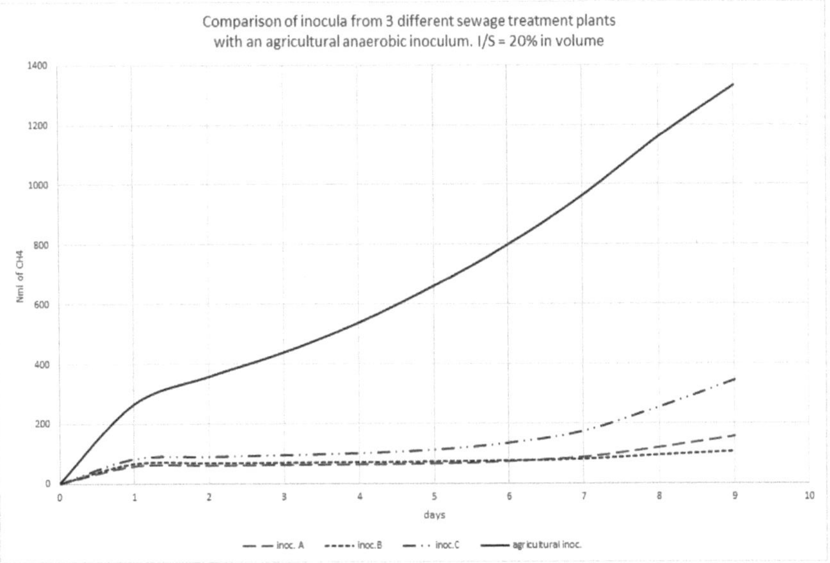

FIGURE 6.4 Comparison of the Endogenous Methane Production of Sewage Sludge and Agricultural Digestate.

of the sewage sludge digestate, one would have expected that the respective curves would only differ in their slopes, the agricultural one being steeper at the beginning but both tending to the same BMP at the end. Figure 6.4 leaves no space for doubts: the digestate from the three sewage plants has not only lower VS concentration but also lower methanogenic activities.

6.3.2 Urban or Industrial Wastewater (UASB or EGSB Digesters)

The usual procedure is as follows:

1. Fill the digester with flocculent or granular sludge to at least 2/3 of its capacity (or three times the daily input volume).
2. Heat the digester to the design temperature.
3. Start operation.

N.B.: Constructors of these types of plants usually include the supply of the inoculum in the contract. They check the biological activity in the lab or source 'fresh' inoculum from a similar plant already in stable operation. Some constructors have facilities for producing highly performant inoculum. Figure 6.5 shows the result of an SMA test on three different inocula: granular inoculum produced at the facility of a constructor specialised in upflow anaerobic sludge blanket (UASB) and expanded granular sludge bed (EGSB) plants, a standard flocculent inoculum, and digestate from a plant that had suffered disruption of the biological process.

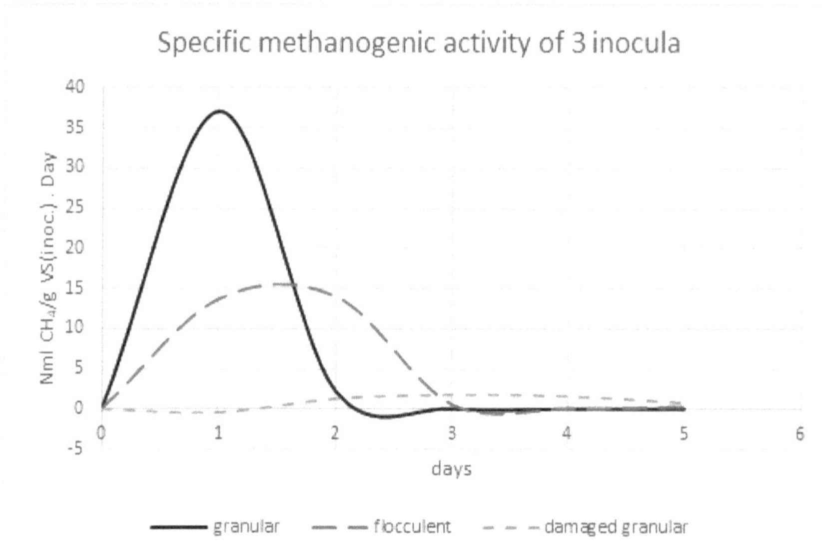

FIGURE 6.5 Comparison of the SMAs of Three Different Inocula.

6.3.3 Agricultural/OFMSW Treatment Plants (CSRT or PFR Digesters)

There are two methods widely employed in Europe. The less frequent one consists in

1. filling the digester with digestate or a mixture of manure and water;
2. heating it to 35–40°C;
3. feeding 1st week at 1/5 of its nominal load, 2nd week at 2/5, and so on.

Those who employ this method argue that feeding a digester at a high OLR during the startup 'will acidify the digester', which, of course, is a false assumption. Acidification happens when acetogenic bacteria produce acetic acid at a higher rate than acetoclastic archaea can convert it to CH_4 and CO_2. Furthermore, organic acids must accumulate until completely neutralising the inoculum's alkalinity, which is usually huge in agricultural plants.

This method has the following drawbacks:

1. It requires at least 1 week to reach the nominal operational temperature plus 3–4 weeks to obtain a gas production stable enough to start running the engine or the boiler, plus 1 month to reach the rated power.
2. Feeding at low OLR results in a small production of biogas, insufficient for the stable operation of the combined heat and power (CHP) plant, so the necessary heating will consume heating oil while slowly collecting biogas in the dome.

A more modern variant of the said method does not feed the digester at a fixed OLR for fixed periods. Instead, the OLR varies according to the following protocol:

1. Fill the digester with digestate or manure diluted with enough water to permit the operation of the pumps and stirrers.
2. Heat it to 35–40°C.
3. Feed the digester at a given OLR, wait for *N* hours and check some parameters (oxidation–reduction potential (ORP), pH, or FOS/TAC), and then increase or decrease the OLR to fit the predefined parameter.

The rationale is always the same: 'avoiding the acidification of the digester'. The author pointed out the drawbacks of the said control parameters in the first volume [1]. As a reminder, we can summarise them as follows:

1. pH and ORP are unreliable parameters. In general, the buffer capacity of manure is so high that when detecting an eventual pH drop, it is usually too late. The ORP has little or no correlation with the SMA.
2. Despite being very popular, the FOS/TAC method has a logical fallacy because it relies on false hypotheses. The basic principle of logic is that false hypotheses cannot lead to true conclusions (although they sometimes may do but only by chance).
3. Furthermore, the 'canonical value', 0.25, is valid for steady-state operation, which contradicts the scope of using the FOS/TAC as the control parameter for the startup phase. Nevertheless, healthy inocula usually work better at much higher FOS/TAC ratios under transient conditions.
4. The method has a logical flaw: assuming that the inoculum already has a good SMA instead of measuring it in advance. There are many reasons why the methanogenic activity of manure or digestate could be low or even inexistent: too much dilution, accidental inhibition from antibiotics or deodorant products employed in the farm, the inoculum was left in contact with the air for too long, etc.

A few European biogas plant constructors adopt a variant of the described protocol: seeding the inoculum with a 'special mixture of selected bacteria' usually lyophilised and packed as a powder. See Section 6.4.2 for a detailed description of the drawbacks of such products and a case study.

6.3.4 The Startup of Anaerobic Digesters in the Academic Literature

Thanks to the Internet, it is relatively easy to download free papers containing guidelines on how to start the anaerobic digestion process. The reader can find some examples in the Bibliography:

a. Ref. 2 mostly focuses on CSRT digesters.
b. Ref. 3 focuses on dry digesters and the optimum recirculation rate.

c. Ref. 4 shows that in a small agricultural single-stage CSRT of 500 m^3, the operating temperature was reached in 42 days and the maximum daily flow of methane was achieved in 123 days. Note that the main focus of the study was on the evolution of microbial communities, not on minimising startup cost.
d. Ref. 5 shows a training for undergraduates probably meant for starting up a small digester in a developing country or a university campus.

The author chose the first reference as an example because it is the most recent among those available in open access. It is a complete summary of previous research, including all the commonplaces already discussed, and some more. Despite being academically correct, it provides little practical information for a biogas plant manager. Some recommendations such as adding lime, sodium bicarbonate, or iron salts are arguable and will be discussed in more detail in Section 6.4.2. According to the paper in question, the startup of a digester would require a campaign of continuous testing in the lab for at least three times the nominal hydraulic retention time (HRT), which in practice means 90 days. Such a long testing time may be acceptable only in special cases. For instance, the author carried out a 3-month continuous tests campaign for a customer needing to feed his digester with residual soap. For more common substrates, such as sludge or agricultural waste, a month of (batch) laboratory tests is enough. The articles cited above are just a small sample of the literature, but they have one thing in common: they focus only on the biological aspects of the digester's startup, assuming *a priori* that it is impossible to shorten it. Wrong hypotheses lead to wrong conclusions. In doing so, they neglect the energy and operational costs of the startup. Hence, the cited literature provides little help to improve the profitability of the biogas plant. Apart from the economic losses, failing to shorten the digester's startup has a negative environmental consequence: heating the sludge with fossil fuels for an unnecessarily long period will emit more CO_2 into the atmosphere.

6.4 PRACTICAL PLANNING OF THE DIGESTER'S STARTUP

The workflow depicted in Figure 6.6 is the outcome of the author's personal experience. The goal is to overcome all the limitations and prejudices of the methods explained so far. Planning the digester's startup requires measuring the biological constants in advance, but keeping the tests to a minimum is also necessary to contain costs. Coupling the data obtained from the laboratory tests with the digester's physical features and the site's expected climatic data at the moment chosen for the startup is mandatory. The size of the digester, insulation thickness, thermal power of the auxiliary boiler, local climate, etc., are all aspects not considered in the literature, at least not as a whole. The method proposed here couples all the said data into a holistic model. Our ultimate goal is to find a strategy for minimising the startup time and cost under given border conditions while maximising the energy production without compromising the process stability.

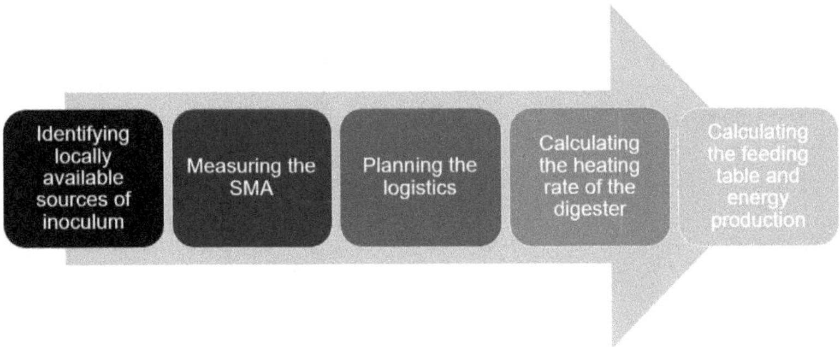

FIGURE 6.6 Workflow of the Activities Necessary to Minimise the Startup Time of the Digester.

6.4.1 Selecting the Inoculum among Locally Available Sources and Measurement of the SMA

The ideal source of inoculum will be the one nearest to the site and must have the maximum volatile solids (VS) concentration to minimise transport costs. Several sources with similar characteristics could be available in the area, so the following rules of thumb are useful to preselect the potentially most convenient one.

1. Mixed agricultural inoculum (dairy, swine, and poultry manure) is suitable for seeding any digestion plant because it usually has higher bacterial biodiversity.
2. Agricultural inoculum from only swine or chicken manure is less suitable for digesters fed with fibrous biomass.
3. Agricultural inoculum from only ruminant manure is the best for digesting cellulosic biomass but less suitable when digesting lipo-proteic waste.
4. Sewage sludge inoculum is not the best option for agricultural or OFMSW plants, but it may be suitable for digesting some industrial effluents.
5. A mixture of inocula from different sources may favour a better performance because of the resulting higher biodiversity of the bacterial consortia.
6. Inocula are living ecosystems. Darwinian evolution makes them suitable for specific substrates, but the adaptation process may require several months. Hence, if you plan to feed your biogas plant with some particular substrate, the ideal inoculum will be that of another plant already running stably with the same feedstock. Figure 6.7 shows data from a case study. The plant was initially running on a mixture of cereal silage with cattle and swine manure. The owner decided to add slaughterhouse residues to the diet, but no plant in the area was already fed with lipo-proteic waste. During the first test, the inoculum was able to digest the slaughterhouse waste, although at a lower efficiency than the theoretical value. After a few months of digesting increasing percentages of animal residues, the inoculum developed a more specialised bacterial flora and reached a biochemical methane potential (BMP) close to the theoretical value.

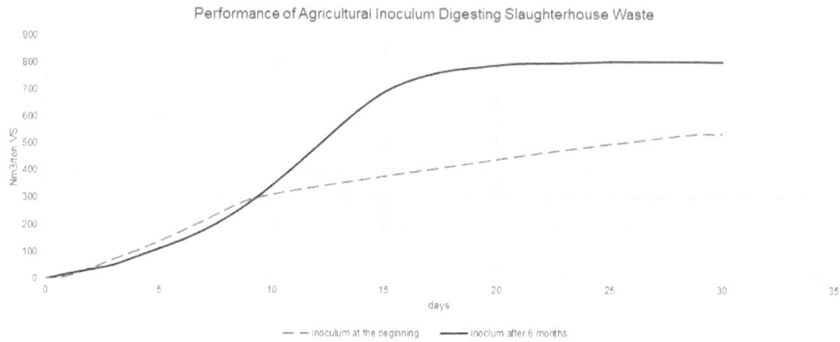

FIGURE 6.7 SMP Curves of Slaughterhouse Waste, Obtained with Inoculum from the Digestion of Vegetable Waste and the Same Inoculum after 6 Months of Adaptation to the New Feedstock.

After shortlisting the potential inoculum sources for the new plant, it is necessary to take samples and test them in the laboratory. The most important test is the SMA, which was already discussed in detail in Chapter 5. The SMA assay is a good indicator of the inoculum's quality but only provides partial information. For instance, granular sludge may have a high SMA but a low hydrolytic capacity. If you intend to start a biogas plant foreseen to run on cellulosic biomass, agricultural inoculum will prove more efficient than granular sludge.

It is advisable to perform also the following tests before deciding the most suitable source of inoculum for your new plant:

1. **Hydrolytic capacity**: The reference substrate for this test is cellulose; the acceptable range is $330 < BMP < 410\ Ncm^3/g$ VS; complete digestion should require about 20 days. This test is recommendable for agricultural digesters fed with vegetal residues and by-products.
2. **Proteolytic capacity**: The reference substrate for this test is neutral (flavour-free) fish or beef gelatine or egg white. The acceptable range is $380 < BMP < 470\ Ncm^3/g$ VS; the digestion time is usually 20–30 days.
3. **Lipolytic capacity**: The reference substrate for this test is vegetable oil; the acceptable range is $900 < BMP < 1{,}000\ Ncm^3/g$ of dry oil; the digestion time is usually 40–60 days. Perform the test with ISR > 5 to avoid substrate inhibition.

Proteolytic and lipolytic capacity tests are highly recommendable if you plan to feed the plant with slaughterhouse waste or fatty industrial waste like soapmaking or biodiesel effluents. For plants treating OFMSW, it is advisable to check all bacterial groups.

6.4.2 DIGRESSION: 'SELECTED BACTERIA', ENZYMES, AND TRACE ELEMENTS: ARE THEY WORTH THEIR COST?

Some biogas plant constructors propose speeding the startup process by adding 'boosters' to the inoculum. Such products are usually expensive, and the hype about their efficacy sometimes transcends the pure marketing and propagates into the peer-reviewed literature. There are four kinds of 'boosters': selected bacteria, trace elements, enzymes, and mixtures of two or all. Selected bacteria are usually lyophilised, but some companies offer concentrated sludge as a semi-solid product. The following considerations summarise the author's experience with some of the products available in the Italian and Spanish markets. The product brands and customers' names are not cited for confidentiality reasons.

6.4.2.1 Lyophilised Bacteria

Increasing the biodiversity of the inoculum by adding more microorganisms is conceptually a good idea. But it also has several drawbacks:

- Seeding a digester with a weak (cheap or free) inoculum and then adding (costly) lyophilised bacteria has no sense. It may prove cheaper (or even free) to check all the locally available inocula in the lab and then choose the one(s) with the best biological activity to seed your digester. Usually, a mixture of different inocula will give the best results because of the higher biodiversity.
- In the case of an existing digester that suffered some biological disruption, adding lyophilised bacteria or digestate from a 'healthy' plant may have the same effect, as shown in Figure 6.8. At this point, the choice is just a

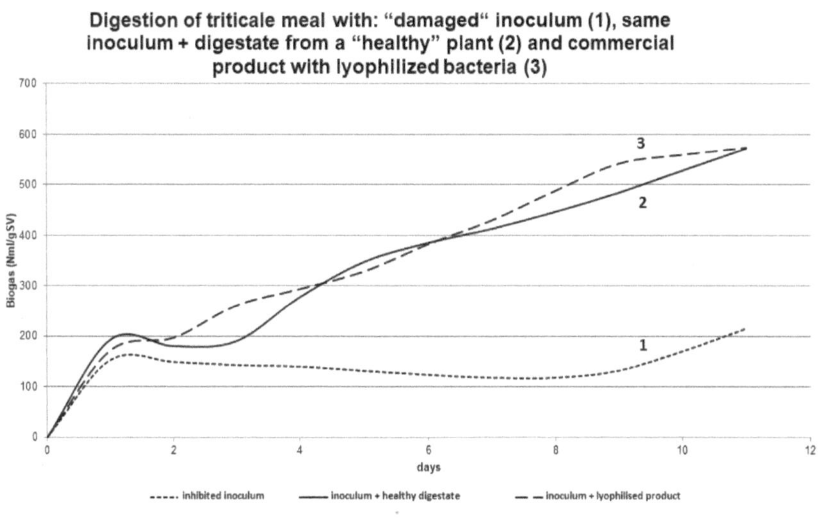

FIGURE 6.8 SMP Curves of a Single Triticale Meal Sample, Obtained with Different Inocula and Additives.

question of cost: is it cheaper to transport *X* m^3 of healthy digestate from another plant placed *N* km away or to purchase *Y* kg of 'a patented special mixture of selected bacteria'? The only way to answer this question is to check both hypotheses in the lab and determine the minimum quantities *X* and *Y*, and their respective costs.

- CAVEAT: Archaea consortia may lose part of their activity because of the lyophilisation process (ref. 6) – another reason to ask the vendor for a sample to test in the lab before deciding whether to purchase such a product.

6.4.2.2 Trace Elements

Trace elements are necessary only when

a. the inoculum has a deficit of them (rare case, must be checked in the lab); or
b. when the feedstock consists of cereal silage (in particular triticale) with little or no cow manure; or
c. the feedstock has a high sulphur content.

In case b, trace elements may be present but are unavailable to the archaea because of phytic acid chelation. Cow manure has bacteria that produce phytase, and it is usually cheaper than commercial mixtures of trace elements.

In case c, S is converted into H_2S, which reacts with metallic salts, precipitating insoluble sulphides. Again, adding manure will cause H_2S to react with NH_4^+. If high levels of H_2S persist, try adding (cheap) ferric salts or consider installing a desulphuriser.

Trace elements are supposed to boost the acetoclastic path because they are biocatalysts for the acetoclastic archaea. This is theoretically true, but in some practical cases, they may indeed increase the biogas yield while lowering the CH_4 concentration. The reason is the stoichiometry of acetic acid:

$$C_2H_4O_2 \rightarrow CH_4 + CO_2$$

Since 1 mole of acetic acid gives one mole of CH_4 and 1 mole of CO_2, boosting biogas production and increasing its quality simultaneously are impossible. At high dosages, trace elements inhibit the hydrogenotrophic archaea, which may reduce the digester's productivity, as shown in Figure 6.9. The difference between both BMP curves is higher than the error margin of the test, so it is licit to infer that, in this particular case of a 'healthy' inoculum, paying for a product that reduces productivity is a bad business.

The simplest and cheapest way to decide if adding trace elements may improve the digester's productivity is to perform several SMA tests with increasing dosages of nutrient solution and observe which performs best.

6.4.2.3 Enzymes

Vegetal biomass consists of two fractions: cellulose and hemicellulose (fully digestible) and lignin (refractory). The rationale for adding enzymes to the inoculum is the belief that they will pre-hydrolyse the biomass, increasing the efficiency of the anaerobic

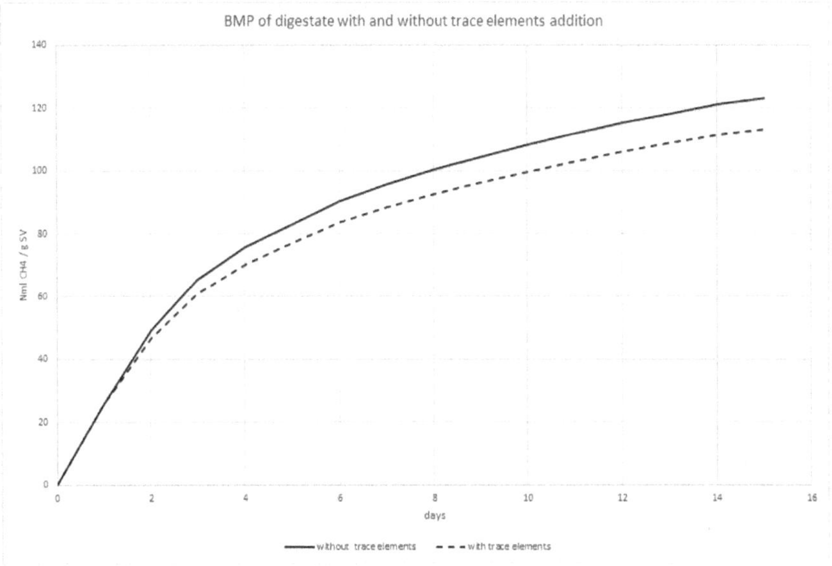

FIGURE 6.9 Digestion of Sodium Acetate with and without Trace Element Addition. An Excess of Trace Elements Decreased the BMP instead of Improving It.

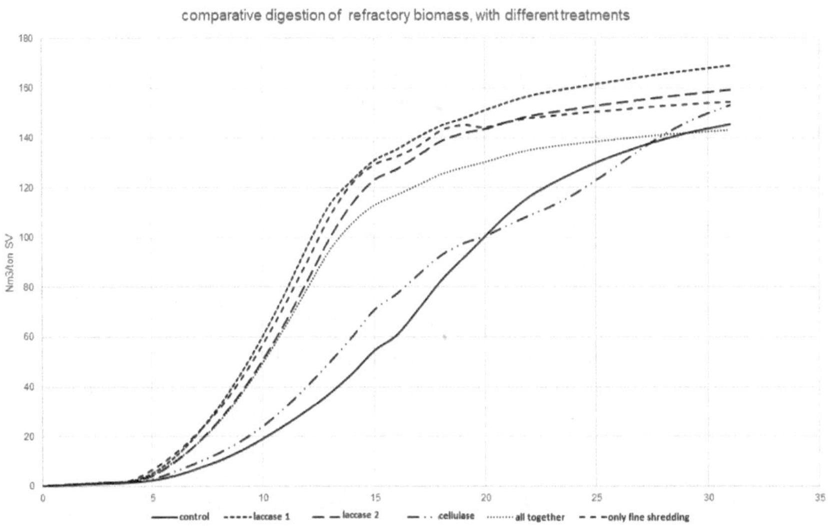

FIGURE 6.10 Effect of Different Enzymatic Treatments on the SMP Curves of Fibrous Biomass.

digestion. Although theoretically true, vendors hype the efficacy of enzymes too much, usually showing as 'proof' the results of the operation of real-scale plants 'before' and 'after', i.e., the post hoc fallacy. When checking the results under controlled laboratory conditions, the crude reality is the one shown in Figure 6.10.

The best results correspond to commercial laccase, a lignin-breaking enzyme. Making an indigestible fraction of vegetal biomass partly digestible increases the BMP, but the cost–benefit balance is negative because of the high price of the enzyme. Adding cellulase brings little benefit because anaerobic bacteria are naturally capable of digesting cellulose. Adding cellulase does not increase the BMP: it just shortens the digestion time. Mixing all the enzymes in a cocktail has no effect. Fine shredding of the biomass increases the BMP by 12% but requires more energy.

6.4.3 Planning the Logistics of the Inoculum and Feedstock

Once we know the SMA, VS concentration, and eventually the hydrolytic capacity of the locally available inocula, we need to organise transport from their sources to our plant. The startup time depends on how quickly we can fill and heat the inoculum. There are two possible cases: digesters that need to be full before heating and feeding them, and digesters that can start the heating and feeding once they reach a minimum level.

6.4.3.1 First Case: The Digester Needs to Be Nearly Full before Starting

Some digesters are built in such a way that reaching a given water level is a necessary condition before starting to heat them. Otherwise, heating the partially filled digester would result in losing the biogas through the digestate output tube. In this case, the bottleneck of the startup process is the time to fill the digester. Once the digester is full with inoculum (or inoculum + substrate), it is possible to start heating it. A second bottleneck could be the available thermal power. For given border conditions (overall thermal losses), higher thermal power will heat the inoculum more quickly, shortening the time to start biogas production.

6.4.3.2 Second Case: The Digester Can Start before Being Full

Other digesters are designed and built so that it is possible to start heating them without losing gas only after reaching a minimum water level. It is then possible to fill and feed them simultaneously, producing biogas from the beginning. In this case, the bottlenecks of the startup process are the maximum quantity of inoculum and feedstock that is possible to feed daily and the heating power. Since the fermentation and methanation rates vary with T, attention must be paid to avoid process disruption.

6.4.4 Calculating the Temperature Rise of the Digester

The temperature in a given point of the digester is a function of the external factors (air temperature, solar radiation, wind speed, and eventual rain), the temperature of the thermal power plant (CHP or boiler), the efficiency of the heat exchanger, the stirring intensity, and the shape and materials of the digester. Heat transfer phenomena and currents in electric circuits have the same mathematical expressions, so it is possible to employ a principle of physics called *system equivalence*. System equivalence allows solving thermal flow problems with an electrical model, a method called *mathematical similarity*. According to the said method, the following equivalences apply:

Electric resistance = thermal resistance [°C/kW] = (thermal conductivity)$^{-1}$
Voltage = temperature [°C]
Current = heat flow [kW]
Electric capacity = thermal capacity = specific heat × mass [kWh/°C]

Caveat!: Pay attention to the units. For our purposes, we will adopt 'technical units'. The formulas discussed below will then allow us to calculate the heating time in hours. If we had adopted SI units, the time would have been determined in seconds.

The mathematical models discussed in the next sections are based on the following simplifying hypotheses:

- The stirring is perfect, so the temperature at any point of the sludge volume is uniform, and the thermal resistance between the liquid and the inner walls of the digester is null.
- The gas in the headspace is at the same temperature as the sludge.
- The heat leaving the digester as sensible and latent heat in the biogas is neglectable.
- The additional heat resistance of the boundary layers of air around the digester is neglectable.
- The soil temperature is constant throughout the year, equal to 10°C. In tropical zones, assume 15°C.
- The net thermal power input is equal to the product of the thermal plant's nominal power and the heat exchanger's efficiency. This implies knowing the said efficiency or at least being able to estimate it with acceptable accuracy.
- Stirring power << heating system's thermal power. Hence, the contribution of the stirrers to the heating (i.e., friction heating, aka Joule heating) is neglectable.
- The slurry inside the digester has the same density, thermal conductivity, and thermal capacity as water. This supposition is fairly true for CSTR digesters usually running at 10%–12% TS concentration.

The aforementioned hypotheses are realistic under most operative conditions and allow the calculations with reasonable accuracy.

6.4.4.1 Dynamic Method

Figure 6.11 shows the electric model of thermal flows in a digester.

T_{ext} is a variable resulting from a semi-deterministic term (average seasonal and daytime temperature), plus a random term representing the weather.

P_{sun} is the total thermal flow of the solar radiation. It is a function of the season and daytime (deterministic signal) and clouding (random signal).

P_{therm} is the thermal power of the boiler (or CHP), assumed to be constant from the beginning of the startup procedure until reaching the design temperature.

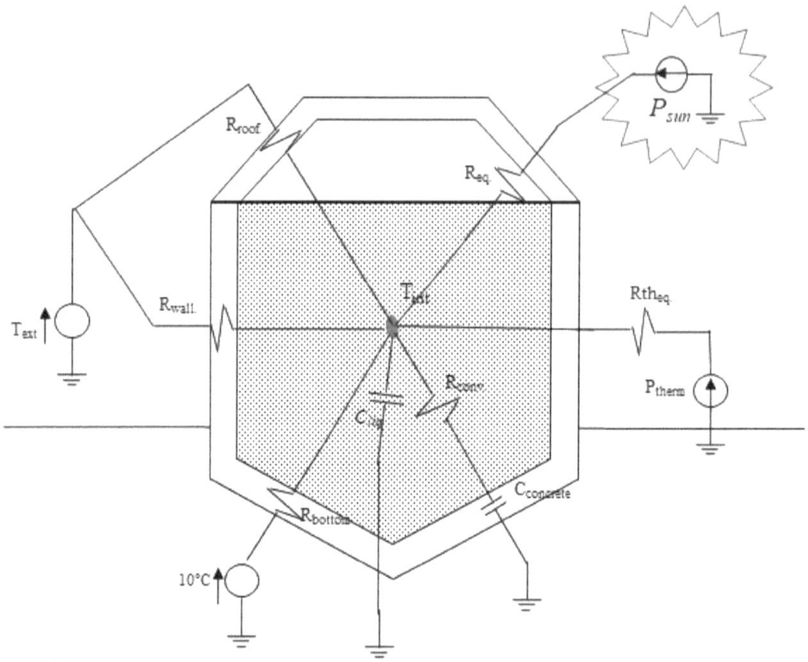

FIGURE 6.11 Electric model of Thermal Flows in a Digester.

10°C is the ground temperature at 1 m depth, assumed to be constant.

R_{eq} is a function of the colour of the digester/dome, its material, and wall thickness.

T_{int} is the unknown variable we want to calculate, starting from room temperature at $t = 0$ until reaching the design temperature (usually around 40°C).

The circuit is relatively simple, but the 'sources' are not pure sinusoids or constants. Modelling them is relatively difficult. Even if a mathematical representation could be found, there are three methods for solving this model circuit:

- Classical theory of circuits: functions in the domain of time (difficult with non-sinusoidal functions).
- More general theory of circuits: Fourier transform and functions in the domain of frequency.
- Numerical simulation with finite elements in the domain of time: The algorithm must include a generator of random noise to simulate clouds and weather variations. Conceptually simple but time-consuming.

Any of the above three methods is a nice argument for a degree thesis or a research paper, but from the viewpoint of an industrial engineer or a biogas plant manager, all of them are too complex and time-consuming.

Fortunately, the principle of system equivalence comes to our rescue. The topology of the equivalent circuit in Figure 6.11 is a passive quadrupole. From the theory of circuits, it is known that capacitive quadrupoles behave as low-pass filters if the capacity is big enough. In other words, quick amplitude variations of the signal (in our case, the external temperature and solar radiation) will produce only tiny variations of our unknown output variable (the slurry's temperature). The digester is full of sludge composed of 90%–95% water, so it has a huge thermal capacity. In general, the digester's mass is so large that the daily variations of *T* and solar irradiation become irrelevant. Then, the digester's thermal inertia is large enough to justify employing the simplified model discussed in the next section.

6.4.4.2 Static Method

We can neglect the daily temperature variations and assume a fixed value, for instance, the average temperature in the month chosen for the digester's startup. The thermal power of the boiler or CHP is usually several orders of magnitude the power of the solar irradiation, so we can neglect the latter. The thermal capacities of the sludge and the digester's envelope are in parallel, so they simply add up. If we know the heat exchanger's efficiency in advance, it is possible to eliminate Rth_{eq} and replace P_{th} with its net value. Since we assume that the temperature of the slurry is uniform because the stirring is perfect, then $R_{conv} = 0$. Figure 6.12 shows the simplified model.

Applying Thevenin's theorem, we can further simplify the model, replacing the series/paralleling with a single series equivalent, as shown in Figure 6.13.

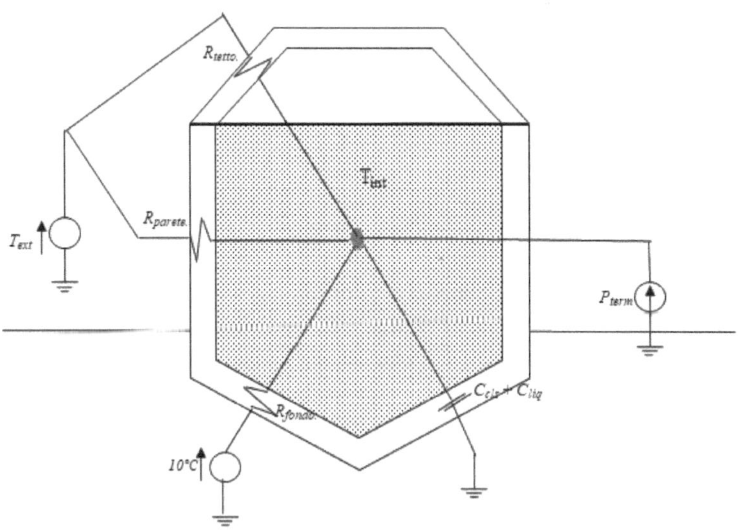

FIGURE 6.12 Simplified Electric Model of Thermal Flows in a Digester.

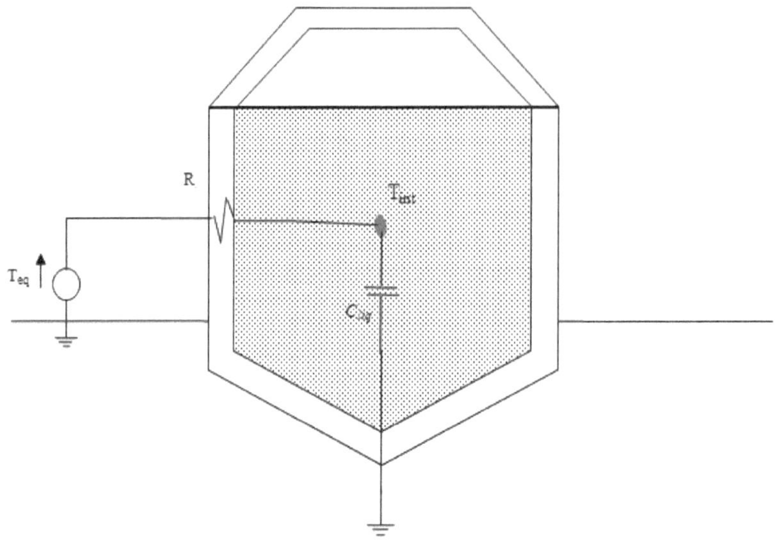

FIGURE 6.13 Equivalent Electric Model of Thermal Flows in the Digester.

Where the equivalent elements are

$$R_1 = \left(\frac{1}{R_{\text{wall}}} + \frac{1}{R_{\text{roof}}} \right)^{-1},$$

$$R = \left(\frac{1}{R_{\text{wall}}} + \frac{1}{R_{\text{roof}}} + \frac{1}{R_{\text{bottom}}} \right)^{-1},$$

$$C = C_{\text{water}} + C_{\text{concrete}},$$

$$T_{\text{eq}} = 10°\text{C} + R_{\text{bottom}} \frac{T_{\text{ext}} - 10°\text{C} + P_{\text{th}} R_1}{R_{\text{bottom}} + R_1}.$$

Knowing C, R_l, R_{eq}, and T_{eq}, the calculation of the digester's temperature at any time is straightforward. The principle of mathematical similarity allows us to employ the classical formula for the charge of a capacitor, replacing voltage by temperature:

$$T_{(t)} = T_0 + T_{\text{eq}} \left(1 - e^{-\frac{t}{RC}} \right),$$

where
$T_{(t)}$ = temperature of the sludge in the instant t,
T_0 = initial temperature of the sludge.

Conversely, it is possible to calculate the time *t* for reaching a given temperature *T*:

$$t = -RC\,ln\left[1 - \frac{T - T_0}{T_{eq}}\right]$$

This formula allows us to calculate the minimum fuel quantity needed for the startup.

Now we have a simple model for calculating $T_{(t)}$ or *t*. We only need to calculate the values of T_{eq}, R_1, *R*, and *C* and introduce them in the above formulas.

Note that T_{eq} is a fictitious value. It represents the temperature of an ideal thermal power source that injects a constant heat flow. Hence, it is also the temperature that the digester could reach in an infinite time (thermal equilibrium) if the external temperature remains constant. In real life, the boiler (or the CHP) heats the sludge indirectly, i.e., the heat flows through a thermal vector (hot water) and across a heat exchanger (usually some pipes coiled around the walls of the digester). The efficiency of a heat exchanger *is not constant*: it is proportional to the temperature difference between the sludge and the hot water and to the water speed at each side of the exchanger's wall. Since the sludge's temperature increases while the temperature of the thermal vector remains constant, the heat flow must decrease. This method tends to overestimate T_{eq}, resulting in slightly higher heating rates than the actual. Note that T_{eq} is a function of P_{therm}, which is equal to the product $P_{boiler} \times \eta_{h.exch}$. The value of $\eta_{h.exch}$ is usually an 'educated guess' because we usually do not know it with certainty. We can estimate $\eta_{h.exch}$ assuming an overall heat exchange coefficient multiplied by the heat exchanger's total area, *A*. We usually know the value *A* from the plant's blueprints, or in the worst case, we can measure it on-site. In any case, we can estimate with better accuracy the value of $\eta_{h.exch}$ with the following formula:

$$\eta_{h.exch.} = \frac{kA\left(T_{boiler} - T_{digester}\right)}{P_{boiler}},$$

where

$k = 150 - 1{,}200$ W/m^{2}°C (tubular heat exchanger; 1,000 is a realistic value for countercurrent exchangers);

$k = 150 - 1{,}200$ W/m^{2}°C (plate heat exchanger, not employed in contact with the sludge);

$k = 700 - 2{,}500$ W/m^{2}°C (for spiral heat exchangers, assume 900);

Empirical data from ref. 7

A = area of heat exchange (m^2).

N.B.: For a realistic estimation of *k*, assume values within the lower half of the ranges provided above. Usual values of $\eta_{h.exch}$ are in the range of 50%–80%.

Calculating the thermal capacity *C* is straightforward:

$$C = C_{water} + C_{concrete}$$

$$C_{water} = \mathrm{V} \times 1{,}000\ \mathrm{kg/m^3} \times 0.00116278\ \mathrm{kWh/kg°C} = \mathrm{V} \times 1.16278\ [\mathrm{kWh/°C}]$$

Here,

V = volume of digestate or sludge assumed as water (m^3),

$C_{concrete} = V_c \times 2{,}300\ kg/m^3 \times 0.0002444\ kWh/kg\ ^oC = V_c \times 0.5612\ [kWh/^oC]$, where

V_c = volume of the concrete wall + bottom (m^3) = thickness × area of the wall and bottom.

Calculating the thermal resistances of the perimetral wall and bottom of the digester is also straightforward: the most common case is a cylindrical tank with a perimetral wall made of concrete, insulated with foam panels. The thermal resistance of the external coating or upholstery panels is neglectable because their only function is protecting the insulant from atmospheric agents. The general formula of the thermal resistance of a single-component wall is

$$r = 1/U = s/k\ [m^2\ ^oC/kW],$$

where

s = thickness of the layer (m),

k = thermal conductivity (usually given in W/m.K, divide by 1,000 for our 'technical' unit system),

U = thermal loss coefficient (usually W/m^2K),

$k_{concrete} = 0.0014\ kW/m^oC$,

$k_{PU\ foam} = 2.8E\text{-}05\ kW/m^oC$,

$k_{PE\ foam} = 3.5E\text{-}05\ kW/m^oC$,

$k_{sandwich\ panel} = 2.2E\text{-}05\ kW/m^oC$.

The thermal resistance of a multilayer wall is the sum of the individual resistances of the layers. Hence,

$$R_{wall} = (r_{concrete} + r_{foam})/A_{wall} = (r_{concrete} + r_{foam})/\pi Dh,$$

where

D = diameter of the digester [m],

h = height of the digester [m].

The bottom of the digester is usually a concrete slab directly in contact with the soil, with no insulation. Hence

$$R_{bottom} = r_{concrete}/A_{bottom} = 4\ r_{concrete}/\pi\ D2.$$

The thermal resistance of the digester's roof is a bit more difficult to calculate, at least in modern digesters. In ancient times, the construction of digesters was fully concrete or sandwich insulant panels, so the formula for the thermal resistance of the top was analogous to that of the bottom, plus the eventual insulation. In modern digesters, the most common cover (aka gasholder) is a pneumatic structure. Its shape can be conical, semi-spherical, or a spherical cap. It can be a single membrane or a membrane with one or two air chambers, and in some cases, the external layer is a membrane of insulant foam with a reflective coating on the inner side. The colour of the external membrane influences the overall thermal dispersion coefficient, U, which is the inverse of r, because of the radiation losses. Furthermore, the shape of the gasholder is kept by controlling the pressure in the gap between the inner and the outer membranes through an air blower, which in turn influences the convection losses. The following values from different sources give an idea of the high variability of the value of R_{roof}:

$r = 244,5\ m^2\ {}^{\circ}C/kW$ (3 mm external insulation membrane, summer, data from a manufacturer 8),
$r = 174,5\ m^2\ {}^{\circ}C/kW$ (3 mm external insulation membrane, winter, idem),
$r = 481\ m^2\ {}^{\circ}C/kW$ (double membrane, winter, from ref. 9),
$r = 1448\ m^2\ {}^{\circ}C/kW$ (double membrane, winter, ref. 10),
$r = 270\ m^2\ {}^{\circ}C/kW$ (air gap, neglecting thin membranes' resistance, ref. 11).

6.4.5 Our Goal: The Digester's Loading Table

Once we can calculate the temperature at any time, it is possible to define the laboratory setup to simulate the alternative strategies for simultaneously heating and feeding the digester. The scope is to reach a biogas production that allows starting the CHP in the shortest time, thus saving fossil fuel. Warming up the digester will proceed until reaching its final design temperature, while the OLR must remain within safe limits to avoid process disruption. There are several ways to achieve such goals, depending on the type of digester. For instance:

a. Fill the digester with inoculum and a given quantity of feedstock, then start heating it.
b. Fill the digester with inoculum, then heat it to a predefined temperature. At this point, start feeding it with growing OLR until reaching a steady state.
c. Load daily a given quantity of inoculum and feedstock while heating, proceeding until the digester is full.

Depending on the instruments and reactors available at the laboratory, the experimental campaign can adopt either dynamic or static simulation techniques.

6.4.5.1 Dynamic Lab Simulation

This technique requires employing continuous or semi-continuous CSRT reactors and an adequate instrument. The author has obtained a good correlation between the data from tests performed with the bioreactor simulator with semi-batch 2-l reactors (Figure 6.14), the AMPTS Light with the same reactors, and a custom continuous CSRT reactor with individual BPC Go instrument and data measured during the startup of industrial-scale plants. Scaling up results from continuous or semi-batch tests is straightforward: OLR in g VS/l in the lab translates directly into kg VS/m^3 in the biogas plant; Nml of methane measured in the lab translate into Nm^3 in the plant;

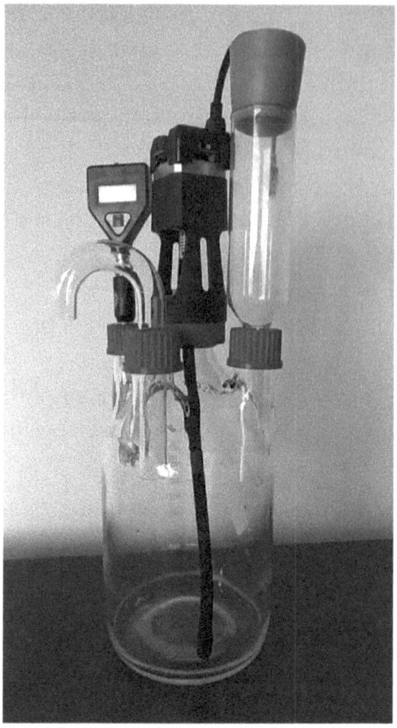

FIGURE 6.14 Two-Litre Semicontinuous Reactor.

and thermostatic baths usually employed in the lab allow programming the heating ramp, matching closely the digester's temperature rise. Semi-batch reactors can provide similar results to continuously fed ones if they are fed daily at the same time.

6.4.5.2 Static Lab Simulation (Batch Tests and Numerical Approximation)

Batch tests are easier, cheaper, and less time-consuming than continuous or semi-batch tests. However, the said advantages come at the expense of interpolation errors. The procedure is as follows:

1. Perform four batch tests with ISR = 1, 2, 3, and 4 at 20 °C.
2. Repeat the same tests at 30 °C.
3. Repeat the same at 40 °C.
4. Since the operation during the startup period is similar to a batch or semi-batch test (transient state), it is possible to estimate the gas volume or flow on any *j*-day at the corresponding T_j by simple interpolation between the SMP curves (Figure 6.15) or the daily flow curves.
5. The ISR that provides the highest flow at a given temperature without causing inhibition allows you to calculate the maximum quantity of feedstock to add to the inoculum to fill the gas holder and start running the CHP in the shortest time. If you are starting with the digester already full of cold inoculum, then you can calculate how much feedstock to add before switching

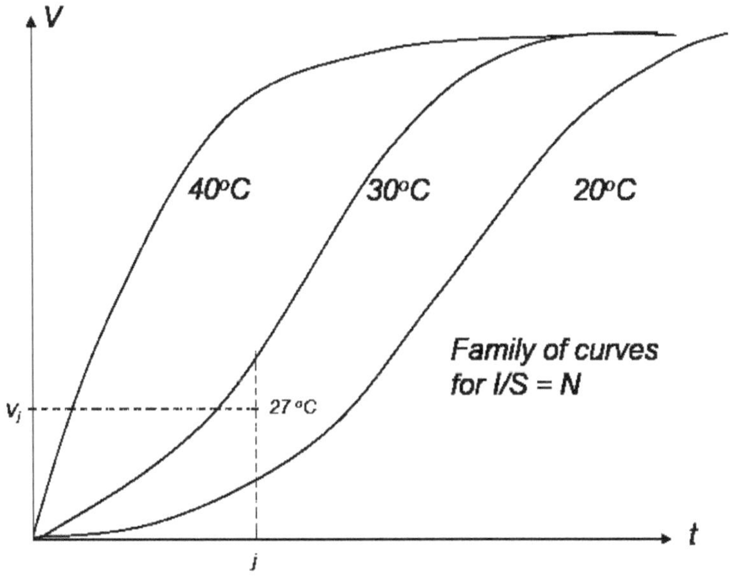

FIGURE 6.15 Family of SMP Curves at Different Temperatures for a Given ISR.

on the heating. If you intend to heat the inoculum while pumping a given daily quantity into the digester, then knowing the most suitable ISR at any temperature will allow you to calculate how much feedstock you can safely add to the daily inoculum quantity.

Measuring the BMP at 20°C may be difficult because most thermostatic baths can only heat and accurately control the temperature if the setpoint is at least 5°C above the room temperature. A thermostatic system like BPC Air can heat, cool, and control the inner temperature independently of the room temperature, but it is much more expensive. The following simplified version of the procedure described above will yield acceptable, although less accurate, predictions.

1. Perform four batch tests with ISR = 1, 2, 3, and 4 at a temperature equal to 90% of the design temperature, T_d. For instance, if $T_d = 40°C$, as in most digestion plants, then perform the tests at 36°C.
2. Assume that the anaerobic activity at 10°C is null and that it will increase linearly in the range 10°C–36°C.
3. Estimate the gas flow on the *j*th day at the corresponding T_j by linear interpolation between the *j*th value of the daily flow curve at 90% T_d and the null case at 10°C. For instance, if $T_d = 40°C$, you performed the tests at 36°C, so you have a set of curves of daily methane flow at each ISR. Suppose that, according to the heating forecast, the digester's temperature on the 7th day

of heating is 27°C. If the methane flow on the 7th day at 36°C was $Q_{7(36°C)}$, then you can calculate the methane flow on the 7th day at 27°C as

$$Q_{7(27°C)} = Q_{7(36°C)} \frac{(27-10)}{(36-10)}.$$

4. The ISR that provides the highest flow at a given temperature without causing inhibition allows the calculation of the maximum quantity of feedstock. The same considerations in point 5 above apply.

CASE STUDY 6.1

Our goal is to start up, in the shortest time and with the lowest possible cost, a digester with the following features:

- $D_{\text{digester}} = 20$ m ; $h_{\text{digester}} = 6$ m, made of concrete
- Bottom not insulated, $A_{\text{bottom}} = 314$ m^2
- $A_{\text{side wall}} = 377$ m^2
- $V_d = 1{,}885$ m^3
- $s_1 = 30$ cm, concrete side wall
- $s_{21} = 10$ cm, polystyrene (PS) foam. Only the side wall is insulated
- $T_{\text{nominal}} = 40°C$
- $T_{\text{ext}} = 15°C$ (daily average foreseen at the moment of the start-up, from historical meteorologic information of the site retrieved from the Internet)
- semispherical gas dome, double membrane
- $A_{\text{dome}} = 2\,\pi\,r^2 = 628$ m^2
- $V_{\text{dome}} = 2/3\,\pi\,r^3 = 1{,}279$ m^3 $V_{\text{dome}} = 2/3\,\pi\,r^3 = 1{,}279$ m^3
- $P_{\text{th}} = \eta_{\text{h.exch}} \times P_{\text{boiler}} = 0.8 \times 300$ kW $= 240$ kW
- Installed combined heat and power (CHP) is 60 kW_{el} and 100 kW_{th}.

First step: Calculate the thermal resistances and capacity

This is straightforward: just replace the input values above in the formulas explained in Section 6.4.4.2.

$$r_{\text{concrete}} = s/k = 0.3 \text{ m}/0.0014 \text{ kW/m°C} = 214 \text{ m}^2\,°\text{C/kW}$$

$$s/k = 0.1\text{m}/3.5\text{E-05 kW/m°C} = 2{,}857\text{m}^2\,°\text{C/kW}$$

$r_{\text{dome}} = 481$ m^2 °C/kW (average value chosen from the literature, unless your supplier provides more accurate data)

$$R_{\text{wall}} = (r_{\text{concrete}} + r_{\text{foam}}) / A_{\text{side wall}} = 8.146 \text{ °C/kW}$$

$$R_{\text{bottom}} = r_{\text{concrete}} / A_{\text{bottom}} = 0.568 \text{ °C/kW}$$

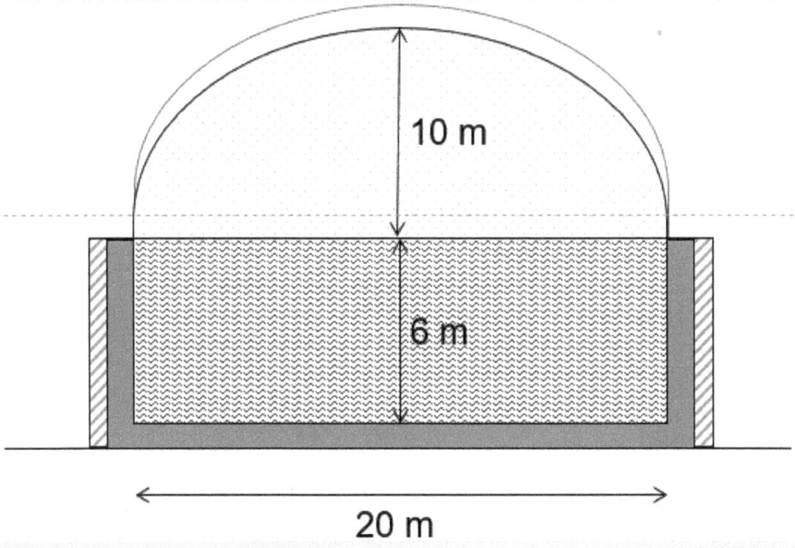

FIGURE 6.16 Sketch of the Digester's Cross-Section.

$$R_{dome} = r_{dome} / A_{dome} = 0.766\ ^\circ C/kW$$

$$R_1 = (1/R_{wall} + 1/R_{roof})^{-1} = 0.700\ ^\circ C/kW$$

$$R = (1/R_{wall} + 1/R_{roof} + 1/R_{bottom})^{-1} = 0.314\ ^\circ C/kW$$

$$C_{water} = 1{,}885 \times 1.16278 = 2{,}192\ kWh/^\circ C$$

$$V_c = \text{volume of the concrete wall + bottom}\left(m^3\right) = \text{thickness} \times \text{area}$$

$$C_{concrete} = \left(377 + 314\right) \times 0.3 \times 0.5612 = 116\ kWh/^\circ C$$

$$C = 2{,}192 + 116 = 2{,}308\ kWh/^\circ C$$

First step: calculate the heating ratio
Knowing *R* and *C*, it is possible to calculate the heating ratio:

$$T_{eq} = 10^\circ C + R_{bottom} \frac{T_{ext} - 10^\circ C + P_{th} R_1}{R_{bottom} + R_1}$$

$$T_{eq} = 10^\circ C + 0.568 \frac{15^\circ C - 10^\circ C + 240 \times 0.7}{0.568 + 0.7} = 88^\circ C$$

$$T_{(t)} = T_0 + T_{eq}\left(1 - e^{-\frac{t}{RC}}\right)$$

$$RC = 0.314 \times 2308 = 724.7\ h$$

$$T_{(t)} = 10 + 88\left(1 - e^{-\frac{t}{724.7}}\right)$$

Time for reaching 40°C:

$$t = -RC\ ln\left[1 - \frac{T - T_o}{T_{eq}}\right] = -724.7\ ln\left[1 - \frac{40 - 10}{88}\right] = 302\,h = 12.6\,days$$

From 302 hours onwards, the temperature is assumed to be constant.

Second step: Calculate the maximum quantity of heating oil necessary for the startup

INPUT DATA:

LHV $_{\text{heating oil}}$ = 11.89 kWh/kg (from EngineeringToolbox, https://www.engineeringtoolbox.com/fuels-higher-calorific-values-d_169.html)

$\rho_{\text{heating oil}}$ = 0.84 kg/l (from the same source)

Assumed efficiency of the boiler's burner = 0.9 (range for standard oil-fuelled boilers: 0.85–0.95)

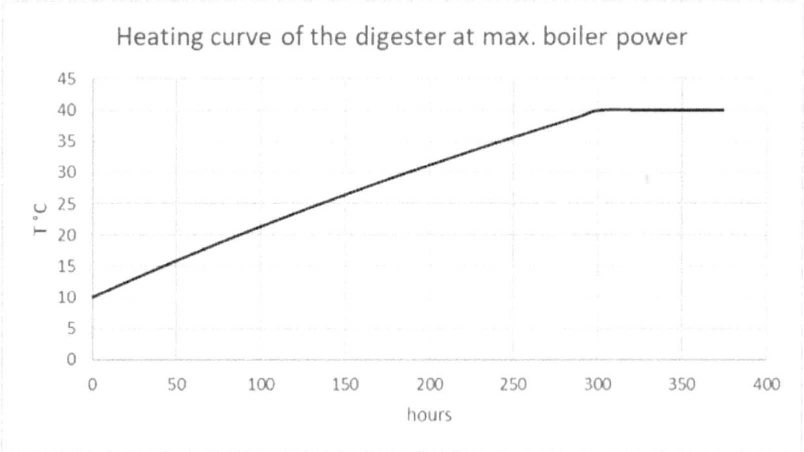

FIGURE 6.17 Temperature vs. Time Curve of the Digester at Maximum Heating Power.

CALCULATIONS:

$$m_{\text{h.o.}} = 300\ \text{kW} / 0.9 \times 302\ \text{h} / 11.89\ \text{kWh/kg} = 8{,}466\ \text{kg}$$

$$V_{\text{h.o.}} = 8{,}466\ \text{kg} / 0.84\ \text{kg/l} = 10{,}079\ \text{l}$$

$V_{\text{h.o}}$ is the maximum theoretical heating oil consumption, assuming all the biogas produced during the startup will be burnt in the torch without any heat recovery.

In practice, the biogas production will be fairly steady when reaching about 30°C, allowing the CHP to run at growing power, thus saving heating oil. Since it is impossible to heat the digester faster, our goal is to find a feeding strategy that allows the production of enough methane to start the CHP before reaching the design temperature.

Third step: using the methane production curves from the laboratory tests
In this case study, the customer was particularly 'price-sensitive' and did not want to pay for replicating BMP tests at different temperatures. The only BMP test on the available substrate (a mixture of agricultural residues) was performed at 40°C. When measuring the BMP with an AMPTS II, the instrument provides two sets of data: the cumulated volume and the daily flow (Figure 6.18). In our case, the second curve turns out to be the most useful. Knowing the daily gas production at 40°C as a function of the digestion time allows estimating the correspondent values at lower temperatures by simple linear interpolation, assuming that the methane flow is null at 10°C.

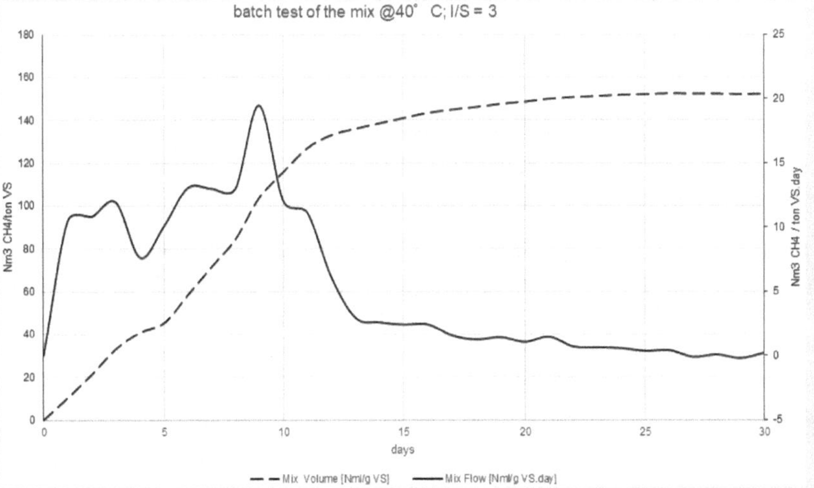

FIGURE 6.18 SMP and Net Methane Flow Curves Obtained from the Test at 40°C.

For instance, suppose we load the digester with inoculum plus a certain quantity of substrate and want to estimate the gas production on the 5th day from the beginning of heating. From the heating curve, we know that the average temperature on the 5th day will be 24°C. The lab test at 40°C demonstrates that the average gas flow on the same day is 90 Nm^3/ton VS day. We can estimate that the gas production on the 5th day will be

$$\frac{Q_5}{(25-10)} = \frac{90}{(40-10)} \rightarrow Q_5 = 45\,\mathrm{Nm}^3\,/\,\mathrm{ton\ VS\,day}.$$

As explained in Section 6.4.5.2, although this method is based on oversimplified hypotheses, its accuracy is acceptable for industrial purposes, and the author found good correspondence between the calculated values and the results observed in the field.

Fourth step: finding the optimum feeding strategy during the startup

The final step is just a question of trial and error. A spreadsheet like the one shown in Figure 6.19 allows us to check several feeding hypotheses and choose the most suitable one for our context. The dilemma is: is it better to overfeed the cold digestate, reaching a conspicuous methane production in a few days to save heating oil, but wasting gas in the torch later? Or should we feed the cold digester with a small quantity of substrate and start the CHP at constant power, but later? There is no simple answer. Everything depends on the cost–benefit analysis for each particular case. Depending on the feedstock's cost, overfeeding the digester and eventually burning the excess gas in the torch may be cheaper than feeding a smaller quantity and waiting longer to turn the CHP on, thus consuming heating oil for a longer time.

6.5 CONCLUSIONS: BEWARE OF MURPHY'S LAW!

The former case study demonstrates that despite 'the literature' and what the 'experts' say, it is possible to safely feed the digester from the very beginning without any 'risk of acidification'. The startup conditions of a real-size digester are the same as those of a laboratory-scale batch test. If keeping a given ISR in the lab prevents acidification, then the same applies to the industrial digester.

A full checkup of the digester is advisable before starting the loading and heating procedure. The following checklist will be a useful guide to avoid delays during the startup phase.

1. Check for liquid and gas leaks.
2. Check all valves. Do not trust the lights on the controller's screen. Walk to your digester and visually check if they are closing/opening correctly!

initial feedstock quantity		max. Feedstock quantity	35,416667	ton
18	ton	V inoc	1700	m3
		VS substrat	32%	
		VS inoc	2%	
		I/S	3	
		V gas dome	1279	m3

Day	Flow @ 40°C [Nm3/ton VS.day]	calculated T (°C)	Calculated CH4 Flow (Nm3)	cumulated V (m3)	V in gas holder	Av. Electric Output (kW)	Avg. Thermal Output (kW)	Boiler Output (kW)	Heating oil consumption (kg)
0	0,00	10,00	0,00	0,00	0,00	0,00	0,00	0,00	0
1	10,49	12,87	60,73	60,73	60,73	0,00		300,00	673
2	10,81	15,64	76,11	136,84	136,84	0,00		300,00	1346
3	11,91	18,32	98,21	235,05	235,05	0,00		300,00	2019
4	7,61	20,92	71,62	306,67	306,67	0,00		300,00	2691
5	10,05	23,43	105,96	412,62	412,62	0,00	0,00	300,00	3364
6	13,04	25,86	151,79	564,41	564,41	0,00	0,00	300,00	4037
7	13,00	28,21	165,02	729,43	639,00	15,01	25,01	274,99	4654
8	13,00	30,48	178,28	907,71	639,00	44,61	74,31	225,69	**5160**
9	19,47	31,34	274,54	1182,25	820,00	60,13	100,18	0,00	
10	12,00	32,40	174,98	1357,23	990,00	60,96	101,56	0,00	
11	11,12	33,43	167,29	1524,53	1160,00	60,51	100,81	0,00	
12	6,10	34,43	94,47	1618,99	1260,00	59,59	99,28	0,00	
13	2,94	35,00	46,24	1665,23	1279,00	64,11	106,81	0,00	
14	2,58	35,94	41,65	1706,88	1279,00	71,03	118,33	0,00	
15	2,41	36,85	39,89	1746,77	1279,00	77,65	129,36	0,00	
16	2,41	37,74	40,85	1787,62	1279,00	84,43	140,66	0,00	
17	1,53	38,59	26,63	1814,25	1279,00	88,85	148,03	0,00	
18	1,22	39,42	21,58	1835,83	1279,00	92,4	154,00	0,00	
19	1,39	40,22	25,19	1861,02	1279,00	96,62			
20	1,03	40,00	18,50	1879,52	1279,00	99			
21	1,42	40,00	25,60	1905,12	1279,00	10			
22	0,67	40,00	12,00	1917,12	1279,00	105,			
23	0,59	40,00	10,65	1927,77	1279,00	107,70			

Output > nominal power
=
gas wasted in the torch

FIGURE 6.19 Screenshot of the Spreadsheet Simulation of a Heating Plus Feeding Startup Strategy.

3. Check the pumps, stirrers, and blowers. Look through the digester's porthole and check if they are effectively turning!
4. Check if the overpressure valves are full with water + glycol mixture in compliance with the manufacturer's specifications.
5. Check the sensors (pH, *T*, *P*, gas composition, etc.). If necessary, perform a calibration.
6. Ensure you have enough fuel oil for the boiler (calculated heating time + 10%).
7. Ensure you have obtained all the licenses to sell energy or gas.
8. Check the grid parallel functionality.
9. Check the inoculum and feedstock stock/transport schedule.
10. Check the weather forecast. If possible, avoid starting your plant during heavy storms or snowfalls.
11. If possible, measure the efficiency of the boiler + heat exchanger assembly for a more accurate prediction of the heating rate.

BIBLIOGRAPHY

[1] MA Rosato, *Managing biogas plants: A practical guide*, CRC Press (September 29, 2017). ISBN 9781138626614.

[2] A Schnürer, I Bohn & J Moestedt, "Protocol for start-up and operation of CSTR biogas processes," in *Hydrocarbon and lipid microbiology protocols, Springer protocols handbooks* (2017). https://doi.org/10.1007/8623_2016_214

[3] Sigrid Kusch-Brandt, Understanding and managing the start-up phase in dry anaerobic digestion, *Technical Paper Presented in the Research Conference In Technical Disciplines* (November 18–22, 2013). http://www.rcitd.com

[4] Xavier Goux, Magdalena Calusinska, Marie Fossépré, Emile Benizri & Philippe Delfosse, Start-up phase of an anaerobic full-scale farm reactor – Appearance of mesophilic anaerobic conditions and establishment of the methanogenic microbial community, *Bioresource Technology*, 212, 217–226 (2016). https://doi.org/10.1016/j.biortech.2016.04.040. ISSN 0960–8524.

[5] Joan Richmond-Hall, *Module 5: Start-up, operations and process monitoring*, Online Handouts of a Course on Biogas Technology. https://richmond-hall.weebly.com/5-start-up--operations.html

[6] Keerti Cherukuri, *Quantitative analysis of methanogenic community in anaerobic digesters and its response to freeze drying and exposure to oxygen*, Marquette University Master's Theses (2009). https://epublications.marquette.edu/cgi/viewcontent.cgi?referer=&httpsredir=1&article=1178&context=theses_open

[7] Data from Engineering Toolbox. https://www.engineeringtoolbox.com/overall-heat-transfer-coefficient-d_434.html. https://www.engineeringtoolbox.com/heat-transfer-coefficients-exchangers-d_450.html

[8] Technical Sheet of the Manufacturer Sattler. https://protex.sattler.com/fileadmin/user_upload/pro-tex/Downloads_ENVIRONMENT/998_POLYPLAN_Biogas_Thermo_Type_II_CD2019_web.pdf

[9] Matteo Bavutti, Luca Guidetti, Giulio Allesina, Antonio Libbra, Alberto Muscio & Simone Pedrazzi, Thermal stabilization of digesters of biogas plants by means of optimization of the surface radiative properties of the gasometer domes, *Energy Procedia*, 45, 1344–1353 (2014). https://core.ac.uk/download/pdf/82056277.pdf

[10] Piotr Rynkowski, Numerical modelling of heat loss through the cover in the anaerobic digester, *Ekonomia I Środowisko*, 2:65 (2018). http://yadda.icm.edu.pl/yadda/element/bwmeta1.element.baztech-2293415c-a060-4f63-975a-069abe4905e0/c/Rynkowski_numerical_2_2018.pdf

[11] Francesco Calise, Francesco Liberato Cappiello, Massimo Dentice d'Accadia, Alessandra Infante & Maria Vicidomini, Modeling of the anaerobic digestion of organic wastes: Integration of heat transfer and biochemical aspects, *Energies*, 13, 2702 (2020). https://doi.org/10.3390/en13112702. www.mdpi.com/journal/energies

7 Selection of Biogas Upgrading Systems

A Thermodynamic Approach

7.1 DEFINING THE PROBLEM: WHAT IS THE 'BEST' UPGRADING SYSTEM?

Natural gas (NG) is a widely diffused commodity and is considered the cleanest of all fossil fuels because its combustion produces small quantities of undesired compounds. Replacing liquid fossil fuels with NG has been a priority in many industrial sectors to reduce air pollution. Nevertheless, the CO_2 resulting from NG combustion contributes to the greenhouse effect because it returns carbon that had been confined in geological formations millions of years ago to the atmosphere. On the other hand, the combustion of biogas is 'greenhouse-neutral' because it sends back to the atmosphere the same carbon that plants absorbed in their biomass and that bacteria converted into CH_4 and CO_2. The carbon neutrality of biogas triggered policies in the European Union and United States to encourage the replacement of NG or fossil fuels with the so-called bio-NG or biomethane. Treating NG at the extraction site produces a standardised commodity consisting of almost pure CH_4 and small amounts of other hydrocarbons. Raw biogas, on the contrary, is a mixture of gases featuring highly variable physico-chemical properties, as shown in Table 7.1.

Converting biogas into biomethane requires conspicuous investments in equipment and consumes large quantities of energy and materials. The 'best' technology is then the one that provides the most economical combination of capital expenditure (CAPEX) and operation expenditure (OPEX), and the said combination depends on the context of each project. Since the profitability of a plant is a direct consequence of the choices made at the moment of its construction and commissioning, this chapter proposes a method based on thermodynamics for selecting the most convenient biogas upgrading technology under given boundary conditions. Thermodynamic laws, and more concretely exergy analysis, allow comparing the OPEX – mainly energy cost – of each technological solution, making the decision-making process more rational. Bias towards a given technology or brand may still result from other factors, but it is not considered in our analysis: relative cost of electricity and heat, availability of components, after-sales service of the different providers, etc.

DOI: 10.1201/9781003609278-7

TABLE 7.1
Physical and Chemical Properties of Biogas from Different Sources

Property	Minimum	Middle Range	Maximum
CH_4 content	40% (landfill gas)	55% (vegetal biomass and dung)	80% (digestion of fats)
SH_2 content	0.3% (cow manure)	0.5% (chicken manure)	>1% (some industrial effluents)
Other impurities	VOC, siloxanes, O_2, N_2, H_2, dust, and moisture, depending on the biomass source and process		
Density at 20°C	0.90 (80% CH_4)	1.06 kg/m³ (60% CH_4)	1.37 (40% CH_4)
Explosivity range in air	12.5%–37.5% (40% CH_4)	8%–25% (60% CH_4)	5%–15% (pure CH_4)
Solubility in water at 20°C	0.023 g/l (pure CH_4)		1.7 g/l (pure CO_2)

7.1.1 Differences between Cleaning and Purification of Biogas

The Italian norm UNI 10458:2011 defines two levels of biogas treatment:

- Cleaning, meaning the removal of SH_2, N_2, O_2, volatile organic compounds (VOCs), moisture, siloxanes, dust, and any other impurity, except CO_2.
- Upgrading or purification, meaning the removal of CO_2 from clean biogas, leaving 'pure' CH_4.

The question is then: how pure must biomethane be? In Europe, biomethane needs to meet two different quality specifications depending on its intended use:

a. The standard EN 16723–1, biomethane quality for injection into the gas grid
b. The standard EN 16723–2, biomethane quality for use in road transport.

Table 7.2 (extracted from ref. 1) summarises the minimum quality requirements of EN 16723–1 for biomethane injection in the NG grid:

7.2 THERMODYNAMICS: THE KEY TO COMPARING APPLES WITH ORANGES

Drafting the investment plan for a biogas upgrading facility is difficult and sometimes confusing because of the large variety of available technologies. At first sight, all upgrading systems look very different, giving the impression that comparisons are impossible. Gas separation is a thermodynamic transformation, subject to the same rules as the most widely studied compression, expansion, or heat exchange. Thermodynamics provides a solid basis for a fair comparison between systems that seem to be incomparable. The classical thermodynamical approach starts by applying the first and second laws to our problem:

TABLE 7.2
Minimum Quality Requirements for Biomethane Injection in the NG Grid, according to EN 16723–1

Quality Parameter	Value
Relative density (air = 1.293 kg/Nm^3)	0.555 ÷ 0.7
Max. H_2S content	5 mg/Sm^3 (3.3 ppm in vol.)
Max. S_{total} content	20 mg/Sm^3 (30 if odourised)
Max. O_2 content	0.02% vol. or 1% in exceptional cases
Max. CO_2 content	2.5% vol (4% in exceptional cases)
Max. H_2 content	0.1% vol
Min. CH_4 content (some national regulations may require higher purity)	90%
Wobbe Index at 20°C	47.02–51.98 MJ/m^3
Maximum heat of combustion (GCV) at a base temperature of 20°C and 101.325 kPa	34.87 MJ/m^3
Minimum heat of combustion (GCV) at a base temperature of 20°C and 101.325 kPa	31.82 MJ/m^3

First law: Conservation of mass and energy.

$$m_{mix}\, h_{mix} + Q_{in} = m_A h_A + m_B\, h_B + Q_{out}$$

where

m_{mix} , m_A and m_B = mass of the gas mixture and of gases A and B, respectively;
h_{mix} , h_A and h_B = enthalpy of the gas mixture and of gases A and B, respectively;
Q_{in} and Q_{out} = heat injected and extracted, respectively, from the system.

Second law: Entropy must always grow; otherwise, transformations are impossible.

$$\Delta S > 0 \leftrightarrow \left(\frac{Q_{out}}{T_{out}} + m_A\, s_A + m_B s_B\right) - \left(\frac{Q_{in}}{T_{in}} + m_{mix}\, s_{mix}\right) > 0$$

The physical meaning of 'entropy growing' is 'altering the equilibrium' of the system. There are three ways to unbalance a gas mixture, causing its entropy to grow:

1. Add mass: For instance, adding an alkali will cause a reaction with CO_2 and precipitation of carbonate salts, leaving pure methane.
2. Add or extract heat: For instance, lowering the temperature of the gas mixture will cause one of the gases to liquefy first.
3. Exert mechanical work: For instance, compressing the gas mixture and forcing it to pass through a selective membrane.

It is possible to envisage an ideal machine that can separate a mixture of two gases, calculating the minimum mechanical work necessary for the said transformation. Since Carnot's theorem defines the equivalence between work and heat, it is possible to expand the concept of 'minimum work of separation' to 'minimum heat of separation'. This will allow fair comparisons between different gas separation systems.

7.2.1 Minimum Work of Gas Separation

Figure 7.1 (adapted by the author from Ref. 2) depicts Van't Hoff's ideal machine, named after the Dutch physical chemist Jacobus Henricus Van't Hoff (1852–1911), the first Nobel Prize laureate in chemistry.

The ideal features of the Van't Hoff machine are as follows:

a. It separates a binary mixture of gases A and B.
b. It performs a reversible isothermal cycle, which, in turn, means that the gas separation process would take infinite time.
c. The process is adiabatic.
d. The cylinders are infinitely rigid, and their surface produces no friction.
e. The pistons are perfectly selective membranes, operating at constant *P*.

The complete mathematical treatment of the ideal gas separation is available in Ref. 2 (already cited). For our practical purposes, the following formula fully represents the process:

$$W_{\min} = -NRT\left[c_A \ln c_A + \left(1-c_A\right)\ln\left(1-c_A\right)\right],$$

where

N = total moles of gas mixture (1 Nm^3 = 44.64 moles);
c_A = molar concentration of one of the components (e.g., CH_4);
R = constant of ideal gas = 8.314 J/mol K;
T = absolute temperature of the ideal isothermal process.

N.B.: The initial volume of the mixture is the sum of the partial volumes of its components (Amagat's law), so the final volumes V_1 and V_2 in the ideal machine will correspond to the initial partial volumes in the gas mixture. The pressure of each of

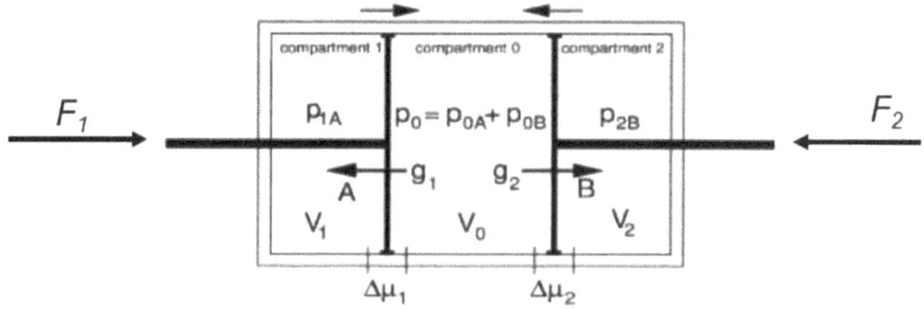

FIGURE 7.1 Concept of the Van't Hoff Ideal Machine.

the separated gas streams will correspond to the initial partial pressure they had in the mixture (Dalton's law).

We can draw three important conceptual conclusions from Van't Hoff's formula:

1. Theoretically, the minimum work necessary for the perfect separation of a binary gas mixture depends only on the process temperature and the initial concentration of each gas.
2. It is impossible to upgrade biogas with less energy than the minimum theoretical value, so the figure is a benchmark for comparing actual processes in an industrial environment.
3. In the ideal separation process, the pressure is constant. Under actual operational conditions, it is necessary to compress the product to the grid's nominal pressure or into cylinders. When benchmarking gas separation processes, pay attention to the final pressure required for the product and consider the compression work separately.

To further refine the Van't Hoff formula to fit the reality better, we must consider the following:

a. Processes are always irreversible, which implies unavoidable energy losses.
b. The separation is always imperfect. The minimum acceptable CH_4 concentration in the product is usually defined by regulations, and the off-gas always contains a small amount of CH_4.
c. In general, accepting the lower purity of the products implies that $W_{imperf} < W_{min}$.
d. Since it is mandatory to compress the product for storage or transport, we can assume

$$W_{real} = W_{ideal} + \Delta W$$

$$W_{ideal} = W_{imp} + W_{comp}$$

where

ΔW is the sum of all the energy losses (irreversibility of the transformation);
W_{imp} is the (ideal) minimum work of an imperfect separation process;
W_{comp} is the ideal adiabatic compression work.

Working out the said hypotheses results in the following formula of the ideal imperfect separation:

$$W_{imp} = -NRT \left\{ \begin{array}{l} c_A^i \, ln\, c_A^i + \left(1 - c_A^i\right) ln\left(1 - c_A^i\right) \\ -\dfrac{c_A^i - c_{2A}^f}{c_{1A}^f - c_{2A}^f}\left[c_{1A}^f \, ln\, c_{1A}^f + \left(1 - c_{1A}^f\right) ln\left(1 - c_{1A}^f\right)\right] \\ -\dfrac{c_{1A}^f - c_{2A}^i}{c_{1A}^f - c_{2A}^f}\left[c_{2A}^f \, ln\, c_{2A}^f + \left(1 - c_{2A}^f\right) ln\left(1 - c_{2A}^f\right)\right] \end{array} \right\},$$

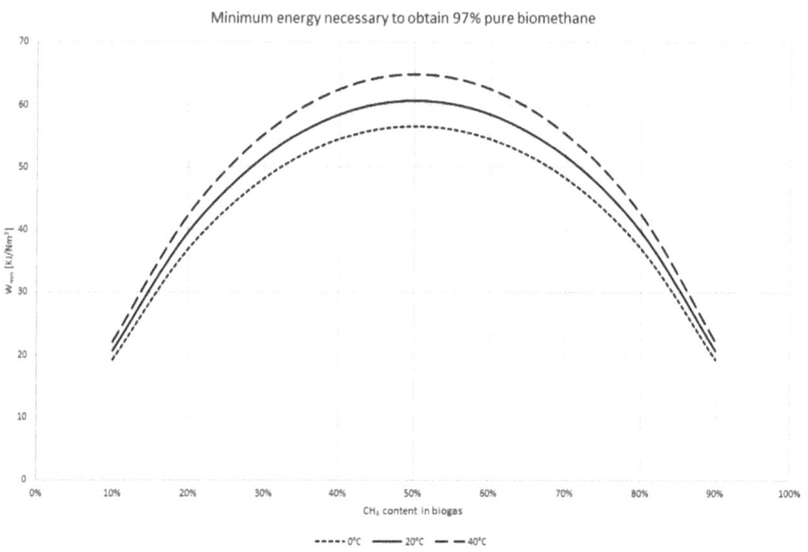

FIGURE 7.2 Ideal Work of Separation as a Function of Methane Concentration in Biogas.

where

c_A^i = initial CH_4 concentration in the feed gas,
c_{1A}^f = final CH_4 concentration in the product,
c_{2A}^f = final CH_4 concentration in the tail gas.

Although the formula may look complicated, plotting it with a spreadsheet is straightforward. Figure 7.2 shows the minimum work of separation for obtaining biomethane with 97% purity, as a function of the process temperature and the initial concentration of CH_4 in biogas.

Since industrial gas separation systems will consume more energy than an ideal machine operating under the same conditions of imperfect separation, it is possible to define a benchmark figure, the efficiency of separation:

$$\eta_{sep} = \frac{W_{imp}}{W_{real}}.$$

7.2.2 Minimum Heat of Gas Separation

What happens when the separation process employs heat instead of mechanical work? From the thermodynamics point of view, Carnot's theorem gives the ideal equivalence between mechanical work and heat.

$$W_{ideal} = \frac{Q_{in} - Q_{out}}{Q_{in}} = Q_{in} \frac{T_h - T_c}{T_h}$$

We can then calculate the equivalent minimum heat required for an imperfect transformation, Q_{imp}; the irreversible heat loss, ΔQ; and the efficiency of thermal separation, $\eta_{t.s}$, using the same criteria already introduced.

$$Q_{imp} = W_{imp} \frac{T_h}{T_h - T_c}$$

Here, W_{imp} is calculated using the Van't Hoff formula; T_h is, in general, the regeneration temperature of the process; and T_c is the process temperature or the refrigeration circuit temperature, if any.

We can hence define the thermal irreversibility and the thermal separation efficiency as

$$\Delta Q = Q_{real} - Q_{imp},$$

$$\eta_{t.s.} = \frac{Q_{imperf}}{Q_{real}}$$

N.B: Converting all Q flows to W, or vice versa, will not change the result, as long as you don't mix heat and work figures.

7.2.3 Minimum Work of Gas Compression

Now we know the minimum energy – work or heat – necessary to produce biomethane with any given purity percentage from biogas with any given initial composition. It is time to calculate the minimum compression energy. Compression is necessary to deliver the product to the desired grid or storage cylinder. From classical thermodynamics, it is known that the ideal compressor must perform an isentropic (adiabatic) transformation. The isentropic work of 1 mole of a perfect gas, $W_{i.c.}$, is given by [3]:

$$W_{i.c.} = \frac{k}{k-1} RT_1 \left[\left(\frac{P_2}{P_1} \right)^{\frac{k-1}{k}} - 1 \right]$$

Here,

$k = c_p/c_v$ = ratio of the heat capacities of the gas at constant pressure and constant volume;

R = universal constant of gases = 8.314 [J/(mol · K), when pressures are in Pa and temperatures in K];

T_1 = initial temperature of the gas;

P_1 = initial pressure of the gas;

P_2 = final pressure of the gas.

Assuming both CO_2 and CH_4 as ideal polyatomic gases, we can state $k = c_p / c_v \approx 1.333$ and then:

$$W_{i.c.} \approx 33.28 T_1 \left[\left(\frac{P_2}{P_1} \right)^{0.2498} - 1 \right]$$

With T_1 in K and P_1 and P_2 in Pa or kPa, the value of $W_{i.c.}$ is given in J/mol. Divide by 3,600,000 to convert the result to kWh/mol and then multiply by 44.64 to further convert to kWh/Nm^3.

7.2.4 Minimum Work of Gas Liquefaction

The demand for liquid biomethane, aka bio-LNG, is skyrocketing because of the decarbonisation policies for the goods transport sector, especially naval and truck transport. Liquefying biogas provides the separation of CH_4 and at the same time the final form of the product.

Gas liquefaction is a process that may employ three types of transformation: compression at a constant temperature, cooling at constant pressure, or adiabatic expansion. In the same way that we calculated the minimum work of separation and compression, it is possible to determine the minimum work of liquefaction. We will analyse two possible paths.

7.2.4.1 Ideal Separation + Liquefaction Process

The simplest ideal biogas upgrading and liquefaction process starts with the perfect separation with a Van't Hoff machine at the initial temperature T_i, followed by liquefaction of the resulting methane mass, m_{CH4}, with a Carnot refrigerator.

In the particular case of biogas upgrading, the gas constants for CH_4 are as follows:

Liquefaction point $= -162°C = 111K$

Average $c_p = 2.156$ kJ/kg K

Latent heat of condensation = 555 kJ/kg (source: https://www.engineeringtoolbox.com/methane-d_1420.html)

The heat extracted to liquefy methane is

$$Q_{ref} = m \Delta H \approx m_{CH4} \left[2.156 (T_i - 111) + 555 \right]$$

The work on the Carnot refrigerator is proportional to the extracted heat and to the efficiency of the cycle, which in turn depends on the initial and final temperatures:

$$W_{ref} = Q_{ref} \frac{111}{T_i - 111}$$

The ideal work of liquefaction is

$$W_{ideal} = W_{min} + W_{ref}.$$

Under our ideal process hypotheses, CO_2 is discarded after the separation step (ideally at room temperature), so most of the energy input is just for the liquefaction of CH_4.

Putting the above formulas in figures, let's calculate the liquefaction work of 1 Nm^3 of biogas with 60% CH_4 at the initial temperature of 35°C:

1. W_{min} = 76.92 kJ/Nm^3 of biogas (just apply the formula of perfect separation)
2. 0.6 Nm^3 CH_4 = 26.78 moles = 428.54 g
3. Refrigeration at constant pressure up to the liquefaction point of methane

$$Q_{ref} = m\ \Delta H \approx 0.428\ \left[2.156(308-111)+555\right] = 419.8\ \text{kJ}$$

4. Energy consumed by the Carnot refrigerator

$$W_{ref} = 419.3\frac{111}{308-111} = 236.6\ \text{kJ}$$

5. $W_{ideal} = W_{sep.min} + W_{ref} = 76.92\,kJ + 236.6\,kJ = 313.5\,kJ = 0.087\,\text{kWh/Nm}^3$ of processed biogas

This is the benchmark figure to evaluate a real process, consisting of separation with membranes followed by cryogenic refrigeration, operating at the same initial conditions.

7.2.4.2 Ideal Two-step Cryogenic Liquefaction

This alternative foresees the refrigeration of clean biogas at constant pressure until the liquefaction and separation of CO_2, followed by the further refrigeration and liquefaction of the remaining CH_4. The refrigerator is a Carnot ideal machine.

First step: Refrigeration of biogas at constant pressure up to the liquefaction point of CO_2 $(195\ \text{K})$ Average $c_{p\,CH4} = 2.156$ kJ/kg K, $c_{p\,CO2} = 0.849$ kJ/kg K, latent heat of condensation of $CO_2 = 591$ kJ/kg (source: https://www.engineeringtoolbox.com/methane-d_1420.html and https://www.engineeringtoolbox.com/CO2-carbon-dioxide-properties-d_2017.html).

$$Q_{ref\,1} = \sum m\,\Delta H \approx \left[m_{CH4}\times 2.156 + m_{CO2}\times 0.849\right](T_i - 195) + 591\times m_{CO2}$$

Second step: Refrigeration at constant pressure up to the liquefaction point of CH_4 $(-162\,°\text{C} = 111\ \text{K})$. Average $c_p = 2.156$ kJ/kg K, latent heat of condensation = 555 kJ/kg (source: https://www.engineeringtoolbox.com/methane-d_1420.html).

$$Q_{ref2} = m_{CH4}\Delta H \approx m_{CH4}\ \left[2.156\ (195-111)+555\right]$$

$$Q_{ref2} = 736.1\,m_{CH4}\left[\text{kJ}\right]$$

Knowing the amount of heat to extract, we can now calculate the energy absorbed by the Carnot refrigerator:

$$W_{ref1} = Q_{ref1} \frac{195}{T_1 - 195},$$

$$W_{ref2} = Q_{ref2} \frac{111}{195 - 111} = 1.3214\, Q_{ref2}.$$

The total ideal work of separation and liquefaction (subindex *s+l*) of both gases is then given as: $W_{s+l} = W_{ref1} + W_{ref2}$

Putting it in figures, under the same initial conditions of the former example – 1 m³ of clean biogas at 35°C with 60% CH_4 (429 g) and 40% CO_2 (786 g):

1. Refrigeration at constant pressure up to the liquefaction point of CO_2 $(195\ K)$.

$$Q_{ref1} \approx [0.429\,\text{kg} \times 2.156 + 0.786\,\text{kg} \times 0.849](308 - 195) + 591 \times 0.786$$
$$= 644.1\ \text{kJ} = 0.179\,\text{kWh}$$

2. Further refrigeration of the remaining CH_4, at constant pressure, up to the liquefaction point of CH_4 $(-162\,°\text{C} = 111\ \text{K})$.

$$Q_{ref2} \approx 0.429\ \text{kg}\ [2.156\ (195 - 111) + 555] = 315.4\ \text{kJ} = 0.876\ \text{kWh}.$$

3. Work on the Carnot refrigerator

$$W_{ref1} = 644.1 \frac{195}{308 - 195} = 362.9\text{kJ} = 1.008\ \text{kWh}$$

$$W_{ref2} = 315.4 \frac{111}{195 - 111} = 416.8\text{kJ} = 1.158\text{kWh}$$

4. Total ideal work of separation and liquefaction of both gases:

$$W_{s+l} = 362.9 + 416.8 = 779.7\,\text{kJ} = 0.217\text{kWh}$$

7.2.5 Conceptual Conclusions from the Thermodynamic Analysis

Summarising the theoretical considerations exposed till now, we can conclude the following:

1. Electricity-driven upgrading processes (e.g., permeable membranes, pressure swing adsorption (PSA), and liquefaction) can be compared to heat-driven ones (e.g., amine scrubbing) by taking the ideal work of separation as a benchmark.

2. The fair comparison of different processes requires converting heat flows into equivalent work, or vice versa. The resulting separation efficiency is the same, regardless of replacing all mechanical work with its equivalent heat flow or conversely.
3. Reaching the final state of the product (compressed or liquefied) is the most energy-consuming stage of the process. The work of compression or liquefaction must be accounted for separately. This allows the comparison of processes with different final states.
4. The underlying theory of gas separation is complex, but the calculation of the separation efficiency and work of compression/liquefaction is a straightforward and reliable benchmark criterion.

Figure 7.3 shows the energy consumed in each stage of the four ideal separation processes + compression and separation + liquefaction. Note that the most energy-consuming process is two-step liquefaction, but the same yields two products simultaneously: liquid CH_4 and CO_2.

7.3 CLASSIFICATION OF UPGRADING TECHNOLOGIES

The IEA Task 37 workgroup focuses on monitoring the diffusion of biogas and biomethane technology. At the moment of writing this chapter (Jan. 2022), its last global report (2019) points out that the most diffused biogas upgrading technologies

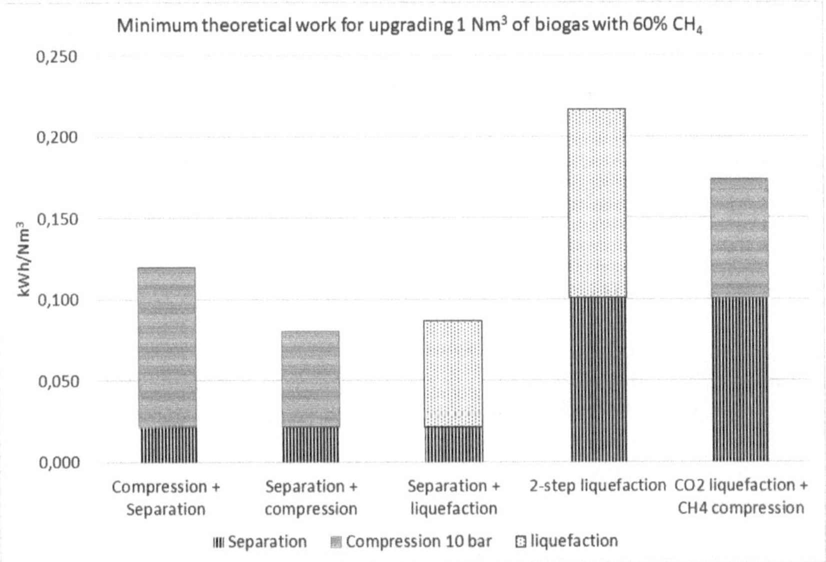

FIGURE 7.3 Minimum Work of Compression + Separation, Separation + Compression, Separation + Liquefaction, Two-Step Liquefaction, and CO_2 Liquefaction + CH_4 Compression.

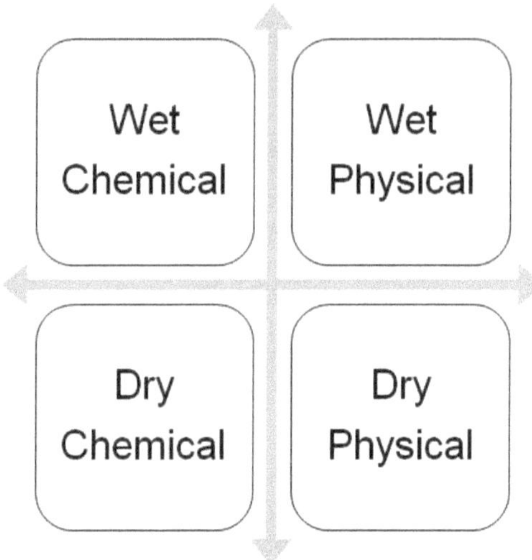

FIGURE 7.4 Classification of Biogas Upgrading Technologies.

are water scrubbing, membrane separation, chemical scrubbing, and PSA (mostly on activated carbon), in this order of abundance. Does this mean that the said are the 'best' upgrading technologies? What is their efficiency compared to their 'ideal' counterparts? This section and the following provide some insights into the state of the art and the improvement margins left.

Biogas upgrading technologies can be classified as wet or dry, physical or chemical, as shown in Figure 7.4.

Dry technologies rely on the different properties of CH_4 and CO_2, which result from their different molecular sizes and weights. Wet technologies rely on the different solubilities of both gases in polar and non-polar solvents. The border between chemical and physical is somehow fuzzy. For instance, the solubility of CO_2 in water is both a physical and a chemical process because the gas reacts with water, forming carbonic acid.

7.4 DRY UPGRADING TECHNOLOGIES

Dry upgrading technologies include several solutions. Some are widely consolidated, while others are considered as 'emerging'. This section presents a short review of their state-of-the-art and market diffusion.

7.4.1 Selective Membranes

This is the most diffused technology among dry ones, with nearly 26% of the market share, according to the IEA 2019 report. The market of membrane upgrading systems was worth US$ 81 Ml in 2022 and is expected to reach US$161 Ml by 2031 [4]. Usually, they operate in the pressure range of 6–20 bar. The most common membrane material is cellulose acetate, having a lifespan of nearly 3 years, but several other polymeric materials are available on the market. The choice of the membrane is a trade-off between performance and selectivity [5]: the higher the CO_2 permeance (i.e., higher flow for a given pressure), the lower the selectivity (i.e., the CO_2 flow will contain higher percentages of

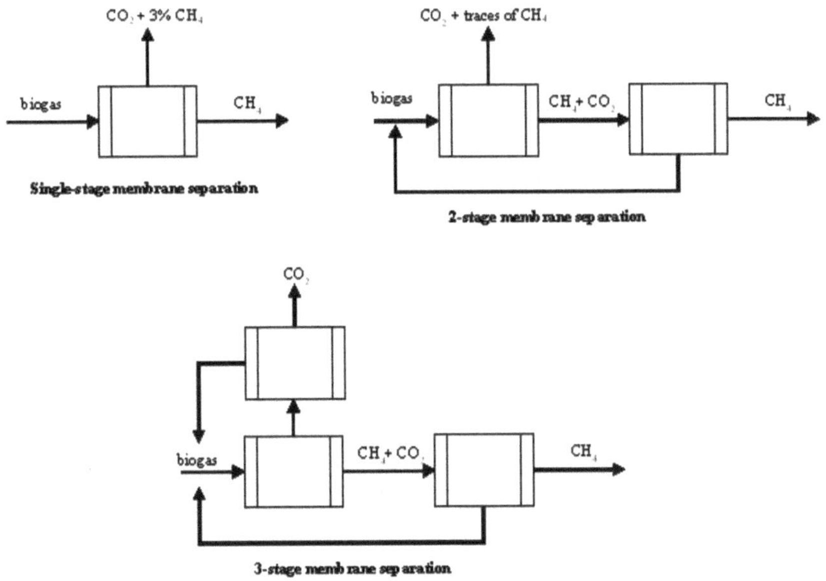

FIGURE 7.5 Different Configurations of Membrane Separation Systems.

residual CH_4). Drying and desulphuration of biogas are mandatory to avoid damage to the membranes. Up to 3% methane is lost with the tail gas in single-stage units, with three-stage arrays providing virtually pure CH_4 and CO_2 streams (Figure 7.5). Since membrane separation requires mechanical work for gas compression, reducing the OPEX requires adopting two- or three-stage compressors driven by IEC Class 4 motors with inverter control for maximum efficiency (see Chapter 2, Section 2.1).

CASE STUDY 7.1: UPGRADING AND GRID INJECTION IN ITALY

The only data available for our analysis are the grid pressure and the specific energy consumption of the plant declared by the manufacturer. The rest are just 'educated guesses', necessary for an approximate estimation.

$$W_{real} = 0.25 \text{ kWh/Nm}^3$$

$$T = 35°\text{C}$$

$$P_1(\text{biogas}) \approx 101 \text{ kPa}$$

$$P\left(\text{product}\right) = 1200 \text{ kPa} (\text{required injection pressure to the gas network})$$

If the pressure of the product is 1,200 kPa, then this must also be the partial pressure of methane at the compressor output. Hence, the total pressure of the biogas at the compressor output must be

P_2 (biogas) $= 1{,}200$ kPa / $0.6 = 2{,}000$ kPa;

$c_A{}^i$ = initial CH_4 concentration in the feed biogas = 60% (assumed value);

$c_{1A}{}^f$ = final CH_4 concentration in the product = 97% (minimum acceptable purity in Italy).

First step: calculation of W_{imp}. This is straightforward: just apply the formula.

$$W_{imp} = 61.53 \text{ kJ/Nm}^3 = 0.017 \text{ kWh/Nm}^3$$

Second step: calculation of $W_{i.c}$. Just apply the formula, considering that

$P_1 = 60\%$ of 101 kPa (pressure at the exit of the Van't Hoff machine) and $P_2 = 1{,}200$ kPa (required network pressure)

$$W_{i.c.} = 0.141 \text{ kWh/Nm}^3$$

The third step is straightforward too:

$$W_{ideal} = 0.0171 + 0.141 = 0.158 \text{ kWh/Nm}^3$$

$$\eta_{sep} = \frac{0.158}{0.25} \equiv 63.2\%$$

This example allows us to draw some interesting conclusions:

1. The compressor absorbs 89% of the total minimum theoretical work of separation and compression.
2. The isentropic efficiency of industrial compressors is in the range of 75–85%. Increasing the final pressure reduces the isentropic efficiency. If we include this factor, the overall efficiency will be 75% × 89% = 66.7%. So, the efficiency calculated with our 'educated guesses' is fairly accurate.
3. The efficiency of electric motors grows with the nominal power. Hence, the efficiency of membrane upgrading plants grows with their capacity: one module of 500 Nm^3/h will be more energy-efficient than two modules of 250 Nm^3/h running in parallel.
4. Open debate on energy policy: the European Commission's Green Deal, the IEA, the US DoE, and several industrialised countries promote replacing NG with 'green hydrogen' obtained from water electrolysis fed with wind and solar power.

The production of 1 Nm^3 of 'green hydrogen' requires at least 5 kWh of electricity, but its lower heating value (LHV) is only 3.54 kWh/Nm^3. Producing 1 Nm^3 of biomethane from biogas and compressing it to 20 bar require just 0.25 kWh, and its LHV is 9.96 kWh/Nm^3.

7.4.2 Molecular Sieves

A molecular sieve is a porous solid that separates particles of molecular dimension according to their sizes and shapes. When two molecules are equally small and fit into the pores, the polarity (charge separation) of the molecule is the dominant separation mechanism. The most polar molecule will be preferentially adsorbed.

Activated carbon (aka activated charcoal) is one of the most common adsorbent materials in the industry. It is very porous and has an extremely high surface area to volume ratio. The pore sizes of activated charcoal can be classified as micropores (width < 2 nm), mesopores (width = 2–50 nm), or macropores (width > 50 nm). Such a wide range of porosity, together with its low cost, makes activated carbon a versatile commodity for many industrial fields at the expense of lower selectivity. Two different grades exist: granular activated carbon (GAC) and powder activated carbon (PAC).

Zeolites are hydrated metal aluminosilicate compounds with well-defined crystalline structures. The silicate and aluminate groups form three-dimensional crystal lattices surrounding cavities that loosely hold the metal ions and the water molecules together. Channels run through the entire crystal, interconnecting the cavities and terminating at the crystal surface. Zeolites lose their bonded moisture when heated, with little or no change in their crystal structure. Dehydrated zeolites can reversibly absorb water, and for this reason, one of the most common uses of zeolite molecular sieves is drying gases. Zeolites can absorb other molecules if the latter are small enough to pass through the channels or pores. Metal ions are also readily replaceable by other ionic units if their charge and size are similar. Engineered zeolites are artificial materials having a very uniform pore size and granulometry. Consequently, they are highly selective, minimise pressure losses in separation columns, and facilitate bulk handling.

Molecular sieves are the heart of PSA upgrading systems. The design of a PSA system is subject to several trade-offs:

1. Size of the columns. CAPEX increases proportionally to the size of the column, while OPEX decreases because of longer cycles (longer life of the adsorbent), lower pressure, and higher selectivity, and less CH_4 lost in the tail gas.
2. It is advisable to redact a supply and after-sales service contract containing very clear specifications. The following aspects require special attention:

    a. Moisture in the biogas feed. This is especially important for engineered zeolites because moisture drastically reduces their lifespan.
    b. Frequency of replacement and cost of the new or regenerated adsorbent.
    c. Minimum guaranteed selectivity.
    d. Frequency of disposal (or regeneration) of the exhausted adsorbent and its cost.

Systems providing methane slip < 2% are more expensive. The acceptable percentage of methane lost with the tail gas depends on a series of considerations: cost–benefit balance, local environmental protection laws (methane is a powerful greenhouse gas), eventual CO_2 recovery as a by-product, etc.

7.4.3 Activated Carbon

It is probably the most diffused adsorbent for PSA biogas upgrading systems. Figure 7.6 shows the operation principle. After compression and drying, the biogas flows through column C1, leaving behind CO_2 and SH_2. After a predefined time, V1A and V1D close, V1C, V2A, and V2D open. Now, the biogas will flow through C2, while the tail gas contained in C1 will be recycled to recover the residual methane. When the pressure in C1 equals the aspiration pressure of the compressor, V1C closes while V1B opens for the vacuum pump to regenerate the column. After a predefined time, the biogas flow shifts to C3, repeating the cycle and going back to C1.

The PSA technology is simple and robust. It has the advantage of upgrading and desulphurating in one single step but has some drawbacks too.

The mesoporosity varies largely, depending on the source of activated carbon (coconut shell, bamboo, woodchips, etc.), which reflects in the selectivity of the product [6].

The selectivity (ratio of the specific absorptances at a given pressure) can be as low as 5 [7]. This means losing up to 20% of CH_4 in the tail gas, requiring a higher recirculation flow and hence higher energy consumption.

7.4.4 Natural and Engineered Zeolites

Zeolites are natural materials. Engineered zeolites for biogas purification allow two alternative methods of regeneration: pressure swing and temperature swing. When

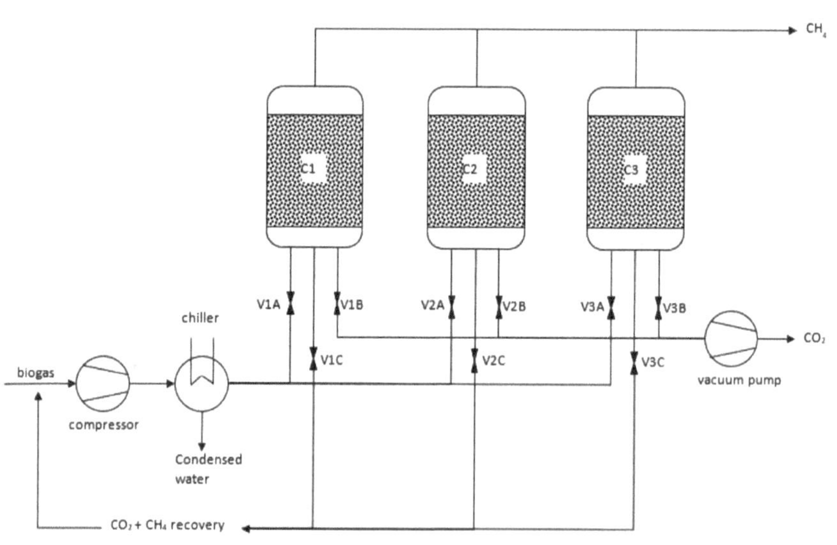

FIGURE 7.6 Operation Principle Sketch of the Activated Carbon PSA System.

hot air is available at low cost, or in countries where electricity is too expensive, temperature swing adsorption (TSA) is an option to consider. The design requirements of both systems are the opposite: PSA units usually require small columns and short cycles in the order of minutes, whereas TSA units require large columns and a cycle duration of several hours or even a day.

The length/diameter (L/D) ratio of zeolite columns is usually L/D < 5. This rule-of-thumb ensures a good compromise between gas turbulence inside the bed (good exchange factor) and mechanical resistance to pressure and vacuum cycles.

There are two types of commercial engineered zeolites.

7.4.4.1 Type A Structure

All molecular sieves are composed of sodium and aluminosilicate, which are grown to form a three-dimensional crystal structure called type A. This crystalline structure allows a molecular sieve to adsorb small molecules, rejecting those too large to fit inside the crystal. The commercial denomination '4A' means that the pore size is 4Å (0.4 nm) in diameter. 4A crystals can adsorb molecules smaller than 4Å in diameter, such as water, while allowing larger molecules, such as methane, to pass along to the desired product stream. When exposing 4A crystals to potassium, larger potassium ions will replace sodium ions, creating pores of about 3Å in diameter, called 3A molecular sieves. 3A molecular sieves absorb water more selectively. Exposing 4A crystals to smaller calcium ions will produce larger pore sizes of about 5Å in diameter. Type 5A zeolite can adsorb certain mercaptans while being still small enough to block larger hydrocarbons from entering the crystal.

These properties allow 5A sieves to be utilised in LNG and biogas refining to remove sulphur compounds while allowing the desired hydrocarbons to pass through to the product stream.

Type 5A is also useful for producing relatively pure streams of hydrogen or oxygen since it can adsorb nitrogen, carbon monoxide, and other contaminants.

7.4.4.2 Type X Structure

Type X crystals are shaped differently from type A, tending to offer much larger pore sizes of about 9Å in diameter. 13X molecular sieve is often used in air treatment processes, such as in the production of high-purity oxygen, because of its high selectivity to CO_2. In addition to oxygen production, type X crystals are commonly used in cryogenic distillation processes to dehydrate LNG and bio-LNG streams. Removing all water from these streams is mandatory to prevent pipelines from clogging and freezing. There are many variations of type X crystals, with most being type 13X, although some products, called type 10X, may have pore openings about 8Å in diameter. The NG industry employs 13X zeolites to eliminate CO_2. It is worth noting that the molecule of CO_2 has a critical diameter of 3.3Å, while CH_4 has 3.6Å. Hence, both components can enter 4A (4Å pore) and 13X (10Å pore) molecular sieves. Hydrocarbons up to propane can enter 4A molecular sieves. However, the sieving effect is not the only parameter playing a role in the removal mechanisms of zeolites. CO_2 is non-polar due to its linear symmetry. An apolar adsorbent like activated carbon can indeed remove CO_2, but it will have to compete with other apolar molecules. On the other hand, due to the difference in electronegativity between C and O, CO_2 can have induced dipoles. This last property is important when adopting polar adsorbents like zeolites and explains why they better adsorb CO_2 than CH_4, despite both gases being able to enter the pore. The electric field gradient of the 4A zeolite is weaker when compared to the 13X zeolite. This is the reason why 13X zeolites are more efficient in removing CO_2.

7.4.5 Alkaline Hydrates

Alkaline hydrate upgrading is based on the following reaction:

$$Na_2CO_3 + nH_2O + CO_2 \leftrightarrow 2NaHCO_3 + (n-1)\,H_2O$$

This technology has many theoretical advantages: desulphuration, dehydration, and CO_2 removal in a single step; low cost; mild operation conditions; lower heat demand for regeneration compared to the wet equivalent because of the minimum quantity of water; and sodium carbonate is inexpensive and does not pose environmental or health hazards [8]. The drawbacks are few but difficult to overcome: high heat demand at temperatures higher than 110°C; the alkaline packing tends to clog after several cycles of hydration and regeneration; high-pressure drop across the packing; and the efficiency depends sharply on the uniformity and correct degree of hydration, which is difficult to achieve in a static packed reactor. Big packed columns are necessary, so the cycle must be long enough to allow heating the whole mass of steel and hydrated carbonate, removing the CO_2, cooling, and rehydrating the alkaline packing before starting a new cycle.

7.4.6 Bed-packed Amines

This is an emerging technology (only one plant in Italy by 9 January 2022) based on the chemical reaction (R_2 represents the amine radical):

$$2R_2NH + CO_2 \rightarrow R_2NH_2^{\;+} + R_2NCO_2^-$$

Like in the case of molecular sieves, this technology consists in the wet impregnation of Al_2O_3 pellets having high superficial area (270 m^2/g) and bimodal distribution of meso- (8–10 nm) and macroporosity (400–500 nm) with diethanolamine (DEA), 36% wt.

The advantages of bed-packed amines over wet amine scrubbing are 63% of the heat demand for the regeneration, lower regeneration temperature (85°C), no foaming and no corrosion.

7.5 WET UPGRADING TECHNOLOGIES

Wet upgrading technologies rely on the solubility difference between CO_2 and CH_4 in a given liquid phase. The efficiency of the interaction between gas and liquid

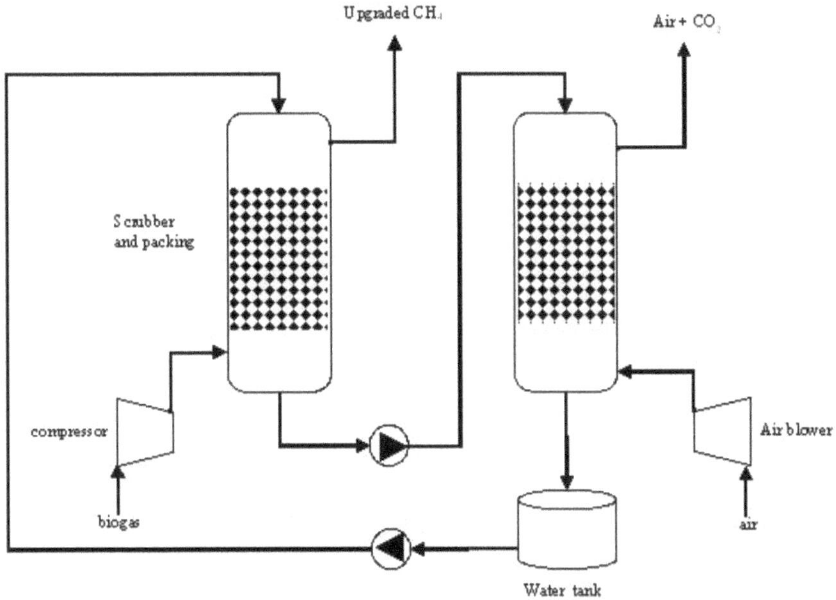

FIGURE 7.7 General Block Diagram of a Wet Separation System Showing the Scrubber and Regenerator.

is proportional to the exchange surface; therefore, packed scrubbers and spray scrubbers with countercurrent flows are preferred to bubbling columns. Figure 7.7 shows a general block diagram of the process.

7.5.1 Water Scrubbing

High-pressure water scrubbing is the most employed technique, accounting for nearly 40% of all biogas upgrading plants worldwide by 2019 [10]. Its main advantages are the capacity to remove CO_2, H_2S, and siloxanes altogether; simple engineering; and widely available equipment (standard pumps, steel carpentry, and standard packing material). Its drawbacks are the high (electric) energy demand for gas compression and pumping, and the exponential decrease of CO_2 solubility with growing water temperature (Figure 7.8, data from [11]). Since the solubility of CH_4 remains almost constant within the practical temperature and pressure range (< 0.05 molar fraction), the selectivity of this separation method decreases with growing water temperature. It is worth noting that this technique is widely diffused in Northern Europe, where cold water is available all year round and the cost of electricity is low. Regeneration of the water saturated with CO_2 can be done by using vacuum or by temperature swing.

7.5.2 Water/PEG Scrubbing (Selexol Process)

Polyethylene glycol (PEG) is a family of polymers of ethylene widely employed in the chemical industry. Water–PEG mixtures have high selectivity for CO_2 absorption

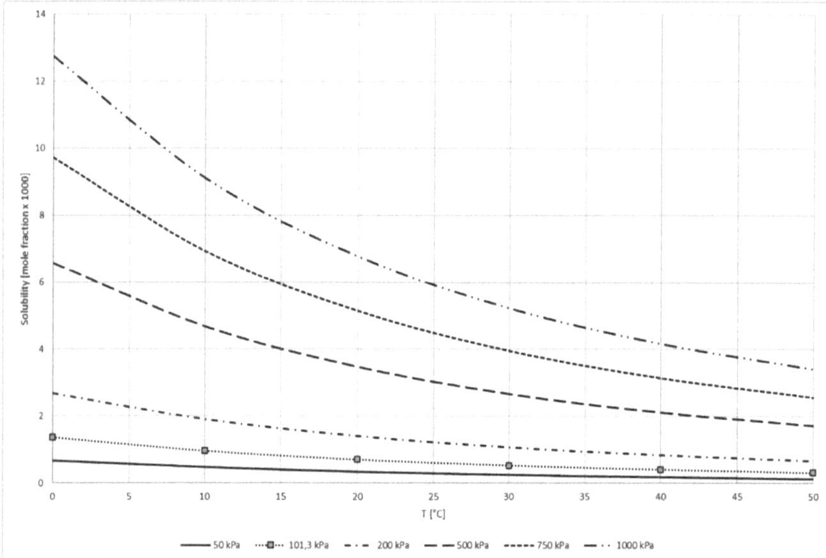

FIGURE 7.8 Solubilities of CO_2 in Water as a Function of Temperature, at Different Pressures.

at atmospheric or moderate pressure. The solubility of CO_2 increases with the molecular weight of the PEG [12]. Regeneration of the solution can be done by vacuum or by temperature swing.

7.5.3 Alkaline Scrubbing (Benfield Process)

The classical Benfield process is widely diffused in the natural gas industry. It employs KOH solutions that react with CO_2 at near room temperature, forming K_2CO_3. Regeneration of the solution requires heating it at about 120°C. It is also possible to employ NaOH solutions.

7.5.4 Amine Scrubbing

Monoethanolamine (MEA; aka 2-aminoethanol) and dimethanolamine (DMEA; aka 2-dimethylaminoethanol) are very selective towards CO_2 (loss of CH_4 < 0.1%). They react with H_2S, but the reaction is not reversible, degrading the solution [13], which needs to be changed periodically. Thermal regeneration of the spent solution requires heating it to 115–126°C. MEA and DMEA are toxic (14 & 15), which indicates high costs of transport, storage, manipulation, and disposal of the waste. Amine scrubbing is a technology widely diffused in the chemical industry but less attractive for agricultural biogas plants because of the cost and safety issues.

7.5.5 Solvent Scrubbing (Rectisol Process)

The Rectisol process is a technology patented by Linde. It works at low temperatures and eliminates H_2S together with CO_2. In this process, the solvent is methanol, and the solute is methane, so it is more suitable for gas mixtures with low CH_4 concentrations, like syngas and landfill gas. Therefore, solvent regeneration with N_2 is mandatory to avoid the risk of explosion.

7.6 EMERGING TECHNOLOGIES

7.6.1 Scrubbing with Liquid Digestate (*In Situ* Enrichment)

This is a special case of wet upgrading in which the absorbent is the liquid fraction of digestate instead of water. Regeneration of the liquid is just standard stripping with air. The concept was tested in Sweden [16], but the percentage of CH_4 in the output biogas was found to be highly variable (60%–90%). The O_2 dissolved in the digestate favours the activity of the facultative hydrolytic bacteria. The faster hydrolysis rate of the substrate is counterbalanced by unavoidable methane emissions to the atmosphere and disturbing the strictly anaerobic microbes with the O_2 dissolved in the digestate.

7.6.2 Cryogenic Distillation

This energy-intensive process exploits the differences in boiling points at a given pressure. Despite its higher CAPEX and OPEX, investors may prefer this solution because of the higher market prices of liquefied biomethane (LBM) compared to compressed biomethane (CBM). LBM has a higher value because it is three times denser than CBM. Its energy density is similar to that of diesel oil; hence, it is attractive for the conversion of heavy trucks.

$$\text{Comparison: } 1\ \text{Nm}^3\ \text{BM} = 5\ \text{dm}^3\ \text{CBM} = 1.7\ \text{dm}^3\ \text{LBM}$$

The high selectivity of the process allows the production of food-grade CO_2, a feature that provides some marginal income, although its price is low. This option is profitable only in places where the said commodity may find a market (biogas plants placed near factories of sparkling beverages or dry ice for food conservation).

There are four different process concepts:

1. Indirect refrigeration (external cryogenic unit or a cryogenic tank containing liquid N produced by a third party).
2. Joule effect (compression of the biogas followed by expansion in a turbine, allowing to recover part of the compression energy).
3. Mixed compression of the biogas and indirect refrigeration with liquid N produced on-site through the Joule effect in a closed-loop circuit [17].

4. Hybrid (upgrading with PSA or membrane or scrubber, followed by liquefaction through a Stirling cryocooler (see, for instance, https://www.bio-lng.it/en/index.html).

Modelling the thermodynamic efficiency of cryogenic distillation plants is a very complex task.

The energy consumption depends strongly on the process configuration. Table 7.3 summarises some data from the literature (16 (already cited above), 18, and 19) and some manufacturers.

7.6.3 Bioupgrading (P2G)

Bioupgrading is better known in Europe as Power to Gas (P2G) technology. It consists of injecting H_2 and CO_2 in an external bioreactor – or in the digester itself – where the hydrogenotrophic archaea will convert the gas mixture into CH_4 and H_2O. The overall stoichiometry is as follows:

$$CO_2 + 4H_2 \rightarrow CH_4 + 2H_2O$$

Despite claims from some European academicians, politicians, and plant builders, P2G is a relatively simple technology. Implementing it requires some preliminary laboratory tests to scale up the process.

The rationale behind the ‘clean hydrogen’ policy of the European Green Deal is that hydrolysing water is a mature technology, suitable to overcome the problem of managing the excess of wind or photovoltaic energy in the short term. This is probably a problem during a few days of a year in Germany, Denmark, or Spain, but it may not be the case elsewhere.

TABLE 7.3
Specific Electric Energy Consumption for Different Biomethane Liquefaction Systems

Manufacturer	Electricity (kWh/ Nm³ Biogas)	Source
Scandinavian GtS	0.8	Johansson N.
Acrion	1.42	Johansson N.
Prometheus Energy	1.54	Johansson N.
Criotec	0.65	Criotec Impianti Spa (www.criotec.com)
Ryan-Holmes Process	0.86	Pellegrini et al.
Dual pressure distillation	0.77	Pellegrini et al.
Anti-sublimation process	1.19	Pellegrini et al.
Ideal separation + liquefaction	0.087	See Section 7.2.4.1
Ideal Carnot refrigerator	0.217	See Section 7.2.4.2

The exergy efficiency of P2G is very low, as will be shown below:

In the best case (perfect stoichiometry), converting 1 Nm^3 of CO_2 into 1 Nm^3 of CH_4 requires producing at least 4 Nm^3 of H_2. Water electrolysis requires a theoretical minimum of 237 kJ/mol; hence, producing 4 Nm^3 of H_2 (178.6 moles) requires 42,328.2 kJ (11.76 kWh) of electricity. Like in any other transformation process, the electrolysis efficiency is always lower than 100%. State-of-the-art electrolysers have efficiencies as low as 46% and up to 80% in the best laboratory prototype. The specific electricity consumptions are in the range from 3.8 kWh/Nm^3 (laboratory-scale solid oxide electrolyser) to 5.5 kWh/Nm^3 (commercial alkaline electrolyser; 20). For now, only alkaline electrolysers have reached commercial maturity. Hence, producing 4 Nm^3 of H_2 with commercially available electrolysers will require 22.2 kWh of electricity. Since 1 Nm^3 of CH_4 has 9.98 kWh lower heating value (LHV), and the best available CHP groups have an efficiency in the range of 40–42%, the maximum quantity of electric energy that one can produce from 1 Nm^3 of CH_4 is just 4.2 kWh. The overall efficiency of the cycle is then:

$$\eta_{elec} = 4.2 / 22.2 \equiv 18.9\%.$$

N.B. The calculations above do not consider the electricity consumed for compressing the biogas, the H_2, and the resulting CH_4, so the real efficiency is even lower than 18.9%.

A close look at the reaction's stoichiometry would have sufficed to show the EC policymakers how inefficient this process is. Half of the energy consumed for producing H_2 from water is wasted in producing residual water.

Just for comparison purposes, the electric (exergy) cycle efficiency of a modern electrochemical Li-ion battery is very close to 100%, and at least 85% for the century-old Pb-acid battery. So, the most efficient way of managing the excess renewable energy generation is storing it in batteries.

The separation efficiency of P2G compared to Van't Hoff minimum work of separation is even lower:

$$\eta_{sep} = \left(69.76 \text{ kJ/Nm}^3 / 3600 \text{ s/h}\right) / 22.21 \text{ kWh/Nm}^3 \approx 0{,}087\%.$$

7.6.4 Electromethanogenesis

Electromethanogenesis seems to be the latest fashion in AD research and peer-reviewed publications. An electrobioreactor is a digester fitted out with electrodes. By applying a small voltage (about 2 V DC) across the electrodes, ions will flow through the sludge. Free ions can easily penetrate the plasmatic membranes of microorganisms, boosting their metabolic paths. This technology is dubious (see Chapter 1, Section 1.2.4).

7.6.5 Scrubbing with NH_3 Solution

Scrubbing biogas with an ammonia solution is a low-cost emerging technology, almost unknown in Europe. The Alstom Group patented the process in 2006 under the commercial name chilled ammonia process (CAP). Its original scope was recovering CO_2 from flue gases. Alstom sold the patent to General Electric, which carried on further improvements [21]. The main advantage of the CAP is the high reactivity of aqueous NH_3 with CO_2, forming ammonium carbamate, carbonate. and bicarbonate, which can be sold as low-cost fertilisers [22] or regenerated to ammonia. The regeneration requires heating the solution at 30–60°C, so the heat demand is low and could be met with thermal solar panels. The reaction is quick because of the multiple possible reaction paths, depending on the reactants being in liquid (l), gaseous (g), or solution (aq) state [23]:

$$CO_2\left(g\right) + 2NH_3\left(g\right) \rightarrow NH_2COONH_4\left(s\right) \tag{7.1}$$

$$NH_2COONH_4\left(s\right) + H\,O\left(g\right) \rightarrow \left(NH_4\right)2\,CO_3\left(s\right) \tag{7.2}$$

$$CO_2\left(g\right) + 2NH_3\left(g\right) \rightarrow CO(NH_2)_2\left(s\right) + H_2O\left(g\right) \tag{7.3}$$

$$CO_2\left(g\right) + 2NH_3\left(aq\right) \rightarrow NH_2COO\text{-}\left(aq\right) + NH_4 + \left(aq\right) \tag{7.4}$$

$$2NH_3\left(g\right) + CO_2\left(g\right) + H_2O\left(g\right) \rightarrow (NH_4)_2\,CO_3\left(s\right) \tag{7.5}$$

$$NH_3\left(g\right) + CO_2\left(g\right) + H_2O\left(g\right) \rightarrow NH_4HCO_3\left(s\right) \tag{7.6}$$

$$2NH_3\left(l\right) + CO_2\left(g\right) + H_2O\left(l\right) \rightarrow (NH_4)_2CO_3\left(s\right) \tag{7.7}$$

$$NH_3\left(l\right) + CO_2\left(g\right) + H_2O\left(l\right) \rightarrow NH_4HCO_3\left(s\right) \tag{7.8}$$

7.7 SELECTION OF THE MOST SUITABLE PROCESS IN EACH CONTEXT

Each site is unique. The optimum technical solution somewhere may be less profitable or even impracticable elsewhere. So, this section summarises the concepts explained throughout the chapter and provides some guidelines for analysing biogas upgrading projects from a multidisciplinary point of view and not just from the economic balance.

1. Always compare apples with apples. When planning to invest conspicuous capital in a biogas upgrading plant, it is logical to ask several plant builders for quotations. Of course, each manufacturer will propose a different solution and provide technical data measured under different operational conditions. The key to comparing different technologies is referring to a benchmark. The thermodynamic analysis is then the best-proven approach:
   a. Calculate the energy consumed for the gas separation and the energy consumed for the compression or liquefaction as two distinct figures. Remember that the energy demand for the compression or liquefaction of the product(s) is usually higher than the energy required to separate CH_4 and CO_2.

b. Compare them to the ideal energies of separation and compression or liquefaction.
c. Always account for the thermodynamic equivalences of heat and mechanical work. If supplier A proposes a process driven by heat and supplier B requires electricity, ask for the technical sheets of the electric motors. Remember that four different IEC efficiency classes of electric motors exist (see Chapter 2, Tables 2.2 and 2.3). If the class efficiency is unknown, assume 90%. Ask supplier A for the highest and lowest temperatures of the thermal cycle. Then convert the heat flows into work (or vice versa), and finally compare the equivalent works (or heats) of each process to the ideal work (or heat) of a perfect thermodynamic conversion. The best solution is the one that provides the best ratio between theoretical and real energy consumption for the process. The result must be the same, regardless of choosing work or heat as the reference energy form.

2. The best solution from the energy point of view is not necessarily the most profitable. For instance, if heat is available for free or at a low cost – from a CHP, a geothermal pit, etc. – then a heat-driven process will turn more profitable, even if it may be thermodynamically less efficient.
3. Recovering CO_2 instead of venting it into the atmosphere may provide some marginal income. This argument deserves a deep analysis (provided in Chapter 8, Section 8.2).
4. Table 7.4 provides some conceptual insights into other data that are sometimes overlooked but may influence the project's profitability under different local conditions. As a rule, performing a sensitivity analysis is always advisable. For instance, if the candidate solution requires electricity, analysing how the net profit may vary in case of cost rise, energy shortage, or market disruption will prove vital to discover eventually hidden criticalities.

TABLE 7.4
Factors Affecting the Profitability of Different Upgrading Systems

	Influence on profitability	
Factor	**CAPEX**	**OPEX**
Water availability and local climate	↔	↓
IEC class of the motors	↑	↓
Multi-stage compressors	↑	↓
Heat recovery	↑	↓
Pressure recovery (expansion turbine)	↑	↓
CO_2 recovery	↑	↓
Plant size and capacity	↑	↓
Consumables, periodical maintenance	↓	↑
Proprietary components or technology	↓	↑
Royalties paid to the technology provider	↓	↑

BIBLIOGRAPHY

[1] J Savickis, Namejs Zeltins, I Bode, Leo Jansons, E Dzelzitis, A Koposovs, A Selickis & A Ansone, The biomethane injection into the natural gas networks: The EU's gas synergy path, *Latvian Journal of Physics and Technical Sciences*, 57, 34–50 (2020). https://doi.org/10.2478/lpts-2020–0020. https://www.researchgate.net/publication/343716958_The_Biomethane_Injection_into_the_Natural_Gas_Networks_The_EU's_Gas_Synergy_Path

[2] SA Amelkin, AM Tsirlin, JM Burzler, S Schubert & KH Hoffmann, Minimal work for separation processes of binary mixtures, *Open Systems & Information Dynamics*, 10:04, 335–349 (2003). https://doi.org/10.1023/B:OPSY.0000009555.63816.86

[3] Jiří Oldřich, Isentropic efficiency of centrifugal compressor working with real gas, *Acta Polytechnica CTU Proceedings*, 20, 65–72 (2018). https://doi.org/10.14311/APP.2018.20.0065. https://www.researchgate.net/publication/330892730_ISENTROPIC_EFFICIENCY_OF_CENTRIFUGAL_COMPRESSOR_WORKING_WITH_REAL_GAS

[4] Dana Havas & Haiqing Lin, Optimal membranes for biogas upgrade by removing CO2: High permeance or high selectivity?, Separation Science and Technology, 52:2, 186–196 (2017). https://doi.org/10.1080/01496395.2016.1218515

[5] Mario Pellerano, Pascaline Pré, Mariem Kacem & Arnaud Delebarre, CO2 capture by adsorption on activated carbons using pressure modulation, *Energy Procedia*, 1:1, 647–653 (2009). https://doi.org/10.1016/j.egypro.2009.01.085. ISSN 1876–6102.

[6] D Peredo-Mancilla et al., CO_2 and CH_4 adsorption behavior of biomass-based activated carbons, *Energies*, 11, 3136 (2018). https://doi.org/10.3390/en11113136

[7] Yuanhao Cai, WeilinWang, Liang Li, Zhaofeng Wang, Suying Wang, Hao Ding, Zhengguo Zhang, Luyi Sun & Weixing Wang, Effective capture of carbon dioxide using hydrated sodium carbonate powders, *Materials*, 11, 183 (2018). https://doi.org/10.3390/ma11020183

[8] M Scagliotti & P Castellazzi, Sistemi di upgrading a confronto, *Proceedings of Biometano Day 2021*. https://agroenergia.eu/biometano-day-2021/?lang=en

[9] IEA Task 37 Country Report (2019). https://www.ieabioenergy.com/wp-content/uploads/2020/03/IEA-Task-37-Country-Report-Summaries-2019-1.pdf

[10] John J Carroll, John D Slupsky & Alan E Mather, The solubility of carbon dioxide in water at low pressure, *Journal of Physical and Chemical Reference Data*, 20, 1201 (1991). https://doi.org/10.1063/1.555900. https://srd.nist.gov/JPCRD/jpcrd427.pdf

[11] Jun Li, Yinmei Ye, Lifang Chen & Zhiwen Qi, Solubilities of CO2 in Poly(ethylene glycols) from (303.15 to 333.15) K, *Journal of Chemical & Engineering Data*, 57, 610–616 (2011). https://doi.org/10.1021/je201197m

[12] MS Islam, R Yusoff, BS Ali, MN Islam & MH Chakrabarti, Degradation studies of amines and alkanolamines during sour gas treatment process, *International Journal of the Physical Sciences*, 6:25, 5877–5890 (October 23, 2011). https://doi.org/10.5897/IJPS11.237. https://academicjournals.org/journal/IJPS/article-full-text-pdf/D7C588425191

[13] ECHA (European Chemical Agency) safety sheet of 2-aminoethanol (MEA). https://echa.europa.eu/it/substance-information/-/substanceinfo/100.004.986

[14] ECHA (European Chemical Agency) safety sheet of 2-dimethylaminoethanol (DMEA). https://echa.europa.eu/it/substance-information/-/substanceinfo/100.003.221

[15] Mikael Hansson, Johan Laurell, Åke Nordberg, Åke Rasmuson, Jing Liu, Mihaela Nistor, Sten Strömberg & João Costa, In-situ methane enrichment of raw biogas in the anaerobic digestion process (Metananrikning av rågasen under rötningsprocessen), *SGC Rapport*, 280 (2013). http://www.sgc.se/ckfinder/userfiles/files/SGC280.pdf

[16] SE Hayemi et al., Cryogenic vs. absorption biogas upgrading in liquefied biomethane production – An energy efficiency analysis, *Fuel*. https://doi.org/10.1016/j.fuel.2019.01.172

[17] N Johansson, *Production of liquid biogas, with cryogenic and conventional upgrading technology*. https://lup.lub.lu.se/luur/download?func=downloadFile&recordOId=4468178&fileOId=4469242
[18] LA Pellegrini et al., Biogas to liquefied biomethane via cryogenic upgrading technologies, *Renewable Energy*. https://doi.org/10.1016/j.renene.2017.08.007
[19] L Vidas & R Castro, Recent developments on hydrogen production technologies: State-of-the-art review with focus on green-electrolysis, *Applied Sciences*, 11, 11363 (2021). https://doi.org/10.3390/app112311363
[20] Gianluca Valenti & Davide Bonalumi, "Chemical absorption by aqueous solution of ammonia," in *Carbon capture, utilization and sequestration* (2018). https://doi.org/10.5772/intechopen.78545. https://www.researchgate.net/publication/327625142_Chemical_Absorption_by_Aqueous_Solution_of_Ammonia
[21] AF Antipchuk, GA Iutinska & OY Davydova, The effect of ammonium carbonate fertilizers on the biological activity of soil, *Acta Agriculturaae Scrbica*, VI:11, 11–18 (2001). http://scindeks-clanci.ceon.rs/data/pdf/0354-9542/2001/0354-95420111011a.pdf
[22] Jinzhao Liu, Shujuan Wang, Bo Zhao, Huiling Tong & Changhe Chen, Absorption of CO2 in aqueous ammonia, *Energy Procedia*, 1, 933–940 (2009). https://doi.org/10.1016/j.egypro.2009.01.124. https://www.researchgate.net/publication/240448337_Absorption_of_CO2_in_aqueous_ammonia

8 Adding Value to the Digestion's By-products

8.1 FERTILISERS FROM DIGESTATE

Despite the many environmental benefits of AD, the main scope of installing a biogas plant is to always make a profit from selling electricity or biomethane and eventually from the tipping fees for waste collection and treatment. In the case of agricultural biogas plants, using the digestate as a replacement for chemical fertilisers provides some extra savings. In the case of organic fraction of municipal solid waste (OFMSW) digesters, producing compost is usually mandatory for environmental regulations, but the income from its sale is often marginal. In the case of wastewater treatment plants (WWTPs), the fate of digested sludge depends on local regulations. In some European countries, digestate from WWTPs usually represents a cost because it has the legal status of 'waste' and is managed accordingly. In other countries, it is employed as a fertiliser, either fresh or after some sanitisation post-treatment, such as composting, pasteurisation, or chemical transformation.

From a purely technical point of view and regardless of the legal status that local regulations assign it, digestate is a substance with high moisture content, containing some organic matter and dissolved salts. It has intermediate agronomical features between a fertiliser and an amendment.

DEFINITIONS:

1. A fertiliser is a substance that improves the nutrient content in the soil, directly affecting plant growth.
2. A soil amendment is a substance that improves the soil's physical condition (e.g., soil structure and water infiltration), indirectly affecting plant growth.

Digestate contains variable concentrations of solids, N, P, K, Ca, S, Fe, and Mg, depending on the digester's feedstock. For preliminary design purposes, the data from Risberg et al. [1] are a good departure point, but a full characterisation in the laboratory of an actual digestate sample is mandatory for the accurate engineering of the treatment system. Analysing samples taken in different months is useful in identifying the range of variability of each test parameter and their seasonal trends, which, in turn, provides useful information about the process's design tolerance and the final product's quality.

Digestate consists of a solid and a liquid fraction, usually treated separately. After separation, most N (as NH_3) and K remain in the liquid fraction, while P, minerals, and organic nitrogen (N_{org}) tend to accumulate in the solid fraction. Tilling with a

DOI: 10.1201/9781003609278-8

plough within 3 hours of direct application of liquid digestate to the soil is mandatory to limit volatilisation of NH_3; otherwise, between 30% and 84% of N will be lost to the atmosphere. In soil, digestate induces N_2O emissions higher than those of mineral fertilisers. These emissions are higher for digestate with a low P/N_{org} ratio. Field trials in France [2] have not shown a strong effect (positive or negative) on crop yields (fertiliser coefficient equivalent of 40%–50%). Phosphoric acid added to liquid digestate will react with the ammonia dissolved in it, forming ammonium phosphate and improving the fertiliser quality. Phosphoric acid is relatively expensive, so a small field trial and a cost–benefit analysis are mandatory before adopting the said solution. Sulphuric acid is a cheaper option for limiting ammonia volatilisation, but ammonium sulphate is not as beneficial to plants as ammonium phosphate.

Apart from nutrients, digestate may contain phytotoxic substances too, such as heavy metals, NaCl, pesticides, or other chemicals. The most straightforward way to determine the optimum dosage of digestate (or compost) is by performing a seedling growth test. The protocol is very simple (see 3 for a detailed description) so anybody can do this test at home. The model species are *Lepidium sativum* L. (garden cress), *Cucumis sativus* L. (cucumber), and *Lolium perenne* L. (winter ryegrass). Seeds of such species are available at any gardening shop. Sow ten seeds for each plant species uniformly in each pot filled with different soil + digestate or soil + compost mixtures. After 17 days, cut the seedling shoots and weigh the fresh biomass. Then report the biomass production as a percentage of an untreated control for each plant species. A quicker variant, more suitable for liquid digestate, consists in placing an inert support (e.g., sand or silt or a cotton dish) in several Petri dishes, then placing 5–10 seeds in each one, and finally adding 10 ml of distilled water to one (control) and adding water + digestate mixtures at different proportions to the others. Cover the dishes, place them in a warm place (ideally in an incubator at 27°C), and weigh the total biomass in each dish after 5 days (Figure 8.1). Compare the total biomass of each dish to the control.

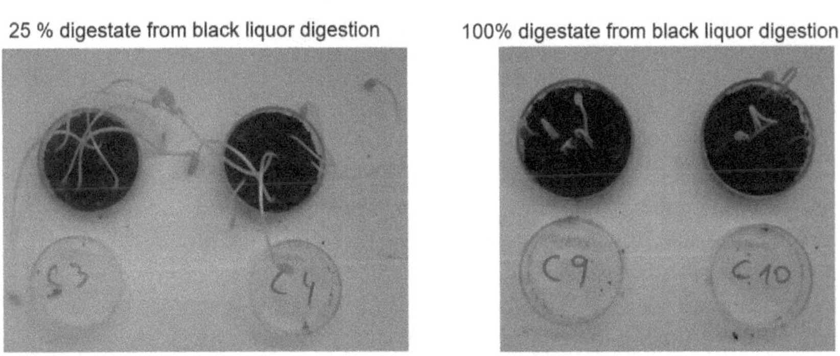

FIGURE 8.1 Germination of Cucumber Seeds with Two Dilution Ratios of Digestate to Identify Possible Toxicity Effects.

There are three methods for managing digestate, depending on local regulations and cost–benefit balance:

a. Application to the soil 'as it is': In Europe, this method is admissible only for agricultural biogas plants. Its cost is near null but provides some marginal savings on chemical fertiliser. Digestate from WWTPs or OFMSW has the legal status of 'waste', so its direct application to the soil is forbidden or at least restricted.
b. Conservative methods allow the recovery of fertilisers, amendments, and eventually other by-products from the digestate, according to the circular economy concept. According to the European legislation on the End of Waste principle [4], treating digestate from WWTPs or OFMSW changes its legal status from 'waste' to 'product'. By definition, a product has a market value; hence, treating digestate with conservative methods allows the waste treatment biogas plant to obtain some additional income.
c. Dissipative methods transform part or all of the digestate in inert effluents discharged into the atmosphere, water bodies, or soil. They follow the logic of linear economy: extract, consume, and discard. The classical nitrification–denitrification wastewater treatment process is a dissipative method widely employed in WWTPs. Dissipative methods usually represent a net cost, so the next sections will deal exclusively with conservative methods capable of bringing extra income to the biogas plant.

8.1.1 Conservative Methods for Treating Digestate

The methods and machinery described in this section reflect the European legislation and trends in the WW and OFMSW biogas plant markets, except for two (black text in Figure 8.2). Presenting such methods and machinery here does not constitute an endorsement or recommendation, nor does it mean that they are the best available solutions, economical, or suitable for a given context. Conversely, the fact that the European Union has not legislated on a given solution, or even forbids or discourages it, does not mean it is ineffective. The author included the next examples only because of his direct professional and academic experience. Figure 8.2 summarises the most common conservative methods for recovering fertiliser and other by-products from the digestate.

8.1.1.1 Solid–liquid Separation Technologies

Solid–liquid separation is a frequent unit operation of chemical and industrial engineering. In general, it is the first step in valorising digestate, providing two streams that will produce fertiliser (liquid fraction) and amendment (compost or biochar from the solid fraction). Table 8.1 summarises the most common technologies. The choice of the machine model is more or less arbitrary, depending mostly on their local availability, capital expenditure (CAPEX) and operation expenditure (OPEX). Although microfiltration and reverse osmosis have also been proposed in some European projects, they are not diffused because of the high energy demand and cost of membrane maintenance and replacement.

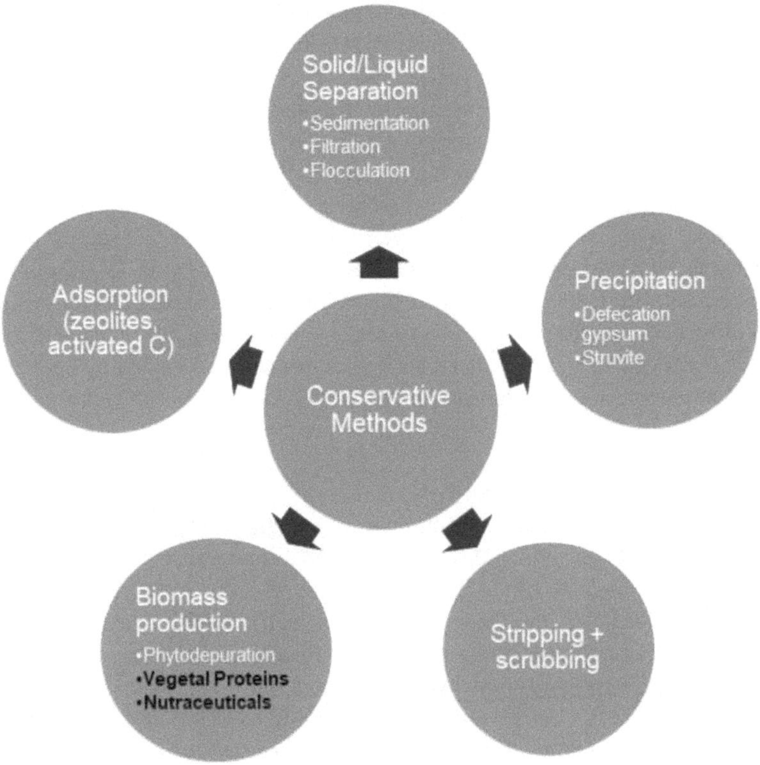

FIGURE 8.2 Conservative Methods for Digestate Treatment.

TABLE 8.1
Solid Removal Efficiencies of Different Solid–Liquid Separation Techniques

Particle Size	Device	Efficiency of Solids Removal
Coarse D > 0.1 mm	Static sieve	20–35%, mesh 0.5–2 mm
	Vibrating sieve	20–35%, mesh 0.5–2 mm
	Rotating sieve	20–35%, mesh 0.5–2 mm
Fine D < 0.1 mm	Sedimentation (adding coagulant)	50%–70%
	Flotation (adding flocculant)	70%–90%
	Ribbon press filter	50%–75%
	Press filter	50%–75%
	Centrifuge	50%–75%
	Auger or hydraulic piston separator	50%–75%
	Coagulation + filtration	80%–90%

8.1.1.2 Precipitation of Defecation Gypsum or Carbonate

Defecation gypsum is produced with WWTP sludge or digestate. It is a registered fertiliser in Italy. The process starts with mild wet oxidation (Fenton reaction of H_2O_2 and $Fe_2(SO_4)_3$) that eliminates organic micropollutants. Adding Cl_3Fe coagulates organic matter, allowing its dewatering. Adding OCa (quicklime) will hydrolyse and flocculate the organic matter, allowing its dehydration with a centrifuge. The resulting mass is strongly alkaline, so it is essential to neutralise its pH. Neutralisation with H_2SO_4 (sulphuric acid) will produce a fertiliser called 'defecation gypsum'. Neutralising with carbonic acid (H_2CO_3), produced on-site with CO_2 from biogas and water, will produce 'defecation carbonate'. Gypsum-based fertiliser is suitable for alkaline soils; carbonate-based fertiliser is suitable for acidic soils. Advantages of defecation gypsum and carbonate: packed as dry pellets, suitable for spreading in the field employing standard top-dresser machines, odourless, slow nutrient release, and low cost.

8.1.1.3 Precipitation of Struvite

Struvite is the mineralogic name of ammonium magnesium phosphate hexahydrate $(NH_4)\,MgPO_4{\cdot}6H_2O$. Struvite is a slow-release fertiliser because it decomposes when in contact with humic and fulvic acids in the soil.

At high pH, ideally in the range of 8–9.5, the NH_4^+ and PO_4^{3-} radicals in the liquid digestate react and precipitate as crystals. Reaching the necessary pH usually requires adding $Ca(OH)_2$ (calcium hydroxide, spent lime). Digestate usually has little Mg and insufficient P, so it is necessary to add OMg or ClMg together with soluble phosphate. Adding lime to increase the pH will form hydroxyapatite (calcium triphosphate, $Ca_5\,(PO_4)_3(OH)$), an insoluble mineral that will precipitate together with struvite. It is then preferable to employ NaOH or KOH to increase the pH. KOH is more expensive than NaOH, but since K is also a nutrient to plants, its use is compatible with the scope of producing a fertiliser.

8.1.1.4 Chemical Recovery by Stripping + Acid Scrubbing

Most N in digestate is dissolved in the liquid fraction as NH_3 and its salts. The solid fraction contains N mainly in the organic form. Chemical recovery of N as ammonium salts has the advantage of producing a 'chemical fertiliser of organic origin'. It has the advantages of chemical fertilisers – purity, easy dosage to the crop, high concentration, and low transport and storage cost – but with a low environmental footprint. The liquid remaining after N extraction is suitable for irrigation or can be employed as process water. Figure 8.3 shows a general block diagram of the process.

There are several options for stripping NH_3 from the liquid digestate: heating it over 60°C, applying vacuum, bubbling N_2 or air, adding an alkali to raise the pH, etc. Similarly, it is possible to precipitate gaseous NH_3 by making it react with any acid. The most common ammonium salts are as follows:

- $SO_4(NH_4)_2$ is the cheapest option, but plants require far less S than N.
- $NO_3(NH_4)$ is the best option from the agronomical point of view because the resulting fertiliser has a double concentration of N. Its trade is severely regulated in the EU and also in some extra-EU countries (explosive product).
- $PO_4(NH_4)_3$ gives a more complete fertiliser, but phosphoric acid is expensive.

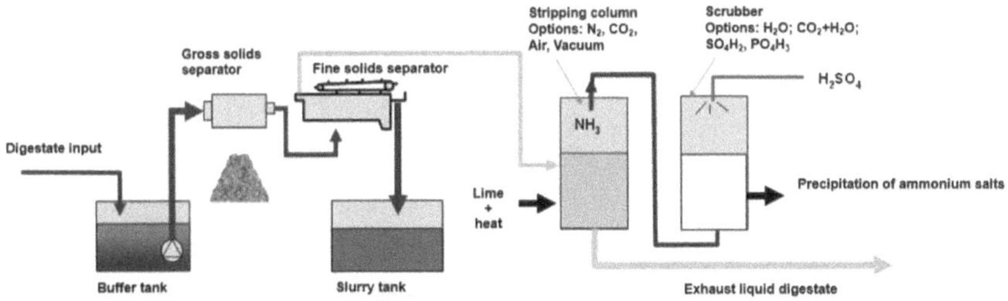

FIGURE 8.3 Block Diagram of a Compound N Separation System.

Ammonium carbonate and carbamate are not widely diffused despite presenting several advantages over the aforementioned ammonium salts: carbonic acid could be easily produced with CO_2 from the biogas while increasing its methane content (see Chapter 7, Section 7.5.5); it has a short reaction time, so the scrubber would be relatively small; and plants will absorb all the resulting carbonate, leaving no traces in the soil. Furthermore, scrubbing biogas directly with liquid digestate would simplify the process and lower both CAPEX and OPEX, producing a liquid fertiliser suitable for fertirrigation and hydroponics.

The efficiency of N removal as a mineral salt is 45%–65%, with the remaining quota contained in the solid fraction of the digestate.

8.1.1.5 Phytodepuration with Industrial Biomass Production

Phytodepuration (aka bioremediation) involves employing plants to remove contaminants or excess nutrients from a wastewater stream. A successful bioremediation project depends on choosing the most suitable plant species for the local climate and the composition of the wastewater to purify. Bioremediation is seldom the primary treatment method for liquid effluents. Its most common use is a tertiary or affination treatment of water before discharging it to a river or the ground. The added value of a phytodepuration system for a biogas plant is producing additional vegetal feedstock for the digester. There are three different concepts of bioremediation systems:

8.1.1.5.1 Artificial Wetland

Figure 8.4 illustrates the concept. The literature has abundance of papers about the use of common reeds (*Phragmites australis*), common cattail (*Typha latifolia*), yellow flag (*Iris pseudacorus*), sedges (*Carex sp.*), rush (*Scirpus lacustris*), and Indian shot or arrowroot (*Canna indica*). *Canna indica* is an interesting plant for biomass production in artificial wetlands in tropical to temperate places, as long as the winter temperature is always higher than –3°C. It produces about 7–10 tonnes/ha of fresh epigeal biomass with low lignin content and 2–4 tonnes/ha of starch-rich rhizomes [5]. The yearly nitrogen abatement capacity ranges from 6,000 to 7,000 kg/ha. These figures refer to the green-leaved variety of *Canna indica*, which produces more epigeal biomass. The violet-leaved cultivar (*Canna edulis* Ker.) produces more rhizomes (claims about 90 tonnes/ha of fresh biomass circulate in non-peer-reviewed publications, but 8 tonnes/ha is a more cautious estimation; 6).

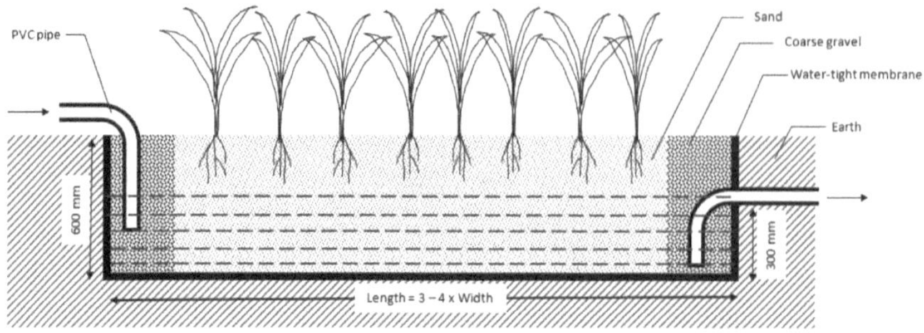

FIGURE 8.4 Sketch of an Artificial Wetland with Subsurface Circulation.

8.1.1.5.2 Macrophytes Growing on Floating Rafts

The rationale of this aquaponics technique is to employ existing drainage channels and ponds, thereby saving the construction cost of an artificial wetland. Some references for further information: results of a LIFE research project [7], an Italian university spin-off that manufactures the rafts [8], and a review of the literature on floating water treatment systems [9].

8.1.1.5.3 Floating Aquatic Plants

Most of the literature concentrates on two species: water hyacinth (*Eichhornia crassipes*) and duckweed (*Lemna minor* and other Lemnaceae). Water hyacinth is invasive and should be kept under control to avoid environmental damage. It is an interesting option only in tropical and warm-temperate locations because employing it as a feedstock for anaerobic digestion is the easiest way to control its proliferation. It grows at a temperature range of 10°C–40°C; at 30°C it doubles its biomass every 11–15 days, yielding 930–2,300 tonnes/ha of fresh biomass [10]. Duckweed is ubiquitous, and its potential has been widely studied, both as high-protein forage (next paragraph) and as feedstock for the production of ethanol or biogas. Growing water hyacinth and duckweed in open ponds full of liquid swine manure digestate was tested in Parma, Italy [11]. The average productivities measured under the Northern Italian climatic conditions were 40 g DM/m^2.day for water hyacinth and 9 g DM/m^2.day for duckweed. The nitrogen and phosphorous uptakes were, respectively, 1.1 g/m^2.day and 0.4 g/m^2.day for water hyacinth and 0.30 g/m^2.day and 0.08 g/m^2.day for duckweed. Duckweed resists winter by sinking and remaining dormant when the temperature drops below 6°C. Water hyacinth will die below 10°C.

In the EU, the current Renewable Energy Directive [12] (called RED III for short because it is the third version) explicitly admits 'aquatic biomass' as a 'double counting' feedstock for the production of advanced biomethane and biofuels. Hence, biogas plants in countries that subsidise 'advanced biomethane' and 'second-generation biofuels' can potentially benefit from higher subsides when feeding the digester with aquatic plants grown on liquid digestate.

8.1.1.6 Production of Vegetable Protein

Extracting protein or animal fodder from water plants grown on liquid digestate is technically possible, and even desirable [13], but is not legal everywhere. For instance, in the EU it is forbidden as of December 2024 because the *Regulation (EU) 2020/741 of the European Parliament and of the Council of 25 May 2020 on minimum requirements for water reuse* applies only to treated urban wastewater and explicitly excludes digestate. On the contrary, growing duckweed in open ponds filled with liquid manure, both to treat the effluent and to produce fodder for pigs and ducks, is a widespread practice in the Far East.

8.1.1.7 Production of Nutraceuticals

The most suitable genera for this scope are microalgae and cyanobacteria. The same considerations in the previous paragraph apply in the EU: liquid digestate is not acceptable as a growth medium. In any case, liquid digestate is opaque, limiting the achievable cellular density and thereby increasing the filtration costs. See Section 8.2.4 for a case study.

8.1.1.8 Adsorption on Zeolites or Activated Carbon

Most of the nitrogen in digestate is present as ammonia or ammonium compounds. Both zeolites and activated carbon can adsorb ammonia from liquid streams, so forcing the liquid digestate to pass through an adsorbent column is a simple treatment method, easy to implement with a relatively low CAPEX. On the other hand, its OPEX is high because the adsorbent material is a consumable that needs continuous replacement. Ammonia-adsorption capacities of the different kinds of zeolites vary between 6 and 14 g N/kg [14]. Activated carbon can adsorb up to 17 g/kg [15]. The spent adsorbent is a slow-release fertiliser and soil amendment, but several factors limit its application to the field. It is possible to regenerate the adsorbent, but in this case, the problem is how to recover the ammonia from the regenerant solution [16].

8.2 AMENDMENT FROM SOLID DIGESTATE

Composting is the most common treatment method for solid organic waste worldwide. In the specific case of the European Union, *Regulation (EU) 2019/1009 of the European Parliament and of the Council of 5 June 2019* recognises several combinations of anaerobic digestion and composting as valid methods to apply the 'end of waste' criterion. The aforementioned regulation is a reliable technical guideline for choosing the most suitable combination of treatments for a specific project, even outside the European territory. Regardless of the digestion process being mesophilic or thermophilic, the following combinations of composting temperature and composting time will transform 'waste' into a stabilised and safe 'product':

- $T > 70$ °C for at least 3 days,
- $T > 65$ °C for at least 5 days,
- $T > 60$ °C for at least 7 days,
- $T > 55$ °C for at least 14 days.

Apart from composting, pyrolysis, hydrothermal treatment (aka wet pyrolysis), and gasification are possible treatments for the solid fraction of digestate, although they are less diffused worldwide. Their final products are pyrolysis gas or syngas, tar or bio-oil, and biochar.

8.2.1 Compost

The product of aerobic oxidation (composting process) of the solid fraction of digestate (or fresh vegetal biomass) is called 'compost'. Respiration of the aerobic bacteria growing on the composting matter increases the temperature to 50°C–60°C (reaching peaks of 75°C in some systems). Composting is a sanitisation method because it kills most pathogens and weed seeds. On applying compost to the soil, bacteria and fungi break cellulose and lignin, a process called 'humification'. The result is a mixture of humic and fulvic acids, both capable of improving the cationic exchange of nutrients with the roots. The cleavage of lignin changes the material's porosity, favouring water retention and nutrient availability to the roots. Table 8.2 shows the average values of 557 samples of compost from vegetal waste (vegetal compost amendment, VCA), compost from mixed OFMSW and vegetal waste (mixed compost amendment, MCA), digestate of OFMSW composted together with vegetal waste (mixed compost amendment digestate, MCA-D), and cow manure (data from the Italian Compost Consortium [17]).

Table 8.3 shows the optimum ranges of the principal control parameters for the composting process [18]. Other parameters influence the process but are difficult to quantify because they depend on the specific composting system or the substrate: for instance, pile size and shape, pile turning frequency, airflow through the pile, time, external air temperature, and eventual presence of heavy metals, antibiotics, or other chemicals.

There are two broad families of composting systems: open-air and closed-reactor plants. Figure 8.5 shows the classification of the different composting systems belonging to each family. Open-air systems are the oldest technology: the Roman writer Lucius Junius Moderatus Columella mentions them in his treatise on agriculture,

TABLE 8.2
Physical and Chemical Characteristics of Varying Compost Qualities

Parameter	Unit	VCA	MCA	MCA-D	Cow Manure
Moisture	%	35.3	26.4	28.5	65
Volatile solids (organic matter)	%WW	33	38.9	33.3	19.2
Organic carbon	%DM	25.5	26.4	23.3	27.4
Humic and fulvic acids	%DM	9.2	9.7	8.5	12
Total N	%DM	1.8	2.3	2.3	1.6
Conductivity	dS/m	1.3	3.5	3.1	N/D
Phosphorous (P)	%DM	0.7	1.6	1.5	1.9
Potassium (K)	%DM	1.3	1.5	1.5	1.7

TABLE 8.3
Optimum Ranges for the Composting Process

Parameter	Optimum Range
Temperature	35°C–60°C
Moisture	50%–65%
Oxygen concentration in the pile	5%–15%
C/N ratio	25–35
pH	6.5–8.5
Particle size	0.5–5 cm
Apparent density	550–650 kg/m^3
Porosity	32–36%

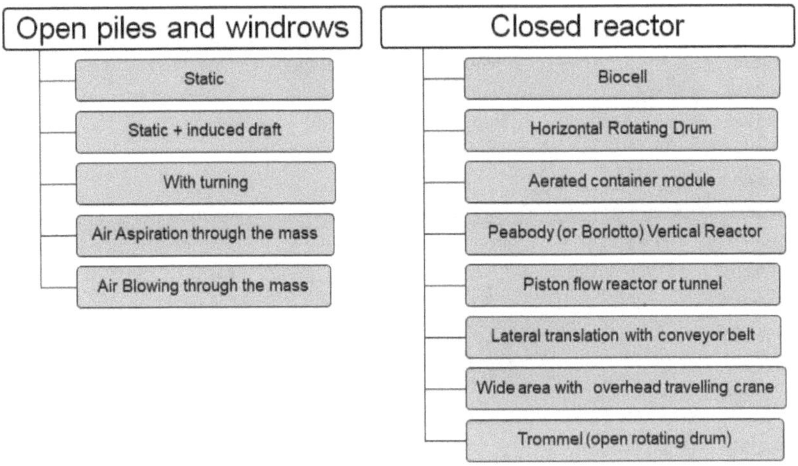

FIGURE 8.5 Classification of Composting Systems.

De re rustica [19], and evidence exists that the Templar Knights employed Columella's agricultural techniques in their feuds. Open-air systems are currently limited to agricultural facilities and small OFMSW treatment plants. Industrial OFMSW treatment plants adopt closed-reactor solutions both because the process is more stable and easier to control and to contain odorous emissions.

Composting is a partially dissipative treatment technique: the intense aerobic activity of bacteria consumes a fraction of the initial C and N. Water partly evaporates and drags P away as leachate. Table 8.4 summarises data about the mass losses [20–24], depending on both the substrate and process. The Washington State University has developed a downloadable spreadsheet [25] to calculate the characteristics of the compost (bulk density, C: N ratio, and moisture) depending on the mixture of feedstocks (manure, hay, garden clippings, etc.)

TABLE 8.4
C, N, P, and Total Mass Losses During Composting of Several Residues

Substrate	C Loss	N Loss	P Loss	Mass Loss	Source
Cattle manure	46%–62%	19%–42%	<2%	15%–20%	Eghball et al.
Pig manure + cornstalks (turned and unturned windrows)	64% (turned), 50% (unturned)	49.8% (turned), 53.8% (unturned)	32% (turned), 29.7% (unturned)	52% (turned), 37% (unturned)	Tiquia-Arashiro et al.
Average of several feedstock	28.9% (C represented 46.4% of total mass loss)	n/a	n/a	11.5%–31.4%	Breitenbeck & Schelliger
Bedding poultry manure composted in open pile, closed chamber, and rotating drum	N.D.	21.2%–27.1% (pile), 13.4–19.8% (chamber), 3.2–8.9% (drum)	3.4%–7% (pile), 0%–1.8% (chamber), 0%–1.6% (drum)	14.8%–19.9% (pile), 15.5%–223% (chamber), 19.4–26.9% (drum)	Uvarov et al.
Mixture of digestate, fresh green waste, and screened compost (DM = 36%, TOC = 34.6% DM)	8.5% DM (20 days) 5.7% DM (63 days)	2.1% DM (64 days)	n/a	Wet mass loss 11.4% (20 days) 13.7% (34 days) 19.2% (63 days)	Franke-Whittle et al.

Dosing fixed yearly quantities of compost to the field is a common agricultural practice among small farmers. Nevertheless, rules of thumb such as 'add 10 tonnes/ha each year' may not be suitable for every combination of crop and soil because different soils and tillage methods require more or less organic matter input to maintain the organic C balance. On the other hand, different crops require different inputs of nutrients. So, the minimum yearly quantity of compost results from a compromise between the environmental requirement of maintaining or increasing the carbon content of the soil and the nutrients for ensuring the yield of a given crop. In general, satisfying the crop's needs for nutrients will result in a net carbon accumulation in the soil because of humification. Humification is the natural process of changing organic matter into humic substances (humus, humate, humic acid, fulvic acid, and humin) by geo-microbiological mechanisms. If all farmers in the world increased the carbon content of the soil by at least 0.4% every year, this would be enough to compensate for 25%–30% of all the anthropic CO_2 emissions [26]. From the point of view of climate change and erosion control, carbon accumulation in the soil is beneficial, but its cost is proportional to the purchase, transport, and application of the compost. On the contrary, adding just the minimum quantity of compost to the soil to keep the carbon balance will result in a nutrient deficit for the crop. Compensating the said deficit requires purchasing chemical fertilisers or employing the liquid fraction of digestate as a liquid fertiliser. The simplest way to find the optimum combination of compost + chemical fertilisers or compost + liquid digestate results from checking the cost with a spreadsheet. The two extreme options are

1. applying the minimum compost quantity for a neutral carbon balance, compensating the deficit of nutrients with chemical fertilisers, and
2. applying the maximum compost quantity to provide all the required N to the crops, resulting in net carbon sequestration in the soil.

The input data to calculate the minimum quantity of compost applied to the soil are as follows:

a. The type of soil (sandy, silty, loamy, clayish, or their combinations) and the percentage of organic matter in it.
b. The humification rate (k_1), i.e., the percentage of VS converted into humic substances in a unit of time (usually 1 year). The value of k_1 depends on the type of organic matter applied: fresh cow or horse manure and OFMSW compost have minimum humification rates of 35% VS, composted manure 30% VS, mature vegetal matter (mature cover crop and straw) 20% of VS, and fresh vegetal matter (young cover crop) 5%. In the case of pig slurry, animal waste (hides, blood, etc.) and liquid digestate, assume 0% [27]. The value of k_1 for the solid fraction of digestate is more variable and depends on the separation method and eventual posttreatment. Indicative values are [28] 70% VS for solids separated by employing a decanter centrifuge, 51%VS for solids separated by screw press, 71% VS for low-phosphorous digestate, 75% VS for solid digestate dried at 60°C, 93%VS for commercial compost, and 91% VS for biochar obtained from solid digestate fraction.
c. The mineralisation rate (k_2), i.e., the quantity of humus lost in 1 year, which is directly proportional to the texture of the soil. Coarse-textured sandy and stony soils present the maximum mineralisation rate (4%–5% per year). Sandy loam has a high mineralisation rate (2.2–2.5% per year). Loam (roughly equal percentages of sand, silt, and clay) loses 2% per year and fine-textured silty and clayish soils lose 1.8% per year. Other factors, such as the intensity of tillage, the slope of the plot, and the rainfall or irrigation may affect the said coefficients.

The following example shows a simplified method to calculate the minimum quantity of compost for a given soil and crop that will keep the organic carbon constant.

1. **Input data**: sandy soil ($k_2 = 2.5\%$) containing 1.5% VS (organic matter) and density 1,250 kg/m^3; the crop (corn) requires 208 kg N/ha, 90 kg P_2O_5/ha, and

90 kg K_2O/ha for standard production of 10 tonnes/ha; and the amendment will be MCA-D, as shown in Table 8.2.

2. The rhizosphere of corn extends nearly 30 cm deep. The total quantity of organic matter in 1 ha of 30 cm topsoil (VS_s) is

$$VS_s = 10{,}000\ \text{m}^2 \times 0.3\ \text{m} \times 1{,}250\ \text{kg/m}^3 \times 1.5\%\text{VS} = 56{,}250\ \text{kg/ha}$$

3. The yearly humus loss (HL) is

$$\text{HL} = k_2 \times \text{VS}_s = 2.5\% \times 56{,}250\ \text{kg/ha} = 1{,}406\ \text{kg VS/ha}.$$

4. The humus lost by mineralisation must be equal to the humus produced by humification, so the yearly quantity of compost necessary to keep the organic matter content constant in the soil (C_{min}) is

$$\text{HL} = k_1 \times C_{min} \times VS_a,$$

where VS_a is the percentage of VS in the amendment. Hence,

$$C_{min} = 1{,}406\ \text{kg/ha} / (0.35 \times 0.333) = 12{,}066\ \text{kg/ha}.$$

We can then condense the above procedure in a single formula:

$$C_{min} = \frac{k_2 \times 10{,}000 \times d \times \rho \times OM_i}{k_1 \times VS_a},$$

where d is the depth of the crop's rhizosphere (assume 0.3–1 m), ρ is the density of the soil (kg/m^3), and OM_i is the initial percentage of organic matter in the soil that must remain constant.

5. The final step is calculating the percentage of nutrients that will be available to the crop, taking N as the reference. The rationale of the calculation is that soil bacteria will decompose the compost, converting a fraction (k_1) into humus and the remaining into soluble N available to the roots (N_1) and CO_2. The mineralisation process decomposes the fraction k_2 of humus, freeing additional nitrogen (N_2). Since humic substances contain 5% nitrogen on average, the total available nitrogen (N_3) will be

$$N_3 = N_1 + N_2,$$

where

N_1 = nitrogen released by the oxidation of organic matter = $C_{min}\ (1-k_1)\ N_{\%}$;
N_2 = nitrogen released by the mineralisation of humus = 5% HL;
$N_{\%}$ = initial percentage of N in the compost.

From Table 8.2, N% of compost type MCA-D is 2.3% of DM, and the moisture is 28.5%, so

$$\text{N\%} = 0.023 \times (1-0.285) \equiv 1.6\%.$$

The total nitrogen available to the roots will then be

$$N_3 = 12{,}066\ (1–0.35)\ 0.016 + 0.05 \times 1{,}406 = 196\ \text{kg/ha},$$

which is almost the necessary N input for the crop's standard productivity. The options are then as follows:

a. Applying just C_{min} and compensating the missing N demand with chemical fertilisers or with the liquid fraction of digestate.
b. Applying a bigger quantity of compost (C) to satisfy the crop's demand of N. From the formula of N_3 above,

$$C = (208–0.05 \times 1{,}406)\ \text{kg/ha}/(0.016 \times 0.65) = 13{,}240\ \text{kg/ha}.$$

Since $C > C_{min}$, the balance of humus will be positive, which is good both for the planet (carbon sequestration) and for the farmer (increased water retention capacity and nutrient uptake efficiency, saving the cost of chemical fertilisers, and reducing hydric stress to the plants).

It is general knowledge that agricultural soils containing between 3% and 6% organic matter are fertile. The said content of organic matter ensures enough water retention capacity, limits erosion, and provides aeration of the roots and the maximum availability of nutrients to the crops [29]. In this example, adopting strategy b will gradually restore the soil's fertility in the subsequent years.

8.2.2 Biochar

In the last 10 years, the EU and many other countries have subsidised the construction of biomass-fuelled power and district heating plants to replace fossil coal. Despite combustion being the predominant technology, some biomass plants adopt either pyrolysis or gasification technologies. Eco-marketing has coined the word 'biochar' to call the main by-product of biomass pyrolysis, although it is nothing else but fine charcoal, too fine to be employed as a fuel. Both the high porosity and the cation exchange capacity of biochar improve water and nutrient retention in the soil. Being mostly mineral carbon, its degradation in the soil is extremely slow, thus representing an effective carbon sink to mitigate climate change. Environmentalist NGOs and producers alike have hyped its fertiliser power, arguing that the high fertility of some areas in Amazonia derives from the high content of microscopic particles of biochar. It is a debated topic whether the Amazonian aboriginal populations deliberately employed the residual charcoal and ashes of their fireplaces as a soil amendment, or it is just the result of ancient wildfires. Some plots in Amazonia, and also in Galicia (Spain) and Japan, contain charcoal dated 7,000–8,000 years old and feature high fertility [30]. Industrial biochar has a potential drawback: it may contain heavy metals and tar, both toxic to plants. Soil microbes will degrade tar during the first year, releasing nearly 10% of the initial carbon content of biochar. The optimum dosage ranges from 10 to 50 tonnes/ha, depending on the initial soil characteristics: on clay soils, it improves the structure (lower tillage cost); on sandy and silty soils, it increases water and nutrient retention; on acidic soils, it increases the pH. During an experiment on durum wheat cultivation in Italy (ref. 27 already cited), the application of 30 tonnes/ha of biochar increased productivity by 20%. Amending the soil with biochar is a potentially effective method to prevent ammonia evaporation when employing liquid digestate as N fertiliser. However, it is difficult to implement because, unlike activated carbon described in Section 8.1.1.8, the adsorption capacity of biochar depends on the biomass source, the pyrolysis temperature, and eventual posttreatments like washing with hot water or acidic or alkaline solutions (some examples in 31).

Mixtures of biochar and compost can provide the ideal balance of carbon and nitrogen for a given combination of soil and crop. Suppose the same soil as in the example in Section 8.1.1, but now the crop will be wheat instead of corn. Wheat requires only 125 kg N/ha. Applying the minimum quantity of composted digestate will preserve the organic matter content in the soil at 1.5%, but the resulting dose of N will be higher than the crop's needs. It is common knowledge that over-fertilising does not improve productivity, causes water pollution because of the excess nitrates leached during rainfall, and may result in partial toxicity for the plants. On the other hand, applying a quantity of compost that provides the exact dose of N will result in a net loss of humus from the already arid soil. A possible strategy is to calculate the quantity of compost that provides all the N and compensate for the missing organic matter with biochar. The procedure is straightforward:

$$N_3 = C_{min}\,(1-k_1)\,N_{\%} + 5\%\ HL,$$

$$125\ \text{kg N/ha} = C_{min}\,(1-0.35)\,0.016 + 0.05 \times 1{,}406,$$

$$C_{min} = 5{,}260\ \text{kg/ha}.$$

Since the humus loss is 1406 kg VS/ha and the humus production is 0.333 × 0.35 × C_{min}, the humus balance (HB) is

$$HB = k_1 \times C_{min} \times VS_a - HL,$$

$$HB = 613\ \text{kg VS/ha} - 1{,}406\ \text{kg VS/ha} = -793\ \text{kg VS}.$$

Biochar from wood pyrolysis at 450°C–500°C has 80%–85% VS (on a DW basis) and 10%–15% moisture (average of the data sheets from several producers). Assuming a minimum of 68% VS for our calculations, the quantity of biochar (*B*) required to integrate the compost amendment in our example is

$$B = 793\ \text{kg VS/ha}/0.68 = 1{,}166\ \text{kg/ha}.$$

8.3 REDIGESTION OF DIGESTATE

Treating the digestate at 70°C for 1 hour allows its redigestion, recovering 0.6 Nm^3/tonne of digestate. See the case study in Chapter 2, Section 2.4.1.

8.4 GROWING EDIBLE MUSHROOMS ON SOLID DIGESTATE

The solid fraction of digestate represents the undigestible portion of vegetal feedstocks: lignin. As already anticipated in Chapter 2, edible mushrooms of the *Pleurotus* family possess the ability to produce enzymes capable of cleaving lignin. Using the residual heat from the CHP, it is possible to sterilise the solid digestate (48 hours at 80°C should be enough). Once sterilised, the solid digestate's moisture must be adjusted to 50%–55%, and then it can be spawned with commercial mycelium of the desired species. *Pleurotus ostreatus* (oyster mushroom) is said to be the easiest species to grow. There is plenty of information on how to grow mushrooms at home, but YouTube tutorials and amateurs' blogs may contain incorrect or hyped information. The author recommends engaging an expert consultant or studying a professional handbook like the one in Ref. [32] before attempting any experience.

Then the investor has four theoretically possible additional profit alternatives that deserve further research and economic analysis:

a. Harvest the mushroom and extract the enzyme-rich juice from it. The juice can be employed for pretreating the feedstock and increasing its BMP. The spent mycelium can be employed or sold as a humus-like amendment.
b. Sell the edible mushroom and extract humates from the exhausted growth substrate (which consists of already hydrolysed lignin and humic-like substances). Humates have a higher market value than compost, but their market is limited.
c. Co-digest both the crushed mushroom biomass and the spent growth medium (but this poses the risk of inhibiting the AD process because of the humates in the spent growth medium).
d. Extract the enzymes from the harvested mushrooms and sell them as a biotech product, employing the extraction residues as additional feedstock for the digester and the spent growth medium as an amendment.

8.5 POSSIBLE MARKETS FOR BIOGENIC CO_2

Carbon dioxide (CO_2) is the main greenhouse gas (GHG) emitted by human activities. CO_2 is an inert gas with a lower GHG power than other gases of anthropic origin, but the volume of its emissions produces the strongest effect on the planet's atmosphere. Despite the many technical applications of CO_2, its low market value hampers its small-scale production and marketing. Currently, only companies specialised in the large-scale production and sales of technical gases can profit from CO_2.

In biogas plants equipped with CHP, the CO_2 derived from the AD process adds to the one produced by the combustion of CH_4, ending up in the atmosphere as flue gas. Biogas upgrading plants usually vent the CO_2 stream, called 'tail gas'. Although AD plants always emit CO_2 into the atmosphere, their impact on the global climate is null; i.e., the GHG emissions are neutral or, under certain conditions, negative. The reason is that residual biomass let to rot in the open air would emit CH_4, a more potent GHG than CO_2. The CO_2 emitted by the combustion of biogas or biomethane is 'neutral' because its C does not derive from fossil sources but from the biomass employed to feed the digester, keeping the short-term atmospheric balance null.

Under favourable local market conditions, recovering part or all of the CO_2 from the AD plant may provide some extra income. Table 8.5 shows some of its possible commercial applications. Performing a thorough market survey and possibly signing a preliminary supply contract with the potential buyer of the CO_2 are mandatory steps to avoid financial failures. The next paragraphs provide more information on each of the said applications.

8.5.1 Cultivation of Vegetables in Greenhouses with Enriched Atmosphere

In Europe, a few farms are already recovering the residual heat from the CHP, and a fraction of CO_2, to increase the productivity of greenhouses. Enriching the atmosphere with CO_2 increases the photosynthetic activity and hence the productivity of plants, but this option has biological and economic limits. Different plant species react in various ways to the increase in CO_2 concentration. For instance, corn, rice, and wheat show signs of stress with concentrations higher than 12,000 ppm (0.10%–0.12%) [33]. Poison ivy (*Toxicodendron radicans*) grows more vigorously with 600 ppm of CO_2 [34]. As a comparison, remember that the record level of atmospheric CO_2 registered by the Mauna Loa Observatory in July 2019 was 411.77 ppm [35].

According to the ASHRAE (American Society of Heating, Refrigerating, and Air-Conditioning Engineers) standards, concentrations of CO_2 in the range of

TABLE 8.5
Possible Applications of Biogenic CO_2

Application	Pros	Cons
Injection in greenhouse	Increases crop productivity, easy	Limited concentration to avoid toxicity, small demand, low margin
Open air diffusion	Increases crop productivity, easy	Higher concentration, more dispersion, low margin
Irrigation with carbonated water	Interesting for water scrubber systems, higher absorption efficiency	Cropland must be attached to the biogas plant. Low margin and pumping cost
Cultivation of microalgae	Product with high added value	Niche market, big investments, low-cost competitors from China and Israel; in Europe it is not possible to employ wastewater as a cheap source of nutrients
Industrial uses (fire extinguishers, dry ice, solvent, etc.)	Higher market value, e.g., dry ice €25–35/kg*	Additional investment cost, need to create a sales and distribution network
Food grade CO_2	Higher market value: €9/kg**	Higher investment cost, niche market

1,000–1,200 ppm (typical of overcrowded rooms) pose no health hazard but can induce headache and a sensation of fatigue. Exposure to concentrations greater than 5,000 ppm for 8 hours can produce toxicity symptoms [36]. Consequently, it is possible to recover only a small fraction of CO_2 for greenhouse cultivation, not only to avoid stressing the plants but also for the safety of the personnel. Finally, the main limiting factor to the recovery of CO_2 for greenhouse cultivation is its high cost. On top of the cost of building the greenhouse itself, a conspicuous investment is necessary for installing the piping, blowers, gas dosing and control systems, heat exchangers, and LED lamps to allow photosynthesis during the night. Both blowers and LED lamps consume part of the electricity produced by the CHP, and the higher growth rate of the plants derived from the enriched atmosphere implies a higher consumption of water and fertilisers.

8.5.2 Cultivation in Open Air

Kimball et al. [37] employed cultivation in open air to simulate the growth of crops under different scenarios of atmospheric CO_2 concentration. In principle, nothing prevents its application to commercial vegetable production on the field, saving the cost of building greenhouses. The only investment required is the installation of porous plastic tubes along the vegetable rows, an air blower, and a gas mixer valve to enrich the air with CO_2. This equipment will create a slightly enriched atmosphere around the plants, with no risk for the personnel or the plants. For now, this technique seems to be limited only to research purposes.

8.5.3 Irrigation with 'Sparkling Water'

CO_2 is highly soluble in water, a feature employed for upgrading biogas using pressurised water scrubbing (see Chapter 7, Section 7.5.1). Irrigating plants with CO_2-saturated water from the scrubber is a straightforward solution because farms already have an irrigation system. It also allows to save the energy for the absorbent's regeneration. According to Enoch & Olesen [38], this technique has been known in the literature since 1866. Irrigating with carbonated water increases plant productivity between 2% and 9% because of five mechanisms:

1. Variation in the nitrification rate and, hence, in N bioavailability;
2. Variation in the soil's texture and pH, i.e., in the availability of other nutrients;
3. Absorption of CO_2 through the roots and transport to the leaves, where it increases the photosynthesis rate;
4. Variation in the plant's hormonal levels;
5. Variation in the speed of decomposition of pesticides in the soil.

Counterintuitively, the said researchers found that the absorption of CO_2 through roots accounts only for 1% of the productivity increment. The strongest contribution derives from the variation in the plant's hormonal levels.

8.5.4 Cultivation of Unicellular Algae

Algae can grow in a pure CO_2 atmosphere, on the condition that the composition of the nutrient broth is perfectly balanced and that all the cells receive the same quantity of light. Otherwise, the algae will not grow because the excess of carbonic acid will cause the pH to drop or, conversely, because of excess O_2. In general, farming unicellular algae and cyanobacteria on a large scale is a very complex process, and the list of European failed projects is long [39]. In Italy, there is only one biogas plant in the province of Oristano that allegedly recovers CO_2 to grow cyanobacteria (*Spirulina platensis*) [40]. A healthy dose of scepticism about the claims in the cited article is advisable. While it is true that *Spirulina* is a food integrator rich in omega-3 and essential amino acids, defining it as 'the food of the future' and 'superfood' is more hype than factual. According to the newspaper article, the plant in Oristano will produce 15 tonnes/year (but it is not clear if they mean dry *Spirulina* or fresh biomass). The exact amount of the investment is not reported; from informal sources, it seems to be higher than €1 million. Assuming that somebody had the scope of producing high-quality proteins in large quantities, for sure it would be more economical to forget about the CO_2 and just cultivate soy, peas, or beans, or farm fish or snails for higher market value. Furthermore, even if spirulina pills have a high value in the European vegan market, it is improbable that a European product can compete against similar Chinese or Israeli products. In China, local laws allow using municipal wastewater as a source of nutrients, while in Europe it is forbidden. Hence, chemical fertilisers and their associated cost are unavoidable when producing *Spirulina* in Europe. The recovery of CO_2 adds consistent CAPEX to such a project, but its contribution to reducing the OPEX is irrelevant. Finally, massive production of *Spirulina* will cause its price to fall because its demand is limited to a small niche of wealthy consumers with a naturist ideology. Every investment requires caution and a thorough sensitivity analysis. Growing algae for the nutraceutical market requires extreme caution, multidisciplinary know-how, and preliminary contracts fixing the price and delivery conditions of the product.

8.5.5 Industrial Use: Fire Extinguishers, Supercritical Solvent, Dry Ice, and Refrigerant Fluid

This option is feasible only if the biogas plant is placed in an industrial region containing one or several buyers within a short distance. The author knows only one example in Italy: the company Montello Spa (https://www.montello-spa.it/riciclo-rifiuti-organici/produzione-biometano/) that runs one of the biggest OFMSW digestion plants in the country. The plant is placed in Lombardy, the centre of the Italian heavy industry, which explains why the CO_2 recovery was a successful business in its special case.

8.5.6 Food Grade CO_2

In Europe, several manufacturers of cryogenic upgrading plants (e.g., Pentair, Haffmans, and Criotec) propose recovering 'bio-CO_2' for the sustainable production of sparkling beverages. In Italy, Caviro Spa has implemented a project with this concept (newspaper article in Italian: https://www.huffingtonpost.it/entry/a-caviro-non-solo-vino-ora-vende-co2-a-chi-produce-bibite-gassate-di-g-lonardi_it_5fe08768c5b60d416343fb9a/). Since Caviro is one of the largest wine and ethanol producers and the owner of one of the biggest biogas plants in Italy, selling the CO_2 to the beverage industry appears as a natural choice for them. Nevertheless, this option seems difficult to implement in the case of a small agricultural biogas plant.

8.6 BIOUPGRADING OF SYNGAS PRODUCED WITH THE SOLID FRACTION OF DIGESTATE

Gaseous fermentation has a theoretical application to biogas plant optimisation too: producing syngas from the high-lignin solid fraction of digestate, or woody residual biomass unfit *for* anaerobic digestion and then upgrading the said syngas into CH_4 in the same manner as the hydrogenotrophic process described in Chapter 9, Section 9.6.

Some bacteria are capable of converting carbon monoxide (CO) dissolved in the aqueous medium of the digester, although the production of the said gas during fermentation is neglectable and usually is not considered in any model of anaerobic digestion. The CO conversion is an adaptive mechanism of such bacterial populations when other substrates are missing. The overall reaction is as follows:

$$4CO + 2H_2O \rightarrow CH_4 + 3CO_2$$

The real process is a complex chain of chemical reactions that we will not discuss here to avoid too much theoretical content. The overall simplified reaction expressed above is enough for our didactical scopes. Readers interested in deeper knowledge of the process will find a detailed description in Ref. 41.

The practical concept to focus on is that different groups of bacteria and archaea constituting the digester's ecosystem can survive if fed with CO. Biogas resulting from the said biological process will contain only 25% of CH_4, and the remaining 75% will be CO_2. But the digester's microbiome contains also hydrogenotrophic archaea able to convert H_2 and CO_2 in methane and water according to the reaction discussed in Section 9.6:

$$4\,H_2 + CO_2 \rightarrow CH_4 + 2H_2O$$

Interestingly, the by-product of the metabolic activity of the hydrogenotrophic population (water) is one of the ingredients necessary to the carboxydotrophic bacteria. Conversely, their exhaust gas (CO_2) is one of the necessary ingredients for the hydrogenotrophic archaea, so both groups can synergically produce methane [42]. At least in theory, bubbling a mixture of 25% CO and 75% H_2 in a reactor full of digestate would produce pure CH_4 and some residual water that will just dilute the digestate. After an adaptation period of the bacterial population, a digestate storage tank equipped with a gas sparger would become a bioreactor performing the following global reaction:

$$CO + 12\,H_2 \rightarrow 4\,CH_4 + 4\,H_2O$$

That reduces to

$$CO + 3\,H_2 \rightarrow CH_4 + H_2O$$

Since the main components of *syngas* are CO_2 and H_2, gasification of the non-digestible fraction of biomass – solid fraction of digestate and lignin-rich agricultural waste – allows one to fully exploit the feedstock with relatively little modifications to the biogas plant. This hypothesis has some pros but also many cons.

ADVANTAGES:

1. Increases the plant's productivity while keeping the feedstock input constant because gasification of the digestate allows the conversion of its residual carbon into biomethane.
2. Possibility of including lignin-rich residual biomass into the plant's process, such as pruning, dry leaves, olive kernels, almond or nut shells, and sawdust.
3. Heat recovery from the gasifier, useful for heating the gas–liquid bioreactor since digestate storage tanks are usually unheated.
4. Since the syngas exits the gasifier at a high temperature, it is theoretically possible to produce electricity by means of an *organic Rankine cycle* (ORC) generator.

DRAWBACKS:

1. Like in any biological process, a fraction of the biomass C (up to 11% in the case of conventional anaerobic digestion, according to 43) will be employed by the microbes to grow and reproduce; hence, the net CH_4 production will always be smaller than the one resulting from stoichiometry.
2. Increase the construction and operational complexity of the plant.
3. Higher CAPEX.
4. Additional electricity will be necessary for the syngas compressor, more intense stirring of the bioreactor, and dry feedstock feeding to the gasifier.
5. Air used for gasification will add N_2 to the resulting gas mixture. Hence, more energy will be necessary to separate N_2 from CH_4.
6. Alternatively, pure O_2 can be employed for gasification, either from commercial suppliers (i.e., no CAPEX but at higher OPEX) or produced on-site (higher CAPEX but requiring additional electricity to work).
7. Syngas from forest residues contains variable proportions of CO and H_2, depending mainly on the gasification technology employed, but always with H_2 much lower than the required 75%: for instance, 20% CO and 40% H_2 in the case of a complex gasification plant like the one in Gothenburg, Sweden [44]; 20% CO and 8% H_2 in the case of a DIY gasifier fed with olive kernels and air [45]. The only gasifier that has yielded syngas with a percentage of H_2 closest to the ideal 75% is called a *sorption enhanced reforming* (SER) reactor, developed by a joint research group of the universities of Vienna and Stuttgart [46]. The said technology requires the simultaneous injection

of pure O_2 and high-temperature steam, so it has a high energy demand for its ancillary equipment.

8. At least for plants based in the overregulated European territory, technical disadvantages are nothing compared to the bureaucratic challenge posed by a process not explicitly mentioned in any national law or EC regulation. Would the national authorities grant the same subsidies to biomethane produced by conventional digestion and gaseous phase fermentation? Or would they pretend to keep a double counting of each product flow and apply different rates of subsidies? Syngas is explosive and toxic – especially if it has a high content of H_2. So what will the additional safety regulations be and what are their associated additional costs? How long will the bureaucratic procedures take to obtain the necessary building and operation licenses?
9. On-site gasification of the solid fraction of digestate is undoubtedly an advantage for wastewater sludge and OFMSW treatment plants since it eliminates the disposal cost of the dry digestate in those countries where its use in agriculture is forbidden or restricted. In the case of agricultural biogas plants, gasification of solid digestate may enhance the production of biomethane at the cost of preventing such organic carbon from returning to the soil (see 47). The environmental benefits of adding the solid fraction of digestate as a soil amendment would be lost, without solving the problem of ammonia, mainly contained in the liquid fraction of the digestate. By the way, adding too much N to the soil without balancing it with enough C results in a higher organic C volatilisation in the soil, reducing its fertility and field capacity. Hence, gasifying the solid fraction of digestate instead of using it as an amendment limits the soil's capacity to accept liquid digestate, resulting in lower agricultural yield because of lower N, P, and K input.

Scientific literature will certainly continue to churn out graduate theses and conference papers on the biomethanation of syngas. However, in the author's opinion, it seems unlikely that we will see plants equipped with this technology in the short term, at least in the EU.

8.7 INSECT FARMING

In 2018, the update to the EU Novel Foods Regulation [48] opened a new business possibility in the old continent: farming insects to produce animal fodder. In 2023, the commission approved partially defatted powder obtained from *Acheta domesticus* (the house cricket) as a novel food on the market. The commission authorised four insects: frozen, dried, and powder forms of *Tenebrio molitor* larva (the yellow mealworm); frozen, dried, and powder forms of *Locusta migratoria* (the migratory locust); and frozen, dried, and powder forms of *Acheta domesticus* and *Alphitobius diaperinus* (the lesser mealworm). The European Food Safety Authority (EFSA) is currently assessing the safety of eight further insect species [49].

The rationale for farming insects in a biogas plant is that insects are *poikilotherms*, i.e., they cannot generate metabolic heat and their metabolic rate depends on the room temperature. In a biogas plant, there is plenty of residual heat to employ for growing and processing the insect biomass, and the leftovers of the growth cycle – called frass – can be easily added to the digester's feedstock or composted and employed as a soil amendment in the farm itself.

Insect-based feed production also has some environmental benefits:

1. Being poikilotherms makes insects very efficient at converting food into biomass. For instance, nutritional conversion rates for meat (i.e., how much feed is needed to produce a unit of increase in body weight in an animal) vary widely depending on the species and the husbandry practices adopted. On average, insects can convert 2 kg of food into 1 kg of mass, whereas a cow requires 8 kg of feed to produce a 1-kg increase in body weight.
2. GHG production by most insects is probably lower than that by conventional livestock. For example, pigs produce between 10 and 100 times more GHGs per kilogram of body weight than mealworms.
3. Insects can feed on organic waste such as food scraps and human faeces, compost, and animal slurry, and can convert it into high-quality protein suitable as animal feed. This practice is permitted in some countries but limited – banned in the case of feed for bovines – in the EU.
4. Insects use less water than conventional livestock. Mealworms, for example, are much more resistant to water shortages than cattle.
5. Finally, insect farming is less dependent on land availability than conventional livestock.

The nutritional content of insects depends on their life stage, habitat, and diet. However, the following are widely accepted:

1. Insects provide high-quality proteins and nutrients comparable to meat and fish.
2. Many insect species have a high concentration of fatty acids. They are also rich in fibre and micronutrients such as copper, iron, magnesium, manganese, phosphorus, selenium, and zinc.
3. Insects present a low risk of transmitting zoonoses (diseases transmitted from animals to humans) such as H1Nr1 (avian influenza) or bovine spongiform encephalopathy (BSE aka mad cow disease). However, European legislation prohibits feeding ruminants with feed containing insects or processed proteins derived from them.

From the consumers' point of view, most of them agree that feeding chickens and fish with feed containing insects is more natural for the animal, and the product's quality is better than those raised with soy-based feed.

Insects represent an interesting source of proteins and amino acids and could therefore integrate the diet of livestock and pets more efficiently than other animals traditionally raised for food. From the point of view of animal nutrition, they could replace with raw materials such as soy (the main source of protein) or more or less synthetic supplements providing amino acids, essential fatty acids (ω-6, ω-3), and mineral salts.

No less important are the homeotherapeutic properties that would make this food source important in supporting especially the early stages of animal development, subject to frequent imbalances with high mortality if not prevented with therapeutic treatments or medicated feed [50].

Besides the production of feed or food, another possibility under study is breeding soldier fly larvae (*Hermetia illucens*) using digestate as substrate. The larvae produced in this way are not suitable as animal feed but useful as industrial ingredients, like bioplastics. The Italian research project *Valoribio* has demonstrated the feasibility of producing biodegradable mulch sheets from *Hermetia illucens* keratin [51].

Commercial experiences of insect farming in Italian biogas plants up to the present (August 2024) have shown that integration with biogas plants is the least of the problems. The author interviewed Dr Federico Frascari, CEO of the Italian company Biological Care srl and one of the pioneers of insect farming in Italy, who kindly provided the following insights about the hurdles that insect farmers must still overcome in the EU.

1. Economies of scale do not justify CAPEX for farms below 10–20,000 boxes. Considering about 1–1.5 kg of live larvae per box per month, the resulting production of about 20,000 kg of live larvae per month is insufficient for industrial projects.
2. Pet or feed factories ask for a defatted and dried product. This means producing 25% of the total fresh weight as meal and 10% as oil.
3. The production cost for 10,000 boxes/month is approximately €1.50–2/kg live larvae. (Author's note: in Italy, a company cannot survive with less than 20% profits, so the price of live larvae should range between €1.80 and €2.40/kg to be profitable. For comparison purposes, on 24 August 2024, an Italian supermarket sold full chicken and chicken legs at €5.60/kg and €3.90/kg, respectively).
4. Investment: the 58 x 38 x 11-cm boxes (Figure 8.6) alone cost from €2 to €3 each.
5. Difficult automation, both for screening (separation of the insects in the various stages of growth) and for feeding (each box must be periodically checked and topped up, if necessary).
6. High feeding costs if the production is oriented towards livestock and humans (author's note: in the EU, organic waste cannot be employed as feed or to produce feed; hence, farming insects for animal or human consumption requires feeding the insects with the same cereals, pulses, or their by-products employed in traditional chicken and pork farming). Furthermore, if you want to operate at low costs, you have to manage trucks of by-products, and here you would need to be close to a biogas plant that consumes the surplus.
7. High cost of the live larvae and extremely high production cost of the larvae meal. Consequently, there is a limited market compared to other animal proteins.

FIGURE 8.6 Plastic Boxes for Growing Larvae in an Italian Insect Farm.

Source: Reproduced with permission of Dr Federico Frascari

8. Relatively simple management but human operations are required for at least 6 out of 7 days with a management limit of 3,000–4,000 boxes per human operator. Automation is possible only for screening and manual handling.
9. Insect farming is an interesting market for hobbyists because the live larvae are sold (as fishing bait, author's note) at €3–6/kg with modest requested volumes. If you can rely on a couple of dealers, then you can make an extra salary with just 3–4,000 boxes in your home's garage.
10. Heat is limitedly interesting on a small to medium scale because the insects survive well even at 5–10°C with a thick litter, albeit without reproducing. In the case of small cold rooms, like those in truck trailers (about 8 x 2.7 x 2.7 m), a heat pump air conditioner easily maintains 25°C or more, as long as the air exchange recovers part of the heat. The investor must decide whether to go on a vacation/carry out maintenance for 3 months and resume production in spring (as almost everyone does) or to spend on heating for year-long production. However, heating is not the most expensive OPEX item.
11. Humidity is a very important factor in saving water but induces mould growth and unwanted fermentations. The same happens when using fruit to hydrate the larvae: midges and other pests already latent in flours will proliferate.
12. Using frass is still little valued despite being potentially interesting: recovering exuviae and faeces could improve profit margins.

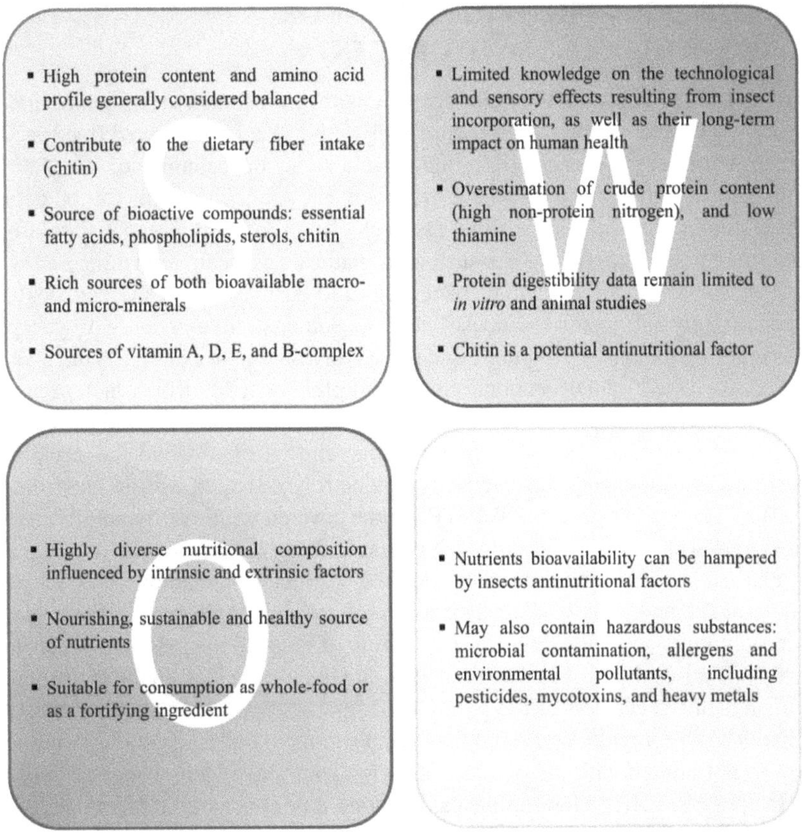

FIGURE 8.7 Strengths, Weaknesses, Opportunities, and Threats (SWOT) Analysis of Insect Protein Consumption.

Source: Ref. 52 already cited, reproduced with permission

13. In my opinion, breeding can be profitable either on a small scale – therefore, without the need for integration into a biogas plant – or on a very large scale – where heating is an important OPEX item but not as relevant as others. Therefore, the only reason to integrate insect breeding in a biogas plant is to optimise the use of biomass and manage the wastewater.

The use of insects in human nutrition is very controversial in the EU, most often based on purely ideological positions of political and consumer groups. A recent study by the University of Padua [52] has highlighted that while it is true that insect proteins have indisputable advantages in terms of environmental sustainability, there is still much to study on the possible negative health effects of their consumption and on the perception that consumers have of them (Figure 8.7).

8.8 HYDROGEN FROM BIOMETHANE OR BIOMETHANE FROM HYDROGEN?

Since the first oil crisis in 1973, the idea of a hydrogen economy has seen periods of turmoil – whenever oil prices hit record highs – but has never gained traction. The International Energy Agency (IEA) published a series of recommendations for policymakers and legislators to promote hydrogen as the energy vector for the ecological transition [53]. At the same time, research efforts focus on alternative hydrogen production methods to the two classical ones: natural gas steam reforming and water hydrolysis powered by photovoltaic panels or wind farms. The EU's Green Deal policy has strongly subsidised the construction of demonstrative green hydrogen projects, triggering the interest of public and private investors. An example is the planned construction of a 'circular economy' hydrogen plant near Madrid, which secured a public funding of €6 million [54]. The planned system consists of producing biogas with 'waste' (not clear if only sludge or sludge and OFMSW from the city), burning it in a CHP and using the electricity to electrolyse treated wastewater from the city's WWTP. Furthermore, the WWTP will be covered with solar panels to produce more electricity – probably because the biogas CHP's power is insufficient to power the treatment process or to reach the desired hydrogen production of 400 kg/day. The fate of the hydrogen is unclear: it will probably be employed in a yet-to-define urban mobility project. But does the availability of €6 million of Brussels's subsidies justify such a tweaky process, without even having a clear destiny of the produced hydrogen? And, regardless of the purely economic aspects, is the overall concept of the Spanish project really beneficial for the environment or the society? An impartial reviewer, not tainted with the political ideology about hydrogen, would quickly dismiss the proposal after a few simple calculations. Just consider the following:

a. The average efficiency of a gas CHP is 40% and the LHV of methane is 50 MJ/kg; hence, 1 kg of methane can produce 20 MJ of electricity.
b. Splitting water with a commercial electrolyser and compressing the resulting hydrogen in cylinders at 25 bar require 54.5 kWh/kg = 196.2 MJ of electricity [55].
c. Hence, producing and compressing 1 kg of hydrogen require the energy equivalent of 9.81 kg of methane.
d. The LHV of hydrogen is 120 MJ/kg.
e. The overall thermal efficiency is then

$$\eta_{th} = \frac{120 \text{ MJ}}{9.81 \text{ kg} \times 50 \text{ MJ / kg}} \times 100 = 24.5\%$$

or 1:9.81, if expressed on a gas mass ratio.
f. From the GHG emissions point of view, both green hydrogen and biomethane are climate neutral. Since 9.81 kg of biomethane is simpler to store than 1 kg of hydrogen and would yield 4 times more energy, the most rational and immediate solution for a future public mobility project would be to retrofit existing diesel vehicles to run on biomethane. Diesel-to-biomethane

conversion kits already exist in the market, and converting the existing fleet is undoubtedly quicker and cheaper than purchasing new hydrogen-fuelled vehicles – vehicles that, by the way, have not reached the mass production stage yet.

g. Producing 400 kg of hydrogen requires electrolysing 3,600 kg of water. Like any politician, the one in charge of the said project wants to show off how 'green' the party at the government is. But honestly who can consider as 'an example of circular economy' a project to replace only 3.6 m^3/day of tap water with the same quantity of treated WW, costing more than €6 million?

Despite the ridiculous scale and efficiency level, Madrid's 'biogas-to-hydrogen circular economy' concept will probably be replicated elsewhere in the EU, and perhaps in other countries that uncritically believe in the predominant ideology of European activists and public opinion. Undoubtedly, a government offering non-refundable money to projects adopting certain technologies or engineering conceptions will strongly condition investors' choices. Most likely, speculators desiring to build the plant, cash public money, and disappear will thrive in such a regulatory context. However, a wise investor seeking a sustainable business must thoroughly check if cashing some public subsidies is worth the cost of adopting an inefficient technology to produce a low-value commodity like hydrogen. The author hopes that the following considerations will guide readers involved in biomethane-to-hydrogen projects through a rational decision-making process.

8.8.1 Which Is Better: Biomethane or Hydrogen?

Such a childish question is usually heard during public debates and results from the hydrogen-based economy being the workhorse of sustainable development 'gurus' and politicians hunting for votes among the ecologist electorate. Attempting to reduce a complex problem to 'one solution better than another' results from most hydrogen economy promoters being either economists – without any clue about thermodynamics – or engineers working for heavy industries that will benefit from a massive deployment of hydrogen projects. In both cases, the preference will fall for hydrogen, regardless of whether its deployment can solve global climate change or not. Their intent is laudable, but the assumptions they base their models upon are rather simplistic. Let's analyse some of them, highlighting their respective weak points:

a. *'The combustion of hydrogen produces energy without emitting CO_2, so hydrogen is an excellent energy carrier to replace fossil fuels'*. True, but it is technically very difficult, if not impossible, to convert a petrol or diesel engine to run on hydrogen. The efficiency of heat engines has a theoretical limit foreseen by Carnot's theorem, which can hardly exceed 50%. The conversion efficiency of thermal engines ranges between 20% (prototype cars with Otto engines) and 52% (triple combined cycle power plant, operating in Siberia). In Marghera (VE, Italy), the national electric company ENEL inaugurated the first hydrogen power plant in the world in 2010, only to close it in January 2018 after spending €50 million on research [56].

b. *'Hydrogen could replace methane, taking advantage of the existing infrastructure'*. Such reasoning may sound true in a political debate (oversimplification bias!) but is inapplicable in practice for precise technical reasons. The LHV of natural gas is 47.7 MJ/kg, that of LPG is 46.1 MJ/kg, that of pure methane is 50 MJ/kg, and that of hydrogen is 120 MJ/kg. Put that way, it seems that H_2 is better because it has double the PCI compared to hydrocarbons. In practice, however, a 10 kg LPG cylinder is made of common steel of at least 1.5 mm thickness, typically 2 mm (*UNI EN 1442:2017 Standard: Equipment and accessories for LPG – Transportable and refillable cylinders of welded steel for liquefied petroleum gas (LPG) – Design and Construction*). A 10-kg capacity natural gas or methane cylinder has approximately the same volume as an equivalent LPG cylinder but requires a more expensive composite material, 6 mm thick, to be able to withstand the 200-bar pressure necessary to contain the said mass in a volume reasonably practical for use (*UNI EN ISO 11439:2013 Standard, Gas cylinders – High-pressure cylinders for the storage of natural gas used as fuel onboard cars*). Considering hydrogen's high LHV, 4.2 kg of this gas would be enough to store and transport the same energy quantity as a 10-kg cylinder of LPG or methane. However, there are two major limitations, one technical and the other physical, which make this objective unachievable:

1. Metals become brittle in contact with hydrogen; therefore, the cylinders must be made of composite materials or special alloys [57].
2. Hydrogen's density is very low, approximately 0.090 kg/Nm^3. For comparison, air density is approximately 1.25 kg/Nm^3 and methane is 0.702 kg/Nm^3. Five cylinders of 50 l each would be needed, at 200 bar pressure, to contain 4.2 kg of hydrogen [58].

Hydrogen's disadvantage compared to that of LPG or methane derives from its low volumetric density: for the same calorific value, hydrogen's storage volume will be approximately seven times greater than that of hydrocarbons. Distributing hydrogen via the current pipeline system would be impossible: they are made of steel, which would become brittle. Furthermore, even if it were somehow possible to convey H_2 safely with the current methane pipelines, as proposed by J. Rifkin and the aforementioned IEA study, the quantity of energy actually delivered to the end user in a unit of time would be much lower than that currently provided by methane. This is because the nominal flow rate of a gas pipeline does not depend on the gas but on its section and length, as well as operating pressure. All things considered, 1 Nm^3 of methane is equivalent to approximately 700 g; therefore, it carries 35 MJ of thermal energy, while 1 Nm^3 of hydrogen is equivalent to 90 g; therefore, it carries 10.8 MJ.

Figure 8.8 clarifies the concept of energy density: this can be expressed either in gravimetric or volumetric terms. Very often, authors who want to enhance the potential of hydrogen as an energy carrier highlight the high energy density of hydrogen in gravimetric terms. Failing, however, to point out the low density in volumetric terms, they hide from the public's attention an important fact: hydrogen is unsuitable for road and aerial transportation applications. This is probably the main reason why the

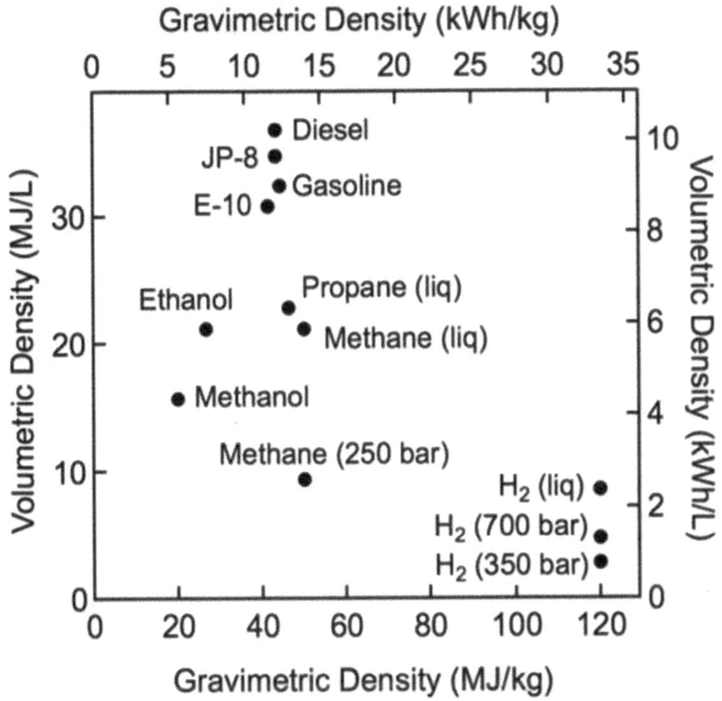

FIGURE 8.8 Energy Density of Different Fuels, Expressed Both in MJ/l and MJ/kg. Notice how hydrogen is at the lower right end: very high density in mass terms but very low in volumetric terms. Source: US Department of Energy, Fuel Cell Office https://www.energy.gov/eere/fuelcells/hydrogen-storage, reproduced with permission

hydrogen economy never became a reality since international bodies, officers, and opinion leaders began promoting it around 1975.

c. *'The hydrogen combustion cell (also called hydrogen stack) is the most efficient electrical generation device possible; therefore, for the same power, much less hydrogen is needed than any other fuel'.* This is partially true because conventional fuels are more suitable for producing heat than generating electricity. However, using hydrogen only for electricity generation with fuel cells is difficult to implement practically. The maximum theoretical efficiency of the hydrogen stack is 85%. The efficiency of the best commercial hydrogen cells built to date ranges between 40% and 60%, which is certainly better than a thermal engine generator set, especially considering that the hydrogen cell is very compact and has no moving parts. However, the limiting factor to the diffusion of this technology on a large scale is logistical: to achieve high efficiencies, it is necessary to use platinum electrodes and other extremely rare materials. Another technology called *solid oxide fuel cell* (SOFC) uses more abundant and economical catalysts, but it only works at high temperatures (500°C–800°C) and, consequently, its conversion efficiency is lower; therefore, its use is justified only for cogeneration, when it is possible to recover the waste heat for some purpose. Finally, the greater efficiency of the hydrogen cell compared to the thermal engine is limited only by electricity generation. Instead, consider the following complete life cycle, applied to a hydrogen cell vehicle: electricity generation (photovoltaic, wind, or any other renewable), water electrolysis, hydrogen compression, transport and storage, and combustion in the cell and inverter to convert the direct current produced by a cell into useful alternating current. If the complete cycle is considered, the overall efficiency is about 35% [59].

For comparison, the fleet of cars produced after 2000 has an average driving efficiency (*tank-to-wheel* in the jargon of life cycle analysis) of around 40% [60], but if we also consider the production and distribution of fuel, the global efficiency of the oil-based cycle (*well-to-wheel*) drops to 25%. Objective data show that the difference between hydrogen cell cars and conventional cars is too small to justify the large-scale implementation of a hydrogen-based energy system.

Finally, SOFCs can run on either hydrogen or methane. Hence, using biomethane directly in an SOFC will be more efficient than converting biomethane into hydrogen.

In general, and considering the current state of the art, the only advantage of hydrogen compared to natural gas is that it does not emit CO_2. Its (theoretically) greater conversion efficiency is nullified by the higher production, transport, and plant costs. On the other hand, if we compare hydrogen with biomethane, the latter is certainly more practical and economically convenient for the following reasons:

a. Although biomethane combustion emits CO_2, these emissions are neutral; i.e., they do not alter the atmospheric carbon balance because they come from short-life biomass and not from fossil fuels.
b. The volumetric energy density of biomethane is greater than that of hydrogen. Methane compressed at 250 bar has an energy density of 10 MJ/l, compared to 5 MJ/l for hydrogen compressed at 700 bar. For comparison, the volumetric energy density of diesel is 38 MJ/l (Figure 8.9), which explains the lower autonomy of methane cars compared to those fuelled with diesel oil or petrol.
c. All sources promoting hydrogen as the 'energy vector of the future' overlook an important factor: the energy necessary to compress it. Since 1 l of methane at 250 bars is equivalent to 2 l of hydrogen at 700 bars, compressing hydrogen will consume 5.6 times more energy (electricity!) for a given quantity of (thermal!) energy stored.
d. Methane does not embrittle metals, so ordinary steel can be used to build tanks and pipelines.
e. The explosive range of methane is limited, between 5% and 15% by volume, while that of hydrogen is very wide, between 4% and 75% by volume. Figure 8.9 shows a comparison between the explosive ranges of H_2 and CH_4 (author's diagram, data from 61).
f. Therefore, the greater danger of hydrogen would entail higher system costs than for methane to guarantee user safety.

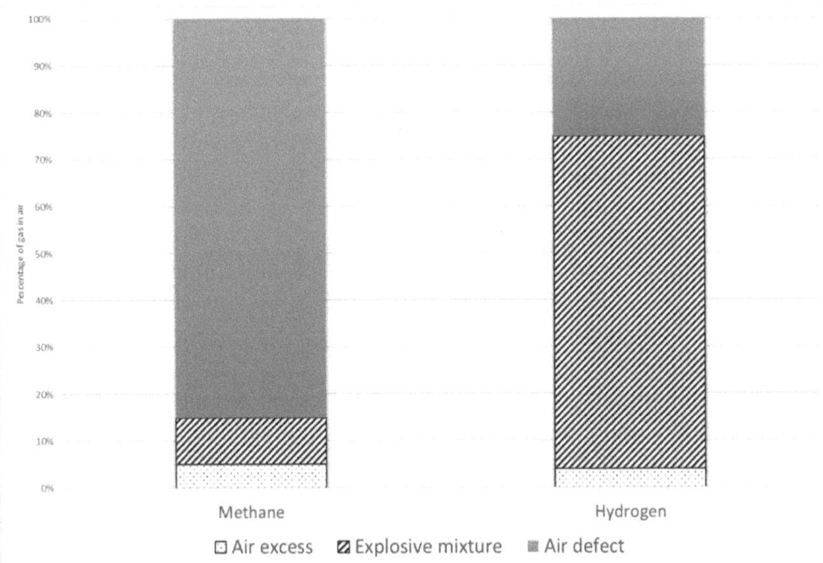

FIGURE 8.9 Comparison of the Explosive Ranges of Methane–Air and Hydrogen–Air Mixtures.

8.8.2 Hydrogen and Bioenergy

Now that we know the lights and shadows of hydrogen technology, we can analyse the different options applicable – in theory – to an AD project and business plan.

1. **Dark fermentation**: It is a fermentation process, very similar to anaerobic digestion – as a matter of fact, it is anaerobic digestion interrupted at the first stage – that produces biohydrogen, i.e., a mixture of H_2 and CO_2. The effluent is an acidic digestate, so it is susceptible to further acetoclastic fermentation to produce CH_4. Despite the claims of many research papers, dark fermentation (DF) has not been commercially successful as, on its own, it does not offer particular advantages. Biohydrogen combustion would indeed emit less CO_2 than biogas, but since the digestate is still fermentable, it can still produce CO_2 and CH_4, exactly as much as a conventional digester. DF is a process that should be seen from the perspective of optimal design of an anaerobic digestion plant, an aspect however overlooked by plant manufacturers, but not as the purpose of the plant.
2. **Hydrogen from biomethane**: The industrial use of hydrogen far exceeds its commercial potential as an energy carrier. Currently, hydrogen is produced on a large scale from natural gas. This method, known as '*steam reforming*', is based on the reaction between water vapour and CH_4 at high temperature, in the presence of a catalyst:

$$CH_4 + H_2O \rightleftharpoons CO + 3\ H_2$$

It is a very energy-intensive process, which ultimately produces as much CO_2 as the direct combustion of CH_4. Research efforts should focus on testing potentially more efficient and environmentally friendly methods. A multidisciplinary group [62] developed the technology known as methane pyrolysis. It involves bubbling methane in a reactor filled with molten tin at 800°C and quartz washers. Under these conditions methane splits, releasing H_2 and leaving a deposit of pure C floating on the molten tin. Another project, coordinated by the Italian Nobel Prize winner Carlo Rubbia and conducted at CIEMAT in Madrid, consists of passing a mixture of CH_4 and a carrier gas (Ar, Ne, He, etc.) through a porous graphite reactor, heated to 1700°C, and obtaining a mixture of H_2 and carbon particles, dragged by the inert gas, at the exit [63].

Yet another group has obtained the thermal decomposition of methane at 700°C–800°C using nanoscale, binary, Fe–M (M = Pd, Mo, or Ni) catalysts supported on alumina [64].

3. **Hydrogen from water and CO (the latter produced by the incomplete combustion of biomass)**: This process is known as the 'water shift reaction'. It is based on an apparently simple reaction, discovered in 1780 by the Italian chemist Felice Fontana:

$$CO + H_2O \rightleftharpoons CO_2 + H_2$$

The resulting mixture, known as *syngas*, is useful for biofuel synthesis through the Fischer–Tropsch process. It works well with (fossil) coal, but the results obtained with biomass at the industrial scale have been unsuccessful [65].

4. **Molten carbonate fuel cell**: This new technology is supposed to be tested in the Drax Plc biomass power plant [66] in England. This is a three-stage technology. Firstly, methane and water are subject to the *steam reforming reaction* explained in point 2. The resulting *syngas* passes to the molten carbonate cell, where it reacts with CO_2 and O_2 coming from the flue gases of the biomass power plant. CO_2 and water are formed inside the fuel cell, producing electricity. The cell's exhaust gas is a mixture of pure CO_2 – to be used for greenhouse cultivation or other purposes – and water vapour, easily separable by simple condensation.
5. **Catalytic reforming of biogas**: The objective of the BIONICO research project (http://www.bionicoproject.eu/) is to produce H_2 for industrial purposes, directly in biogas plants. This is a special type of catalytic reactor, producing *syngas* from biogas and subsequently separating H_2 using special selective membranes. The waste gas is a mixture of CO, CO_2, and CH_4, possibly usable in a conventional CHP unit or boiler.

8.9 CONCLUSIONS

The reactions between hydrogen and carbon are reversible, so processes are possible that tend to obtain CH_4 starting from CO_2 and H_2, or vice versa, to obtain H_2 and C or H_2 and CO starting from CH_4. Despite IEA's policy recommendations and the ideologies of green economy 'gurus', physical laws make it difficult for H_2 to replace fossil fuels because its low volumetric energy density makes its storage and transport impractical and the overall life-cycle efficiency is low. On the contrary, the transformation of biogas into H_2 for industrial purposes could become a possible future business because it is more environmentally friendly than the current *steam reforming process*. The feasibility of this *business* is limited to the existence of an industrial user not too far from the digester, to minimise the cost of hydrogen storage and transport. Furthermore, the technology for converting methane into hydrogen at a small scale is not yet mature.

BIBLIOGRAPHY

[1] Kajsa Risberg, Harald Cederlund, Mikael Pell, Veronica Arthurson & Anna Schnürer, Comparative characterization of digestate versus pig slurry and cow manure – Chemical composition and effects on soil microbial activity, *Waste Management*, 61, 529–538 (2017). https://doi.org/10.1016/j.wasman.2016.12.016

[2] Patrick Dabert et al., *Characterization of DIgestate and evaluation of their agronomic VAlorisation – DIVA; Agricultural recycling of digestate from anaerobic digestion, summary of the DIVA project*, Agence nationale de Recherche. Permanent link: https://anr.fr/Project-ANR-10-BIOE-0007

[3] Isabella Pecorini, Eleonora Peruzzi, Elena Albini, Serena Doni, Cristina Macci, Grazia Masciandaro & Renato Iannelli, Evaluation of MSW compost and digestate mixtures for a circular economy application, *Sustainability*, 12:7, 3042 (2020). https://doi.org/10.3390/su12073042

[4] Directive 2008/98/EC of the European Parliament and of the Council of 19 November 2008 on waste and repealing certain Directives. Summary and full text: https://ec.europa.eu/environment/topics/waste-and-recycling/waste-framework-directive_it and full text https://eur-lex.europa.eu/legal-content/EN/TXT/?uri=CELEX%3A02008L0098-20180705

[5] MA Rosato, *(article in Italian) La canna indica, una coltura dai mille utilizzi*. https://agronotizie.imagelinenetwork.com/bio-energie-rinnovabili/2020/08/24/la-canna-indica-l-una-coltura-dai-mille-utilizzi/67633

[6] G Caicedo Díaz, U Bonilla Ramírez & L Rozo Wilches, *La achira: su producción y beneficio*. http://bibliotecadigital.agronet.gov.co/bitstream/11348/4079/1/La%20achira%2C%20su%20producci%C3%B3n%20y%20beneficio.pdf

[7] FATA Project, *VenetoAgricoltura (in Italian)*. https://www.venetoagricoltura.org/upload/pubblicazioni/Risultati%20finali%20FATA/FATA%20x%20web.pdf

[8] P.A.N spin-off. https://sites.google.com/site/panspinoff/

[9] GS Colares, N Dell'Osbel, PG Wiesel, GA Oliveira, PHZ Lemos, FP da Silva, CA Lutterbeck, LT Kist & ÊL Machado, Floating treatment wetlands: A review and bibliometric analysis, *Science of the Total Environment*, 714, 136776 (April 20, 2020). https://doi.org/10.1016/j.scitotenv.2020.136776. Epub 2020 Jan 17. PMID: 31991269.

[10] Leonardo Lareo and Ricardo Bressani, Possible utilization of the water hyacinth in nutrition and industry, *Food and Nutrition Bulletin*, 4:4 (October 1982). The United Nations University Press (unu.edu).

[11] Alberto Bonomi, Prospettive circa l' impiego di alcune piante acquatiche nell'alimentazione animale, *L'Avvenire Agricolo*, Anno 920 – Numero 11 – 1984

[12] Directive (EU) 2023/2413 of the European Parliament and of the Council of 18 October 2023 amending Directive (EU) 2018/2001, Regulation (EU) 2018/1999 and Directive 98/70/EC as regards the promotion of energy from renewable sources, and repealing Council Directive (EU) 2015/652. https://eur-lex.europa.eu/legal-content/EN/TXT/?uri=CELEX%3A32023L2413&qid=1699364355105

[13] ECS Little, Handbook of utilization of aquatic plants, *FAO Fisheries Technical Papers*, 187, 176 (1979). https://www.fao.org/3/X6862E/X6862E00.htm. ISBN 92-5-100825-6.

[14] MP Bernal & JM Lopez-Real, Natural zeolites and sepiolite as ammonium and ammonia adsorbent materials, *Bioresource Technology*, 43:1, 27–33 (1993). https://doi.org/10.1016/0960-8524(93)90078-P. ISSN 0960–8524.
[15] X-L Long, H Cheng, Z-L Xin, W-D Xiao, W Li & W-K Yuan, Adsorption of ammonia on activated carbon from aqueous solutions, *Environmental Progress*, 27, 225–233 (2008). https://doi.org/10.1002/ep.10252
[16] CJ Castro, HY Shyu, L Xaba, R Bair & DH Yeh, Performance and onsite regeneration of natural zeolite for ammonium removal in a field-scale non-sewered sanitation system, *Science of The Total Environment*, 776, 145938 (2021). https://doi.org/10.1016/j.scitotenv.2021.145938. ISSN 0048–9697.
[17] Italian Compost Consortium, *Biowaste, I DATI DEL SETTORE DEL RICICLO DEL RIFIUTO ORGANICO* (2017). https://www.compost.it/rapporti-cic/rapporto-annuale-2017/
[18] Khalid Azim, Composting parameters and compost quality: A literature review, *Organic Agriculture*, 8, 1–18 (2018). https://doi.org/10.1007/s13165–017–0180-z
[19] Grazia Cacciola, *L'orto naturale for dummies*, Hoepli Editori (2017).
[20] B Eghball, JF Power, JE Gilley & JW Doran, Nutrient, carbon, and mass loss during composting of beef cattle feedlot manure, *Journal of Environmental Quality*, 26, 189–193 (1997). https://doi.org/10.2134/jeq1997.00472425002600010027x
[21] Sonia Tiquia-Arashiro, Tom Richard & Mark Honeyman, Carbon, nutrient, and mass loss during composting, *Nutrient Cycling in Agroecosystems*, 62, 15–24 (2002). https://doi.org/10.1023/A:1015137922816
[22] Gary A Breitenbeck & David Schellinger, Calculating the reduction in material mass and volume during composting, *Compost Science & Utilization*, 12:4, 365–371 (2004). https://doi.org/10.1080/1065657X.2004.10702206
[23] Roman Uvarov, Alexander Briukhanov & Ekaterina Shalavina, Study results of mass and nutrient loss in technologies of different composting rate: Case of bedding poultry manure, *Engineering for Rural Development Jelgava*, 25 (May 27, 2016). https://www.tf.llu.lv/conference/proceedings2016/Papers/N161.pdf
[24] Ingrid H Franke-Whittle, Alberto Confalonieri, Heribert Insam, Mirko Schlegelmilch & Ina Körner, Changes in the microbial communities during co-composting of digestates, *Waste Management*, 34:3 (2014). https://doi.org/10.1016/j.wasman.2013.12.009
[25] Washington State University. *Compost calculator.* https://puyallup.wsu.edu/soils/compost-mix-calculator/
[26] Budiman Minasny, Brendan P Malone, Alex B McBratney, Denis A Angers, Dominique Arrouays, Adam Chambers, Vincent Chaplot, Zueng-Sang Chen, Kun Cheng, Bhabani S Das, Damien J Field, Alessandro Gimona, Carolyn B Hedley, Suk Young Hong, Biswapati Mandal, Ben P Marchant, Manuel Martin, Brian G McConkey, Vera Leatitia Mulder, Sharon O'Rourke, Anne C Richer-de-Forges, Inakwu Odeh, José Padarian, Keith Paustian, Genxing Pan, Laura Poggio, Igor Savin, Vladimir Stolbovoy, Uta Stockmann, Yiyi Sulaeman, Chun-Chih Tsui, Tor-Gunnar Vågen, Bas van Wesemael & Leigh Winowiecki, Soil carbon 4 per mille, *Geoderma*, 292, 59–86 (2017). https://doi.org/10.1016/j.geoderma.2017.01.002. ISSN 0016–7061.
[27] Luca Conte, *Course on biological agriculture*, Veneto Agricoltura (October 2021).
[28] CE Egene et al., Solid fraction of separated digestate as soil improver: Implications for soil fertility and carbon sequestration, *Journal of Soils and Sediments*, 21, 678–688 (2021). https://doi.org/10.1007/s11368-020-02792-z
[29] Fred Magdoff & Harold Van Es, Building soils for better crops: Ecological management for healthy soils, *Sustainable Agriculture Research & Education Program* (2021). National Institute of Food and Agriculture (U.S.). Free download from https://www.sare.org/publications/building-soils-for-better-crops/authors-and-preface/
[30] MA Rosato, Risparmiare acqua e fertilizzanti con il biochar, *AgroNotizie* (March 17, 2017). https://agronotizie.imagelinenetwork.com/fertilizzanti/2017/03/17/risparmiare-acqua-e-fertilizzanti-con-il-biochar/53350

[31] David Arturo Munar-Florez, Darlis Adriana Varón-Cardenas, Nidia Elizabeth Ramírez-Contreras & Jesús Alberto García-Núñez, Adsorption of ammonium and phosphates by biochar produced from oil palm shells: Effects of production conditions, *Results in Chemistry*, 3, 100119 (2021). https://doi.org/10.1016/j.rechem.2021.100119. ISSN 2211–7156.

[32] Philip G Miles & Shu-Ting Chang, *Mushrooms cultivation, nutritional value, medicinal effect, and environmental impact*, 8, CRC Press (March 29, 2004). 478, Color & 98 B/W Illustrations. https://www.routledge.com/Mushrooms-Cultivation-Nutritional-Value-Medicinal-Effect-and-Environmental-Impact/Miles-Chang/p/book/9780849310430. ISBN 9780849310430.

[33] B Bugbee, B Spanarkel, S Johnson, O Monje & G Koerner, *CO2 crop growth enhancement and toxicity in wheat and rice*. https://www.ncbi.nlm.nih.gov/pubmed/11540191

[34] S Milius, *Pumped-up poison ivy: Carbon dioxide boosts plant's size, toxicity*, The Free Library. 2006 Science Service, Inc. (August 31, 2019). https://www.thefreelibrary.com/Pumped-up+poison+ivy%3a+carbon+dioxide+boosts+plant%27s+size%2c+toxicity.-a0147259109

[35] https://www.co2.earth/

[36] Safety sheet of Linde (in Italian) RACCOMANDAZIONE DI SICUREZZA RS04 –Uso anidride carbonica – CO2. http://www.linde-gas.it/it/images/Uso%20anidride%20carbonica%20%E2%80%93%20CO2_tcm335-169727.pdf

[37] Bruce Arnold Kimball et al., Data from the Arizona FACE (free-air CO2 enrichment) experiments on wheat at ample and limiting levels of water and nitrogen, *Open Data Journal for Agricultural Research*, 3, 29–38 (June 2017).

[38] HZ Enoch & JM Olesen, Plant response to irrigation with water enriched with carbon dioxide, *New Phytologist*, 125, 249–258 (1993). https://nph.onlinelibrary.wiley.com/doi/pdf/10.1111/j.1469-8137.1993.tb03880.x

[39] MA Rosato, *(article in Italian) Biocarburanti da alghe in Europa*. http://agronotizie.imagelinenetwork.com/bio-energie-rinnovabili/2016/12/01/i-biocarburanti-da-alghe-in-europa/52069

[40] *Newspaper article in Italian*. https://www.ilsole24ore.com/art/spirulina-parte-sardegna-produzione-superfood-ABKwJ5oB?refresh_ce=1

[41] Sancho Navarro Silvia, Cimpoia Ruxandra, Bruant Guillaume, Guiot Serge R.; Biomethanation of Syngas Using Anaerobic Sludge: Shift in the Catabolic Routes with the CO Partial Pressure Increase, *Frontiers in Microbiology*, 7 (2016). https://www.frontiersin.org/articles/10.3389/fmicb.2016.01188

[42] Kohlmayer Matthias, Huber Robert, Brotsack Raimund & Mayer Wolfgang, Simultaneous CO2 and CO methanation using microbes, *bioRxiv*, 326561. https://doi.org/10.1101/326561

[43] A Cohen, RJ Zoetemeyer, A van Deursen & JG van Andel, Anaerobic digestion of glucose with separated acid production and methane formation, *Water Research*, 13:7, 571–580 (1979). https://doi.org/10.1016/0043-1354(79)90003-4. ISSN 0043–1354.

[44] Hailong Li, Daheem Mehmooda, Eva Thorina & Zhixin Yub, Biomethane production via anaerobic digestion and biomass gasification, *Energy Procedia*, 105, 1172–1177 (2017). https://www.diva-portal.org/smash/get/diva2:1169148/FULLTEXT01.pdf

[45] L Garrido, M Cueva & S Sánchez, Start-up procedure for a downdraft biomass gasifier using olive stones, *S.31st European Biomass Conference and Exhibition*. Free download upon registration to http://www.etaflorence.it/proceedings/index.asp?detail=20742&mode=topic&categories=T83&items=4AO%2E5%2E1. ISBN 978-88-89407-23-3.

[46] https://agronotizie.imagelinenetwork.com/bio-energie-rinnovabili/2020/07/20/idrogeno-da-biomasse-e-green-deal/67467

[47] Vania Scarlet Chavez-Rico, Paul LE Bodelier, Miriam van Eekert, Valentina Sechi, Adrie Veeken & Cees Buisman, Producing organic amendments: Physicochemical changes in biowaste used in anaerobic digestion, composting, and fermentation, *Waste Management*, 149, 177–185 (2022). ISSN 0956–053X.

[48] Regulation (EU) 2015/2283 of the European Parliament and of the Council of 25 November 2015 on novel foods, amending Regulation (EU) No 1169/2011 of the European Parliament and of the Council and repealing Regulation (EC) No 258/97 of the European Parliament and of the Council and Commission Regulation (EC) No 1852/2001. Texts in all EU official languages here: https://eur-lex.europa.eu/legal-content/en/TXT/?uri=CELEX%3A32015R2283
[49] European Parliament, Question for written answer E-000581/2023 to the Commission Rule 138 Charlie Weimers (ECR), (February 22, 2023). https://www.europarl.europa.eu/doceo/document/E-9-2023-000581_EN.html#
[50] Antonella Dalle Zotte, Yazavinder Singh, Bianca Palumbo, Barbara Contiero & Marco Cullere, Live yellow mealworm (*Tenebrio molitor*) larvae: A promising nutritional enrichment for laying quails, *Poultry Science*, 103:7, 103759 (2024). https://doi.org/10.1016/j.psj.2024.103759. ISSN 0032–5791.
[51] Mario A Rosato, Teli di pacciamatura prodotti da scarti agricoli – L'allevamento di insetti e le bioraffinerie del futuro, online article in Italian, *AgroNotizie* (March 8, 2019). https://agronotizie.imagelinenetwork.com/bio-energie-rinnovabili/2019/03/08/teli-di-pacciamatura-prodotti-da-scarti-agricoli/62117
[52] E Pontalti, M Cullere & A Dalle Zotte, Meat alternatives and their impact on human health: A comprehensive review, *Meat and Muscle Biology*, 8:1, 17711, 1–19 (2024). https://doi.org/10.22175/mmb.17711
[53] IEA, *The future of hydrogen*. https://webstore.iea.org/the-future-of-hydrogen
[54] RETEMA, *Online article (in Spanish)*. https://www.retema.es/actualidad/la-comunidad-de-madrid-sera-pionera-en-producir-hidrogeno-verde-partir-de-agua
[55] NREL, *Technology brief: Analysis of current-day commercial electrolyzers*, U.S. Department of Energy. https://www.nrel.gov/docs/fy04osti/36705.pdf
[56] Wikipedia in Italian, *Centrale a Idrogeno di Fusina*. https://it.wikipedia.org/wiki/Centrale_a_idrogeno_di_Fusina
[57] Norm: *UNI EN ISO 11114–4:2017 Transportable gas cylinders – Compatibility of the cylinder and valve materials with the gases contained – Part 4: Test methods for the selection of metallic materials resistant to hydrogen embrittlement.*
[58] Technical data sheet for hydrogen in commercial cylinders from **Air Liquide**. https://industria.airliquide.it/sites/industry_it/files/2016/05/17/hydrogen_italia_it_v6_0.pdf
[59] Rittmar von Helmolt & Ulrich Eberle, Fuel cell vehicles: Status 2007, *Journal of Power Sources*, 165:2, 833–843 (March 20, 2007). https://doi.org/10.1016/j.jpowsour.2006.12.073
[60] David Hafemeister, Review of the 2008 APS energy study, energy future: Think efficiency, *Physics and Society*, 38:2 (April 20, 2009). https://digitalcommons.calpoly.edu/cgi/viewcontent.cgi?article=1313&context=phy_fac
[61] Chrysalis Scientific Inc. https://www.chrysalisscientific.com/pg443-Lower-LEL-Upper-UEL-Explosive-Limits.pdf
[62] T Geißler et al., Hydrogen production via methane pyrolysis in a liquid metal bubble column reactor with a packed bed, *Chemical Engineering Journal*, 299, 192–200 (September 1, 2016).
[63] A Abánades et al., Experimental analysis of direct thermal methane cracking, *Fuel and Energy Abstracts*, 36:20, 12877–12886 (October 2011).
[64] Naresh Shah, Devadas Panjala & Gerald P Huffman, Hydrogen production by catalytic decomposition of methane, *Energy & Fuels*, 15:6, 1528–1534 (2001). https://doi.org/10.1021/ef0101964
[65] MA Rosato, *Article in Italian about Choren's miscanthus-to-fuels project failure*. https://agronotizie.imagelinenetwork.com/bio-energie-rinnovabili/2019/05/10/che-fine-ha-fatto-il-miscanto/62873
[66] The University of Edinburgh, *Project data summary*. https://www.geos.ed.ac.uk/sccs/project-info/2561

9 Applications of Respirometric Tests beyond Anaerobic Digestion

9.1 CHAPTER OVERVIEW

The biochemical methane potential (BMP) test protocol and its derivative, the specific methanogenic activity (SMA) test discussed in Chapter 5, are just two special cases of a broader category of laboratory techniques, called *respirometric tests*. All living things share seven characteristics: motion, sensitivity to the environment, respiration, growth, reproduction, nutrition, and excretion. Measuring the respiration rate is an easy way to obtain indirect information about a microbial population's growth, nutrition, and reproduction rates, or the conversion of a given feedstock into a desired product. For instance:

1. Under anaerobic conditions, yeasts will convert 1 mole of glucose (180 g) into 2 moles of ethanol (92 g) and 2 moles of CO_2 (44.8 Ndm^3).
2. In the case of composting organic waste, unless the organic matter contains highly oxidised and reduced chemical compounds, each mole of consumed O_2 (22.4 Ndm^3) corresponds to 1 mole of biodegraded organic C (12 g), both becoming 1 mole of CO_2. In the case of acetic (aerobic) fermentation, *Acetobacter sp.* will convert 1 mole of ethanol (46 g) and 1 mole of O_2 (22.4 Ndm^3) into 1 mole of acetic acid and 1 mole of water (18 g).

The stoichiometric relationships allow us to employ respirometric tests for various industrial applications, some related directly or indirectly to the biogas business. This chapter will deal with some of them, providing some indications about how to perform the tests correctly and, when the test is standardised, providing the reference norm.

9.1.1 CAVEAT

Regardless of the test protocol, it is important to remember that stoichiometric results are a theoretical model that does not consider the living beings involved in the reaction. As stated in the first paragraph, living beings do not limit their biological activity to eating a feedstock, excreting some metabolite, and giving off some gas as their respiration product. Bacteria will also convert part of the feedstock into

DOI: 10.1201/9781003609278-9

living biomass. Hence, the gas quantity emitted will always be slightly smaller than the stoichiometric expected value. As a rule of thumb, a gas production of about 95% of the theoretical stoichiometric value is perfectly normal. Getting a smaller BMP than the theoretical value does not invalidate the test's correctness, nor does it mean that the instrument underestimates the measurements. When CO_2 is one of the respiration products, always remember that it is highly soluble in water. Hence, the apparent underestimation error of the measured biogas volume is just the result of CO_2 remaining dissolved in the fermentation broth and not being counted in the flow cell.

9.2 THE *DUO* FAMILY OF PRODUCTS: HOW TO MEASURE THE BIOGAS COMPOSITION CORRECTLY

Volumetric flow meters equipped with sealed measurement cells, such as the Gas Endeavour II, the BPC Go, and all the third-generation instruments (AMPTS III, GE III, BRS III, and BPC Blue) offer an interesting possibility to expand the BMP test protocol: measuring simultaneously the biogas flow, its cumulated volume, and the approximate biogas composition. The word 'approximate' is underlined because the method proposed here is based on assuming biogas to be just a binary mixture of CO_2 and CH_4, neglecting all the other trace gases that anaerobic digestion can eventually produce in relevant quantities: H_2, H_2S, NH_3, and siloxanes. This assumption is true in most cases because the concentration of trace gases is usually in the order of a few ppm. But remember that it is not a universal truth because there are cases (e.g., inhibited inoculum and dark fermentation conditions) in which the quantity of H_2 may be higher than some percentual points. When in doubt, checking the gas composition with a gas chromatographer is mandatory.

9.2.1 AMPTS III Duo and GE III Duo: Principle of Operation and Error Margin

The BMP test in Duo configuration employs two measurement cells for each reactor: the first one, connected directly to the batch reactor, will measure the total biogas flow and volume. The exit tube of the first measurement cell is then connected to the CO_2 trap and this latter to the second flow measurement cell (Figure 9.1). In the spreadsheet report of the test results, each reactor will correspond to two columns, one with the total biogas production and the other with the net CH_4 production. The difference between both will then be the CO_2 production.

Like any other laboratory method, the operation of the Duo setup is based on assumptions or simplification hypotheses:

a. The first cell will measure all the biogas produced during the anaerobic digestion of the feedstock.
b. The second cell will measure all the CH_4.

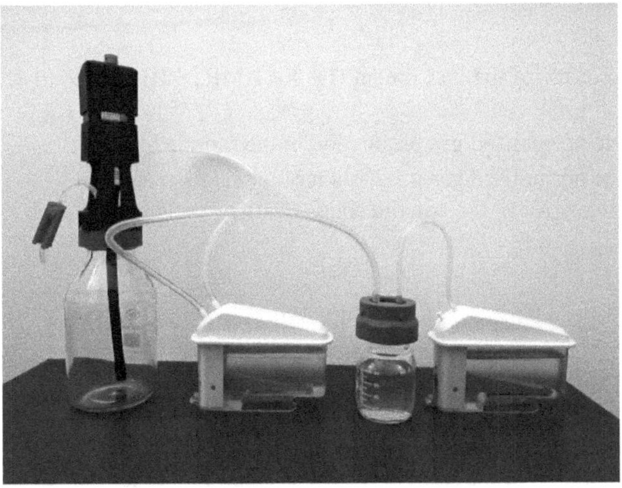

FIGURE 9.1 Setup of the Duo Configuration. Biogas exiting the reactor passes through the first flow cell unit (FCU), then through a flask containing NaOH solution that captures CO_2, and the remaining CH_4 is then conveyed to the second FCU.

While the second assumption will always be true because CH_4 is practically insoluble in water, the first one will often be false. It is unavoidable that a fraction of the CO_2 dissolves in the inoculum, and a further (small) fraction dissolves in the water of the first flow cell. Since only one false hypothesis is enough to lead to false conclusions, the inexperienced laboratory technician will often get confusing results, namely, a correct BMP value associated with an unrealistically high CH_4 percentage (in some cases, as high as 95%). To avoid such a confusing test result, we must ensure that all the hypotheses on which the method is based are as true as possible; i.e., the percentage of CO_2 dissolved in the aqueous medium must be neglectable compared to the total volume of biogas produced.

In this variant of the BMP test, it is necessary to check two error components separately:

a. The metrological error of the calculated BMP.

The formula is the same as already explained in the first volume (9.1):

$$e_{\text{BMP}} = \frac{e_{\text{instr}}\left(\bar{C}+\bar{B}\right)}{\left(\bar{C}-\bar{B}\right)} + e_{\text{VS}},$$

where

e_{instr} = relative error of the instrument (1% for AMPTS III with 9 ml cells, 3% for 2 ml cells);

$\bar{C}$ = average normalised gas production of the sample reactors;

$\bar{B}$ = average normalised gas production of the blank reactors;

e_{VS} = relative error of the volatile solids (usually neglectable) given by:

$$e_{VS} = \frac{2E_{scale}}{DW - Ash} + \frac{E_{scale}}{WW},$$

E_{scale} = absolute error of the scale;

DW = dry weight of the sample;

Ash = weight of the residue after calcination at 550°C;

WW = wet weight of the sample (if $WW > 5{,}000\ E_{scale}$, then $e_{VS} \approx 0$).

b. The error of the calculated percentage of CH_4.

By definition, the concentration of CH_4 in the total biogas is given by

$$X_{ms} = \frac{V_{ms} - V_{mbk}}{V_{bs} - V_{bbk}} \times 100, \qquad (9.1)$$

where

X_{ms} = concentration of methane in the sample biogas [%];

V_{ms} = total volume of methane produced by the sample;

V_{bs} = total volume of biogas produced by the sample;

V_{mbk} = volume of methane produced by the blank;

V_{bbk} = total volume of biogas produced by the blank.

By applying the theorems of error propagation (9.2), the relative error of the calculated methane concentration of the sample, e_{Xms}, will be

$$e_{Xms} = e_{(V_{ms} - V_{mbk})} + e_{(V_{bs} - V_{bbk})},$$

where

$$e_{(V_{ms} - V_{mbk})} = \frac{e_{inst}\left(V_{ms} + V_{mbk}\right)}{\left(V_{ms} - V_{mbk}\right)},$$

$$e_{(V_{bs} - V_{bbk})} = \frac{e_{inst}\left(V_{bs} + V_{bbk}\right)}{\left(V_{bs} - V_{bbk}\right)}.$$

Hence,

$$e_{Xms} = e_{inst}\left[\frac{(V_{ms}+V_{mbk})}{(V_{ms}-V_{mbk})}+\frac{(V_{bs}+V_{bbk})}{(V_{bs}-V_{bbk})}\right] \quad (9.2)$$

Formula (9.2) confirms, once again, the conceptual importance of preincubating the inoculum to minimise the endogenous gas production of the blank. Indeed, if $V_{mbk} \approx V_{bbk} \approx 0$, the minimum uncertainty of the CH_4 concentration reduces to

$$min.\ e_{Xms} = 2e_{inst}.$$

Since CH_4 is almost insoluble in water, we can assume that the measurements V_{ms} and V_{mbk} correspond to the whole methane production of the digestion process, so the first term within square brackets is always valid. On the contrary, the second term is valid only if all the biogas produced during the digestion will flow through the measurement cell, which is usually a false assumption because a fraction of the CO_2 in the biogas will remain dissolved in the inoculum and the water contained in the flow measurement cell, so the instrument will not detect it. Formula (9.2) is then only theoretical because it is based on a simplifying supposition. In general, the uncertainty will be higher. How much higher? This will depend on the test conditions: how much CO_2 will the substrate and blank produce? How much of it will dissolve in the inoculum and flow measurement cell?

We can rewrite Formula (9.2) to account for the biogas underestimation error. By definition

$$X_{ms} = \frac{V_{ms}-V_{mbk}}{(V_{bs}+E_s)-(V_{bbk}+E_{bk})}\times 100,$$

where

E_s = fraction of the CO_2 produced by the substrate, captured in the liquid phase of the sample;

E_{bk} = fraction of the CO_2 produced by the substrate, captured in the liquid phase of the blank.

And rewrite it defining $E = E_{s-}\ Ebk$:

$$X_{ms} = \frac{V_{ms}-V_{mbk}}{(V_{bs}-V_{bbk}+E)}\times 100$$

By definition:

$$e_{Xms} = e_{(V_{ms}-V_{mbk})} + e_{(V_{bs}-V_{bbk}+E)},$$

where

$$e_{(V_{ms}-V_{mbk})} = \frac{e_{inst}\left(V_{ms}+V_{mbk}\right)}{\left(V_{ms}-V_{mbk}\right)}$$

like in the theoretical case, but the relative error of the biogas measurement will be

$$e_{(V_{bs}-V_{bbk}+E)} = \frac{e_{inst}\left(V_{bs}+V_{bbk}\right)+E}{\left(V_{bs}-V_{bbk}+E\right)} = \frac{e_{inst}\left(V_{bs}+V_{bbk}\right)}{\left(V_{bs}-V_{bbk}+E\right)} + \frac{E}{\left(V_{bs}-V_{bbk}+E\right)},$$

$$e_{Xms} = e_{inst}\left[\frac{\left(V_{ms}+V_{mbk}\right)}{\left(V_{ms}-V_{mbk}\right)} + \frac{\left(V_{bs}+V_{bbk}\right))}{\left(V_{bs}-V_{bbk}+E\right)}\right] + \frac{E}{\left(V_{bs}-V_{bbk}+E\right)}. \tag{9.3}$$

Remark: Note the different conceptual nature of both terms. The first one is metrological uncertainty purely related to the instrument's accuracy (or to the dispersion among replicates, whichever is higher). The second component of the uncertainty is related to the very nature of the batch method employed to measure V_{bs}. It depends only on how much of the produced CO_2 will dissolve in the liquid medium during the test, which in turn depends on how much CO_2 absorption capacity the inoculum and the water in the measurement cell had at the beginning of the test, plus the eventual content of ash in the substrate that will add alkalinity to the sample reactors but not to the blank reactors. For a given instrument setup, i.e., for a given volume of inoculum, the volume of biogas produced will depend on the biodegradability of the substrate and on the amount of substrate added to the reactor, which depends on the VS concentration of the inoculum and the inoculum/substrate ratio (ISR) chosen.

From a practical point of view, we cannot calculate the uncertainty of the calculated value X_{ms} unless we find a method to determine E. The possible alternatives are as follows:

1. Calculate the maximum value of E knowing the solubility of CO_2 at standard atmospheric pressure and the incubation temperature, $S_{(Ti)}$, the incubation temperature T_i, and the total volume of the inoculum, v_i, plus the volume of water contained in the flow measurement cell (120 ml in the case of AMPTS III and GE III) and the solubility at room temperature (assumed to be 20°C).

$$E = k\left(S_{(Ti)} \times v_i + 859,7 \times 0,120\right)$$

Table 9.1 provides the solubility of CO_2 in water at 110 kPa of CO_2 partial pressure within the temperature range of interest for anaerobic digestion tests. The figures in bold are from Table 2–120 of [3]; the rest are the author's conversions and interpolations to facilitate the reader's calculations and the comprehension of the examples

TABLE 9.1
Solubility of CO_2 in Water at 110 kPa of CO_2 Partial Pressure as a Function of Temperature

Temperature [°C]	Solubility [g CO_2/100 cm^3 H_2O]	Solubility [Ncm^3 CO_2/dm^3 H_2O]
20	**0.1688**	859.7
30	**0.1257**	640.2
32	0.12002	611.3
34	0.11434	582.3
36	0.10866	553.4
38	0.10298	524.5
40	**0.0973**	495.5
41	0.09518	484.7
44	0.08882	452.4
46	0.08458	430.8
48	0.08034	409.2
50	**0.0761**	387.6
52	0.0724	368.7
54	0.0687	349.9
56	0.065	331.0
58	0.0613	312.2
60	**0.0576**	293.4

included in this chapter. The coefficient k is necessary to correct the solubility values from Table 9.1 because biogas is a mixture of gases and, hence, the partial pressure of CO_2 will be between 40% and 50% of the reference pressure. For simplicity, assume $k = 0.5$.

So far, we have assumed that the only mechanism of CO_2 volume underestimation is its absorption in the inoculum and the water inside the flow measurement cell. Under real-life test conditions, we must consider another factor that increases the uncertainty of the *Xm* value measured with the Duo setup: the inoculum's alkalinity and the ash and proteins in the substrate. The presence of carbonates and NH_4^+ in the inoculum and the release of Na, K, Ca, and Mg salts together with NH_4^+ during the substrate's degradation will increase the overall CO_2 sequestration capacity beyond the mere solubility in the aqueous medium. It could be possible to measure the initial and final alkalinity of the inoculum, and their difference represents the volume of CO_2 absorbed during the digestion.

To account for the additional CO_2 absorption of the ashes, we can assume them to be pure potassium hydroxide, which is fairly approximate in the case of vegetal substrates. The stoichiometry of the reaction is as follows:

$$2\ KOH + CO_2 \rightarrow K_2CO_3 + H_2O$$

Potassium carbonate is still capable of absorbing CO_2, becoming potassium bicarbonate according to the following reaction:

$$K_2CO_3 + CO_2 + H_2O \rightarrow 2KHCO_3$$

The two steps can be summarised in the overall reaction:

$$2\ KOH + 2CO_2 \rightarrow 2KHCO_3$$

So, 112 g of KOH will capture 44,800 Ncm^3 of CO_2, i.e., the theoretical absorption capacity of ash is

$$CO_{2\,\text{ash}} = m_{\text{ash}} \times 400\,\text{Ncm}^3 / \text{g}_{\text{ash}}$$

We can then estimate the maximum theoretical absolute error in the measurement of CO_2 with the AMPTS III Duo system as

$$E_{\text{th}} = 0.5\left(S_{(\text{Ti})} \times v_{\text{i}} + 859{,}7 \times 0{,}120\right) + m_{\text{ash}} \times 400\ \text{Ncm}^3 / \text{g}_{\text{ash}} \qquad (9.4)$$

Here,

$S_{(\text{Ti})}$ = CO_2 solubility at the incubation temperature, from Table 9.1;

v_{i} = volume of inoculum (usually 1,800 ml for 2 l reactors and 900 ml for 1 l reactors);

m_{ash} = ash mass of the sample = % ash × *WW* of the sample added to the reactor.

Knowing the E_{th} of both the sample and blank reactors, we can then calculate *E* as the difference.

The theoretical estimation of *E* has a logical flaw: it is based on two suppositions that we cannot prove true (most probably at least one is false):

a. There is no CO_2 already dissolved in the inoculum and water contained in the system, so these will absorb the gas to their maximum physical capacity.
b. All the ash in the substrate is KOH and will dissolve in the inoculum during digestion.

2. It is also possible to measure the total alkalinity of the inoculum before and after the test, both in the blank and sample reactors; calculate the difference of calcium carbonate equivalent and then the CO_2 captured; and finally add the resulting values of dissolved CO_2 to the correspondent biogas volumes. This alternative relies on a standard laboratory procedure [4]. It is easy to perform either manually or with an automatic titrator, but it is not accurate enough because calculating a value as the difference of two measurements – each

one with its uncertainty – implies that the overall relative error tends to infinity when the measured difference is of the same order of magnitude of the instrument's uncertainty.

3. A third alternative would be to measure E directly. To do this, we could employ a reference substrate like cellulose. We know by stoichiometry that cellulose will always produce biogas with 50% CH_4 and 50% CO_2. E is then

$$E = 2\left(V_{\text{ms}} - V_{\text{mbk}}\right) - \left(V_{\text{bs}} - B_{\text{bbk}}\right).$$

Since cellulose has little or no ash, we should eventually correct E by adding the fraction of CO_2 absorbed by the ash:

$$CO_{2\,\text{ash}} = m_{\text{ash}} \times 400\,\text{Ncm}^3 / \text{g}_{\text{ash}}$$

The drawback of this method is that it requires occupying at least one additional reactor and two cells for the reference substrate. Furthermore, it relies on a further supposition: that the quantity of absorbed CO_2 is the same in all reactors and measurement cells, which is not necessarily true.

9.2.1.1 Practical Example: Estimating the Range of e_{Xm} in the BMP Test with the AMPTS III Duo Setup

To assess the influence of the different error sources in the calculation of X_m from biogas and methane volumes measured with the AMPTS III Duo, the author performed a test with two similar substrates, both known to yield biogas with 50% methane and allowing thus to easily calculate the true value of E. The first substrate was cellulose (toilet paper) having a neglectable ash content and 91.1% VS. The second was sugarcane bagasse containing 6.6% ash and 86.4% VS. We know from the literature that the approximate composition of the VS of bagasse is 45% cellulose, 28% hemicellulose, 20% lignin, and 5% sucrose [5]. Since cellulose, hemicellulose, and sucrose are all saccharides, sugarcane bagasse will certainly yield biogas containing 50% methane. The high content of ash will allow us to analyse its influence on the total CO_2 absorption. Checking the error margin of the calculated X_m in both cases will allow us to find out if the application of a systematic error correction can be acceptable and eventually provide some guidelines to increase the reliability of the test.

9.2.2 Experimental Data

The VS concentration of the inoculum was 3.0%, and the inoculum quantity in all reactors was 1,800 g. It is generally recognised that solubility and diffusion of CO_2 influence the volumetric measurements of biogas by liquid displacement. For instance, a gas containing 60% CH_4 and 40% CO_2 generates an underestimation of about 14% at a flow rate close to 1 Nl/day when measured with AMPTS II [6]. To minimise this effect, the liquid in the FCU has to be flushed with a standard gas having a similar composition to the one that is going to be measured. Since producing a mixture of gases may be a problem for small laboratories, and flushing for long times to saturate the inoculum and the cells is time-consuming, the author deliberately avoided saturating the inoculum and the water in the flow measurement cells with CO_2 and chose ISR 3 for both bagasse and toilet paper. The scope was checking the efficacy of the correction method under suboptimal test conditions (dilute inoculum, ISR = 3 instead of 2, hence little gas production for a big volume of water). To eliminate air and create an anaerobic environment before starting the test, all headspaces were flushed with propane (see Chapter 4, Section 4.7). Under the said test conditions, the resulting quantity of bagasse was 20.8 g, corresponding to 18 g of VS and 1.37 g of ashes. Bagasse was tested in duplicate, while toilet paper and blank were tested in single reactors because of capacity reasons (the instrument was an AMPTS III Light having only 9 cells, so only four reactors were available in duo configuration, and the left reactor was employed in conventional configuration to check the inoculum's SMA). The control reactor was loaded with 19.8 g of toilet paper, corresponding to 18 g VS but without ashes. The quantities of gas measured with the Duo setup after 35 days (overestimation function activated for the methane measurement, deactivated for total biogas according to the manufacturer's indications) were the following:

V_{bbk} = biogas production of the blank =819.2 Ncm^3;
V_{mbk} = methane production of the blank =680.1 Ncm^3;
V_b = biogas production of the sample (average of two replicates) =7,931.3 Ncm^3;
V_m = methane production of the sample (average of two replicates) =4,656.3 Ncm^3;
V_{bc} = biogas production of cellulose =15,623.1 Ncm^3;
V_{mc} = methane production of cellulose =7,964.7 Ncm^3.

The concentration of CH_4 resulting from the uncorrected measurement of the sample was

$$X_m = \frac{V_m - V_{mbk}}{V_b - V_{bbk}} = \frac{4,656.3 - 680.1}{7,931.3 - 819.2} = 55.9\%.$$

By definition, $X_{m\ cellulose} = 50\ \%$; hence,

$$X_m = 0.5 = \frac{V_{mc} - V_{mbk}}{V_{bc} - V_{bbk} + E} = \frac{7,964.7 - 680.1}{15,623.1 - 819.2 + E},$$

$$E = 2\ (7,964.7 - 680.1) - 15,623.1 + 819.2 = -234.7\,Ncm^3.$$

Note that in this case, E is negative, meaning that the quantity of CO_2 absorbed in the blank was higher than the one absorbed in the control. By definition, E is always the absolute value of a difference between the actual biogas production (unknown) and the (averaged) measured values, so we must ignore the minus sign.

Cellulose had no ash while bagasse did, so it is reasonable to assume that the systematic error in the measurement of the biogas volume is the sum of E and the CO_2 captured by the bagasse's ash released during the digestion. The estimated quantity of the latter is

$$CO_{2\,\text{capt.}} = 0.066 \times 20.8\ g \times 200\,\text{Ncm}^3 / \text{g} = 274.6\,\text{Ncm}^3.$$

If we suppose that E has the same absolute value for both cellulose and bagasse and add $CO_{2\,\text{capt}}$ to the biogas volume produced by the bagasse sample, the corrected methane concentration of the bagasse is

$$X_{\text{m corrected}} = \frac{V_{\text{m}} - V_{\text{mbk}}}{E + CO_{2\,\text{ash}} + V_{\text{b}} - V_{\text{bbk}}} \times 100 = \frac{4{,}656.3 - 680.1}{234.7 + 274.6 + 7{,}931.3 - 819.2} \times 100,$$

$$X_{\text{m corrected}} = 52.2\ \%.$$

Since in this particular case we know in advance that bagasse must yield biogas with 50% CH_4, the absolute and relative errors of corrected X_{m} are

$$E_{\text{corrected Xm}} = 52.2\% - 50\% = 2.2\%,$$

$$e_{\text{Xm}} = \frac{2.2\%}{52.2\%} = 4.2\%.$$

Observe that the uncorrected measured X_{m} was 55.9%, which means an absolute error of 5.9% and a relative error of 10.5%.

Let's calculate the metrological error of the corrected value of X_{m} with Formula 9.3.

$$e_{\text{Xms}} = e_{\text{inst}} \left[\frac{(V_{\text{ms}} + V_{\text{mbk}})}{(V_{\text{ms}} - V_{\text{mbk}})} + \frac{(V_{\text{bs}} + V_{\text{bbk}}))}{(V_{\text{bs}} - V_{\text{bbk}} + E)} \right] + \frac{E}{(V_{\text{bs}} - V_{\text{bbk}} + E)}$$

$$e_{\text{Xm}} = 0.01 \left[\frac{(4{,}656.3 + 680.1)}{(4{,}656.3 - 680.1)} + \frac{(7{,}931.3 + 819.2)}{(7{,}931.3 - 819.2)} \right] + \frac{234.7}{(7{,}931.3 - 819.2 + 234.7)}$$

$$e_{\text{Xm}} = (0.026 + 0.032) \times 100 = 5.8\%$$

This means that the corrected X_{m} should be expressed with its uncertainty margin as follows:

$$X_{\text{m}} = 52.2\% \pm (5.8\% \times 52.2\%) = 52\ \% \pm 3\%\,.$$

So, any value in the range of 49.2% and 55.2% is within the range of accuracy of this method.

9.3 CONCLUSIONS

From this practical example, we can deduce what follows:

1. Introducing a correction of the underestimated biogas volumes actually increases the accuracy of the calculated X_m, reducing the relative error margin. In this example, the relative error decreased from 10.5% to 4.2%; i.e., the correction produced a 6.3% improvement in accuracy.
2. The real value of X_m (50%) falls within the calculated range of uncertainty (49.2%–55.2%).
3. The uncorrected value of X_m (55.9%) is almost coincident with the upper limit of the uncertainty range.
4. Applying a correction factor to account for the CO_2 dissolved in the liquid and eventually captured by the ash in the substrate cannot replace a correct test protocol. The best option to minimise the uncertainty derived from the fraction of biogas captured in the liquid is to saturate the inoculum and water in the measurement cell with CO_2 before starting the test. A small overestimation is still possible for substrates containing relevant percentages of ash, but it is easy to correct this systematic error.

9.3.1 Guidelines for the Correct Application of the BMP Protocol with the AMPTS III or GE III Duo Configuration

Obtaining reliable values of X_m when employing the Duo setup requires bearing in mind some conditions:

a. Results are only estimations and do not account for trace gases. An accurate determination of the biogas composition requires using a gas chromatographer (GC) as mandatory. In any case, the GC is unable to determine how much CO_2 eventually remained dissolved in the liquid phase. Using an expensive and highly accurate instrument like a GC will not ensure reliable results if the total quantity of CO_2 produced is in the same order as the absorption capacity of the inoculum.
b. Regardless of using the Duo configuration or the GC, minimising the fraction of CO_2 dissolved in the inoculum is mandatory. Alternatively, calculating a correction factor as explained in Section 9.2.2 is also possible but less accurate.
c. The average gas composition results from the difference between two independent measurements: the accumulated volumes of biogas and methane. Calculating the uncertainty margins of the CO_2 and CH_4 percentages, as explained in Section 9.2.2, is mandatory because it provides an index of the test's overall quality and reliability.
d. As in the case of the BMP test, an adequate preincubation of the inoculum is mandatory to minimise the uncertainty of the methane concentration measurement because the error tends to become just twice the instrumental error when the gas production of the blank tends to 0.

e. The gas composition in the headspace of the reactor at the beginning of the test has a high impact on the final result. Therefore, the 'eliminate overestimation' function of the AMPTS III/GE III should be deactivated in the first unit (measuring the total biogas production) and activated in the second unit (measuring the net methane production).
f. Flushing the headspace of the reactors with a gas mixture having a similar composition to the gas that is going to be produced (e.g., 60% N_2 and 40% CO_2) will help reduce the uncertainty. The said practice is widely diffused in the literature and recommended by VDI4630, IWA BMP Protocol 2016, and UNI/TS 11703. But, in the author's opinion, it does not guarantee the highest accuracy because the quantity of CO_2 employed for flushing the headspace may not be enough to saturate the medium, so some of the CO_2 produced by the digestion may still remain dissolved in the inoculum anyway.
g. The quality of the inoculum has a high influence too. A very diluted inoculum will admit very little substrate; hence, the overall gas production will be low. Maximising gas production, and hence minimising the measurement error, requires keeping the reactor headspace as small as possible and the ISR low enough by adding as much substrate as possible. A high gas production will ensure a high flow rate. The amount of CO_2 dissolved in the inoculum, and to a lesser extent in the water of the flow measurement cell, is directly proportional to the contact time between water and gas. The higher the gas production and the smaller the headspace, the shorter the contact time – hence, the lower the underestimation error because less CO_2 will dissolve.
h. Experience shows that for the gas composition calculation to be accurate, the gas flow rate should fall in the range of 25–300 Ncm^3/h.
i. Conceptually, the gas composition measured with the dual-cell method under thermophilic conditions will be more accurate because the water solubility decreases as temperature increases. The solubility of CO_2 at 38°C is 0.1026 g/100 ml water, while at 50°C, it is 0.0761 g/100 ml water, i.e., 25.8% less.

There is no simple way to simultaneously measure the gas flow and composition using batch tests. Shortcuts and simplifications will reduce the accuracy, especially when testing low-degradability materials. The following general recommendations [7] will help minimise the underestimation error caused by the gas mixture's solubility:

1. For creating anaerobic conditions in the bioreactors at the beginning of the test, several gases can be used: nitrogen or a standard gas having a similar composition to the one that is going to be produced, e.g., 60% CH_4 and 40% CO_2 or 60% N_2 and 40% CO_2.
2. When a gas mixture containing CO_2 is used for flushing the reactors and CO_2 is removed before detection, it is very important to apply a *constant time* for flushing for all the reactors.
3. When one gas component (e.g., CO_2) is removed before detection and when the gas composition at the beginning and the end of the test is different, then the 'eliminate overestimation' in AMPTS II/GE II or 'correction' function

in AMPTS III/GE III should be activated. In this case, it is mandatory to measure the final gas composition in the headspace for the function to produce accurate results.

4. The headspace (abbreviated HS in the instrument's software) volume of the reactor includes the space until reaching the CO_2 absorption unit. For example, it is 200 ml for 500 ml reactors with 100 ml absorption units and about 600 ml for 2 l reactors with 250 ml absorption units.
5. For DUO units where the first unit is measuring the total volume of biogas and the second one is measuring the total volume of biomethane after removing CO_2 by alkaline solution, the correction function should be activated only for the second measuring unit. The HS volume for the second unit also includes the HS of the first FCU in the line.

9.3.2 A Possible Protocol for Minimising the CO_2 Underestimation Error

Note that the guidelines above refer to the instrument's setup. However, the reactors that are full of inoculum will capture part of the CO_2 produced during the fermentation, because of both physical dissolution in water and chemical reactions with the inoculum's alkalinity and substrate's ashes. On top of the general guidelines listed above, flushing with CO_2 before launching the test requires some special attention. This preliminary operation provides the operation conditions most similar to the ideal case exposed in Section 9.2.1. The idea is to convert all carbonates present in the inoculum into bicarbonates. The recommended procedure is as follows:

a. Prepare the blank and the sample reactors, adopting the minimum ISR compatible with the feedstock you intend to test (ideally 2 but 3 may still be acceptable), close them, and start the stirrers at 80%.
b. Connect each reactor directly to the input tube of its corresponding biogas flow measurement cell. Check that the flow measurement cell's output is not connected to the CO_2 trap.
c. Connect the flush valve of each reactor to a manifold and this one to a CO_2 bottle having an adequate pressure reducer.
d. Check that all the flush valves are open and then open the valve of the CO_2 bottle. Flush for at least 20 minutes to ensure saturation of both the inoculum and water in the FCU. Bubbling CO_2 through a short piece of flexible tube inserted in the inner connector of the stirrer cap (Figure 9.2, reactor filled with water for more clarity) will improve the absorption in the inoculum.
e. Close the main CO_2 valve, close the flush valves of the single reactors, and disconnect them from the manifold.
f. Connect the output of the biogas flow measurement cells to their corresponding CO_2 traps and the output of these latter to their corresponding methane flow measurement cells.
g. Go to the 'control' menu of your instrument and start the test as usual.

The procedure described above may still yield a slightly underestimated V_b and hence a slightly overestimated X_m because it does not account for the mineral fraction

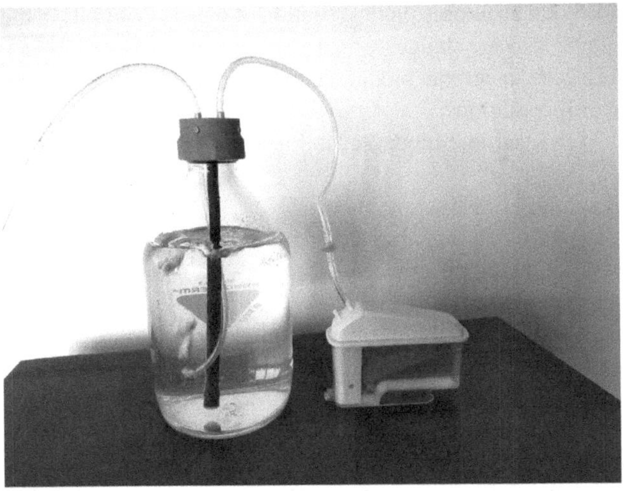

FIGURE 9.2 Bubbling CO_2 in the Reactor and First FCU Will Saturate the Water in Them and Reduce the Underestimation Error.

of the substrate (ashes) that will add alkalinity to the inoculum during the digestion. Unless the substrate has a high ash content (reeds, rice straw, and similar biomass), the said effect is neglectable. It is still possible to estimate the volume of CO_2 absorbed by the substrate's ashes and add it to the total biogas volume as a systematic error.

9.4 ANAEROBIC BIODEGRADABILITY OF PLASTICS AND OTHER POLLUTANTS

Plastic litter polluting seas, soil, and rivers is becoming a general concern of modern society. Many governments have banned or strongly limited the sales of single-use plastic items in favour of biodegradable products. But 'biodegradable' is a meaningless word unless it is associated with a benchmark and a context. Measuring the biodegradability of a material in an anaerobic environment requires defining what biodegradability is and what kind of anaerobic environment it is referred to. At the moment of writing this chapter (June 2024), the applicable standards are

- ISO 14853: ultimate anaerobic biodegradation of plastic materials in an aqueous system;
- ISO 13975: ultimate anaerobic biodegradation of plastic materials in controlled slurry digestion systems;
- ISO 15985: ultimate anaerobic biodegradation and disintegration under high-solid anaerobic digestion conditions;
- ISO 11734: ultimate anaerobic biodegradability of organic compounds in digested sludge;
- ASTM D5511: anaerobic biodegradation of plastic materials under high-solid anaerobic digestion;

- ASTM D5210: anaerobic biodegradation of plastic materials in the presence of municipal sewage sludge
- ASTM D5526: anaerobic biodegradation of plastic materials under accelerated landfill conditions;
- OECD 311: anaerobic biodegradability of organic compounds in digested sludge.

The general layout of anaerobic biodegradability assays is similar to that of a BMP test but with the following differences:

- the source and preparation of the inoculum;
- the minimum volume of inoculum required and the quantity of substrate;
- the temperature;
- the focus of the measurement is the total gas production, regardless of its composition;
- the higher sensitivity of the instrument required for testing highly refractory or very diluted substrates;
- the presentation of the results, usually a curve of consumed carbon (g consumed / g sample. day or % C/day), which requires measuring (or calculating) the total organic carbon (TOC).

From a practical point of view, operators skilled in performing BMP tests should not find any difficulties in measuring the anaerobic biodegradability of materials. Suitable instruments are GE II, GE III, BPC Blue (specific model for measuring biodegradability), and AMPTS III because they all offer the possibility of changing the sensibility from the standard 9 ml to 2 ml when necessary. AMPTS II may be employed, but in some cases, its sensibility may not be enough.

9.5 RUMINAL SIMULATION

In vitro ruminal assays are a special case of respirometric anaerobic tests in which the inoculum is ruminal fluid. Ruminal fluid may be extracted from living cows or sheep or sourced directly from a slaughterhouse. In any case, it must be transported to the laboratory in a thermal container flushed with CO_2 or N_2, to preserve the living anaerobic bacteria from contact with O_2 and from cooling down. Ruminal fluid is then filtered and mixed with standard solutions that mimic the buffer effect of saliva. Another characteristic of *in vitro* ruminal tests is their duration, usually 24 hours and rarely 48 hours. The high dilution of ruminal fluid implies that the quantity of substrate added to the reactor must be small to avoid inhibition effects; hence, the gas production will be small, usually in the 100–200 ml range. It is then preferable to perform these tests using 2 ml FCUs to obtain more detailed digestion curves.

Gas production (GP) has become a routine test in feed evaluation since the early 1980s because of the high correlation between metabolisable energy (ME) measured in live animals and that predicted from *in vitro* GP. The said technique has several advantages over other *in vitro* methods (e.g., measuring residues after enzymatic treatment):

1. GP reflects all nutrients fermented, both soluble and insoluble. Fractions that are not fermentable do not contribute to gas production.
2. Furthermore, the kinetics of fermentation can be obtained from a single incubation, allowing the rate of fermentation to be calculated.
3. GP measurement is a direct indicator of microbial activity and can be a better index of forage ME content than an indirect *in vitro* measurement based on fermented nutrients.

The *in vitro* GP method is effective for a wide range of feed evaluations:

1. **Organic matter digestibility**: The digestibility of organic matter is closely correlated with that predicted by GP and the crude protein and ash contents of feeds. Therefore, Gas Endeavour can be used to predict the extent of digestion for various feeds.
2. **Energy contents of feeds**: The *in vitro* GP method can predict the ME content of feeds with a high correlation with results obtained *in vivo*. The gas measurement provides a better estimate of the ME level of feeds, when combined with some chemical constituents, compared with calculations based on chemical constituents only. Gas Endeavour's uncertainty margin is ±1%, which means a higher accuracy of the ME calculated value than using any other method. The instrument can detect gas volumes as low as 2 ml.
3. **Kinetics of fermentation**: The ruminal digestion rate is as important as the extent of digestion when assessing the nutritive values of fodder. The kinetics of fermentation is one of several factors that influence voluntary feed intake by ruminants. Gas Endeavour provides kinetic information as Excel tables, both of instantaneous gas flow and of total cumulative volume. Selectable time scales are 5 minutes, 15 minutes, and 60 minutes.
4. **Effects of added fat on feed degradation**: Fats are common supplements used in the diets of lactating dairy cows. Gas Endeavour's high accuracy and repeatability ensure that any difference between the fat-added sample and the control sample is exclusively caused by the effect of fat addition, and not by instrumental uncertainties (e.g., variations in the room temperature and pressure when using syringes, calibration offsets, weak batteries, and thermal drift when using pressure sensors).
5. **Anti-nutritive factors**: Some feeds, such as forage legumes and cottonseed, contain phenols, alkaloids, and saponins that have negative biological effects on microbes and reduce microbial growth in the rumen. Gas Endeavour's high repeatability, low detection limit, and independence from external factors like room temperature and pressure ensure the measurement of anti-nutritive levels that syringes and barometric instruments are not able to detect.

6. **Ruminal fluid modifiers**: Monasin, yeast, and yeast fermentation by-products are added often to the diet to modify the bacterial populations in the rumen. Gas Endeavour makes it easy to study their effects, by simply incubating feeds in the presence or absence of such compounds. The data from such tests are crucial for optimising milk production costs and overall productivity in commercial dairy farms.
7. **Feed associative effects**: Rations are mixtures of individual feeds, with a multitude of possible combinations. Because of positive associative effects on *in vitro* GP, it is not correct to sum the individual ME of the ration's components. Gas Endeavour allows performing triplicate tests on five different rations at a time.
8. **Monitoring rumen microbial change**: In addition to rates and extents of digestion, Gas Endeavour can be used to study substrate-related factors that influence microbial populations in the rumen. This enables manipulation of rumen microflora to increase the utilisation of feeds through the degradation of fibre and lignin, increasing the efficiency of nitrogen utilisation or allowing the degradation of anti-nutritional and toxic components of feeds.
9. **Nutrient synchronisation**: Carbohydrate and nitrogen sources must be available simultaneously to maximise microbial growth. The degradation rates of carbohydrates and nitrogen-containing compounds can influence ruminal ammonia concentrations. Gas Endeavour offers an opportunity to easily study microbial requirements for nitrogen and carbohydrates to enable efficient fermentative activity and accumulation in the ruminal fluid.
10. **Plant breeding, biotechnology**: Several forage and cereal crops have been genetically modified to increase yield or produce chemical constituents normally deficient in a particular plant. Forage plants are selected for rapid fibre digestibility. Plants have also been genetically engineered to produce human lysozyme, but it is unclear what effect lysozyme has on the rumen's microbiome. Although many genetically engineered plants are intended for human consumption, their by-products will be fed to animals as a means of disposal. Gas Endeavour III, having 18 reactors and measurement channels, allows the evaluation of large sample quantities in a short time, to select those with the highest feeding values.
11. **Environmental degradation**: More than half of the nutrients consumed by ruminants leave the animal unutilised and undigested and are excreted in faeces, urine, and gases. This increases animal production costs and environmental impacts as well, by contaminating surface- and groundwater and contributing to air pollution. Gas Endeavour allows the quick and easy selection of the rations featuring the highest efficiency of feed utilisation, reducing the amount of unutilised nutrients leaving the animal and hence lowering feed costs and environmental burden.

Although *in vitro* GP is an effective method for analysing ruminant feed, most of the literature was written on tests performed with traditional measurement systems, like syringes and manometers, which yield highly variable results. Furthermore, most – if not all – of the literature on GP contains non-normalised data, making it difficult to compare results from different authors. Measures performed with Gas Endeavour II and III are far more precise and accurate than those obtained with other instruments. Despite higher precision and accuracy, some users point out that the values measured with the said instruments 'are underestimated' when compared to 'literature averages'. The following example shows that there is no underestimation at all but just a difference in conventions to express gas volumes and a bit of cultural prejudice too.

Despite the existence of specific norms [8] on assessing the uncertainty margin of laboratory measures, a constant in all the literature about ruminal *in vitro* tests is the lack of such kind of analysis. Most of the papers on gas production from ruminal fluids provide tables reporting the individual values of several measures and their dispersion. On the one side, such values represent the 'raw' gas production, which is unrealistic because it includes moisture. From a metrological point of view, checking only the standard deviation of a set of values is not fully correct, since it is a measure of the precision (i.e., repeatability and reproducibility) of the test but not of its accuracy. Accuracy is especially important when the measured values must be employed for further elaboration, e.g., the energy value of the feedstock as a function of the gas production, since calculations amplify the error of the final result, according to rules derived from the error propagation theorems. Furthermore, it is common knowledge that gasses vary their volume with temperature and pressure. While the reference temperature is always 39°C in all ruminal fermentation tests, the atmospheric pressure is highly variable and not controllable. Let's see how to elaborate correctly 'raw' data from the literature, to compare them with normalized data obtained with Gas Endeavour II and III. As an example of literature data, we took the average of ten measures of gas production from corn silage, 228.4 ml gas/g DM, having a standard deviation equal to ± 20 ml. The said value was published in [9]. The paper states that the tests were performed according to Menke and Steingass (1988), using 100 ml glass syringes filled with 30 ml of ruminal fluid and 200 mg DM of sample. The paper does not state what the atmospheric pressure was when the test was performed. We will assume $P_{atm} = 101.3$ kPa. The following procedure allows us to calculate the error or uncertainty of the said measurements.

a. Uncertainty of measurement induced by the gas moisture at 39°C (312K).

From the thermodynamic tables of water vapour,

$$\text{P sat.vap.@310K} = 6.230 \text{ kPa}$$

$$\text{P sat.vap.@315K} = 8.143 \text{ kPa}$$

By linear interpolation to 312K:

$$\Delta P = (8.143 - 6.230)/5 \times 2 = 0.765 \text{ kPa}$$

$$P \text{ sat.vap@312K} = 6.230 + 0.765 = 6.995 \text{ kPa}$$

The percentage of the gas volume occupied by water vapour, $G\%$, is equal to the quotient of the partial pressures:

$$G\,\% = 6.995\text{ kPa}/101.3\text{ kPa} \times 100 = 6.9\,\% \approx 7\%.$$

N.B.: According to the ISO guidelines, the error or uncertainty of a measure must be expressed by rounding to the first significant digit. Hence, 7% of the volume measured by the syringe (or by any other method) does not come from the substrate's fermentation but is just water vapour. Note that this is a systematic error; i.e., we know not only its magnitude but also its sign, so it is easy to correct as follows:

$$V_{\text{dry gas}} = 228\text{ ml/g DM} \times 0.93 = 212\text{ ml/g DM}$$

b. Intrinsic measure error of the syringe

Graduated laboratory glassware can be labelled as class A (error = 0.5%) or class B (error = 1%), according to ISO 4788 (equivalent to DIN 12600). The paper does not state what class of syringe was employed, so we will assume the best case is class A. Since the syringes have 100 ml, their absolute error margin is ± 0.5 ml. Since they were filled with 30 ml of ruminal fluid and 200 mg feedstock, we can calculate that the piston's displacement was of the order of 50 ml. By definition, the relative error of such a measure is

$$e_{\text{gas}} = \frac{\pm 0.5\text{ ml}}{50\text{ ml}} \times 100 = 1\%.$$

Remark: Letting apart the systematic error induced by the gas moisture, from a formal point of view, the authors of the paper in question should have stated the results according to ISO GUIDE 98–3:2008 as follows:

$$V_{\text{gas}} = 228 \pm 2\text{ ml/g DM}.$$

From a practical point of view, the syringe is not able to distinguish between a test yielding 226 ml/g DM and a test yielding 230 ml/g DM. Hence, expressing the average as 228.4 ml/g DM has no physical sense because the error margin is bigger than the decimal order of magnitude. Such misconception is a constant in all the literature. Furthermore, all the authors assume that the error source is the dispersion among replicates, which is true in most cases but should be checked case by case.

c. Normalising to 0°C (273.16 K)

According to ISO 10780, gas volumes must always be expressed in 'normal conditions', i.e., at 273.16 K and 101.3 kPa. Note that the reference temperature stated in the paper, 39°C, is also a measured value subject to uncertainty because the Hohenheim incubator has a temperature tolerance equal to ±0.5°C [10]. Furthermore, it is necessary to take the syringes out of the incubator to read the piston's displacement, which unavoidably will vary their temperature. Assuming the best case, i.e., just the tolerance of the incubator's thermostat, the relative error of the temperature determination is

$$e_T = \left(\pm\frac{0.5\ ^\circ\mathrm{C}}{39^\circ\mathrm{C}}\right)\times 100 = \pm 1.28\% = \pm 1\%.$$

d. Error induced by the uncertainty of the atmospheric pressure

The paper chosen for this example does not state if the ten tests were performed simultaneously or on different days, nor is the atmospheric pressure stated. The atmospheric pressure in a given locality can vary as much as 3%. In the case under study, data from the San Francisco airport weather station show that in 2007 the maximum pressure was 101.93 kPa and the minimum was 98.91 kPa. This represents an asymmetric error margin on the reference pressure, stated as

$$e_{P1} = \left(101.93/101.32 - 1\right)\times 100 = +0.6\ \% = +1\ \%,$$

$$e_{P2} = \left(98.91/101.32 - 1\right)\times 100 = -2.35\ \% = -2\ \%\cdot$$

Summarising: The average 'raw' gas production of corn silage reported in our example, 228 ml/g DM, corresponds to a 'net and normalised' gas production comprised in the following range:

$$G_{P\max} = \left(228\ \mathrm{ml/g\ DM} - 7\% + 1\%\right)\left(273{,}16\ \mathrm{K}\right)/\left(313\ \mathrm{K}\right) + 1\% + 1\% = 191\ \mathrm{Nml/g\ DM},$$

$$G_{P\min} = \left(228\ \mathrm{ml/g\ DM} - 7\% - 1\%\right)\left(273{,}16\ \mathrm{K}\right)/\left(313\ \mathrm{K}\right) - 1\% - 2\ \% = 178\ \mathrm{Nml/g\ DM},$$

Ruminal gas production tests, performed with a prototype of the Gas Endeavour instrument on different samples of corn silage from various sources at the Laboratory of Animal Nutrition of the University of Parma (Italy) by Prof. Afro Quarantelli and his team, yielded values between 145 and 192 Nml/g DM, depending on the origin of the sample. The instrumental error of every single measure is just 1% because the instrument can normalise in real time. The representativeness of the sample is higher when employing Gas Endeavour because the 500-ml reactors require adding 6 g DM of feedstock, against just 0.2 g in the standard syringe method. The representativeness of the sample can induce large variability of the results, either arising from the sampling procedure or the substrate's preparation. It is probably the main cause of the standard deviation of the tests reported in the literature. In the paper taken as an example, the standard deviation of the ten individual measures was 22 ml/g DM, while in the tests at the University of Parma, some sets of values had standard deviations as low as 4 Nml/g DM [11].

The former example shows that measures obtained with Gas Endeavour II and with 'traditional' instruments are fully comparable, but it is necessary to refer all values to normalised conditions, 0°C or 39°C and 101.3 kPa. Furthermore, reviewers should carry out correctly the analysis of the instrumental uncertainties and not just check the dispersion among replicates. A test in which the calculated error is of the same order of magnitude as the coefficient of variation of single replicates is at the same time accurate and precise. Most traditional methods are acceptably precise but not fully accurate. Comparing results from different publications is time-consuming because the test conditions (water vapour content and atmospheric pressure) are seldom stated, and the analysis of the uncertainty of the measures is always missing. On the other hand, results obtained with Gas Endeavour II and III in different times and places can become directly comparable with only a 1% uncertainty margin: just clicking on the 'normalisation' check box, choosing 'wet' (including water vapour) or 'dry' (discounting water vapour), and then downloading a table of automatically normalised values. If you want to compare the normalised results to 'literature' values, referred to as 39°C instead of 0°C, then it is necessary to multiply the normalised gas volumes by 1.14.

The applications of the GE III Animal Nutrition instrument are as follows:

- Continuous monitoring of gas production for the extraction of process kinetic information;
- Determination of feed digestibility and its metabolisable energy content;
- Optimisation of feed composition and nutrient content for livestock;
- Investigation of feed additives or supplements to optimise ruminal fermentation;
- Investigation of functional components that may have beneficial effects on methane emissions and/or organic matter digestibility *in vivo*;
- Screening a large range of feeds or additives before testing these *in vivo*;
- Comparing different pretreatments of feed compounds.

9.6 COMPOSTING AND AEROBIC BIODEGRADABILITY OF PLASTICS

9.6.1 Overview of Aerobic Tests

When it comes to aerobic degradation processes, any organic compound can be degraded to CO_2 with the consumption of oxygen. Some methods measure the degradation rate and biodegradability as a function of the produced CO_2, but such a criterion is not reliable. The ratio of the number of CO_2 molecules formed per number of O_2 molecules consumed depends on the oxidation state of the organic substrate. The following examples clarify this concept.

a. Organic molecules with an oxidation state of *C* equal to zero: for each O_2 molecule that is consumed, a CO_2 molecule is formed. For example, the degradation of glucose will consume 6 moles of O_2, yielding 6 moles of CO_2, i.e., the molar ratio is 1:1:

$$C_6H_{12}O_6 + 6O_2 \rightarrow 6\ CO_2 + 6\ H_2O$$

b. Highly *oxidised* organic compounds: In this case, 2 moles of CO_2 are formed for each mole of oxygen consumed. For example, the oxidation of formic acid will consume 1 mole of O_2 and yield two moles of CO_2, i.e., the molar ratio is 1:2:

$$2\,HCOOH + O_2 \rightarrow 2\,CO_2 + 2\,H_2O$$

c. Heavily *reduced* organic compounds: In this case, CO_2 formation is lower for the same oxygen consumed. For example, the oxidation of methane will consume 2 moles of O_2 and yield 1 mole of CO_2, i.e., the molar ratio is 2:1:

$$CH_4 + 2\,O_2 \rightarrow CO_2 + 2\,H_2O$$

Since plastics and organic waste are mixtures of such three molecule species in unknown proportions, measuring the aerobic degradability as a function of the CO_2 produced by microbial respiration is not accurate. This is the reason why most protocols are designed to measure oxygen consumption, like in the classical biological oxygen demand (BOD) test.

For most respiration activity tests (e.g., soil respiration), a parameter named *respiratory coefficient* is calculated using two parameters: volume of O_2 consumed and volume of CO_2 produced. For measuring the CO_2 formed, it is necessary to transfer a given quantity of the alkaline solution to an external container, measure the volume of the solution and titrate it with a strong acid (e.g., HCl 1 N). Using a volumetric displacement instrument like the Gas Endeavour to measure O_2 consumption, instead of CO_2 evolution to determine the biodegradation rate of organic materials, has a great advantage over conventional methods based on infrared sensors: there is no need for a large sample to generate enough CO_2 for accurate quantification. The operator can maintain the same ratio of inoculum to test material as recommended by the protocol, but the sample should occupy a small volume of the reactor. This allows for a larger headspace, which is crucial for proper mass transfer during the tests. The small sample volume also ensures good oxygen distribution throughout the sample, directly improving the biodegradation process. In other words, the test can be performed without agitation, which is a big advantage over systems based on CO_2 evolution requiring weekly agitation based on the sample size. Furthermore, opening the cabinet at 58°C to take the gas sample causes a significant temperature drop, and since the microorganisms used in the test are very sensitive to temperature fluctuations, the said procedure harms the test.

9.6.2 The Biodegradability of Plastics

Plastics are an unavoidable constituent of urban waste. They end up in the composting facilities where they may decompose during the process if they are 'compostable' or remain more or less unaltered and, if not properly separated, finish in the compost and the soil (much literature is available on this subject, see for instance 12). Some plastics sold as 'biodegradable' are not such: they will just break into small particles, called microplastics, that can alter the soil's complex biological equilibrium and enter into the trophic chains, eventually reaching humans. It is then necessary to measure how much 'biodegradable' such products are under aerobic conditions, either in a solid matrix (soil) or in a liquid environment (shallow waters). Since all plastics are carbon-based molecules, their aerobic degradation will consume O_2 and yield CO_2. Since the chemical formula of the plastic product is usually known, it is possible to calculate the theoretical O_2 demand and analyse the dynamics of the process, expressing the degradation as a percentage of the initial C content or as a percentage of the theoretical O_2 demand. Test protocols are available based on the measurement of oxygen demand in a closed respirometer, which is the recommended method for BPC® Blue with a simplified system configuration, providing higher measurement precision and accuracy. Measuring the quantity of O_2 consumed during microbial respiration provides an accurate indication of the quantity of C effectively degraded. At the moment of writing this book (2024), the applicable standards for such respirometric assays were as follows:

- ISO 14851: ultimate aerobic biodegradation of plastic materials in an aqueous system;
- ISO 17556: ultimate aerobic biodegradation of plastic materials in soil;
- ISO 18830: aerobic biodegradation of non-floating plastic materials in a seawater–sandy sediment interface;
- ISO 23977–2: aerobic biodegradation of plastic materials exposed to seawater (part 2);
- OECD 301: ready biodegradability of chemical materials in an aerobic aqueous medium (public access downloadable PDF in 13);
- ISO 14852: ultimate aerobic biodegradability of plastic materials in an aqueous medium;
- ISO 22404: aerobic biodegradation of non-floating materials exposed to marine sediment;
- ISO 19679: aerobic biodegradation of non-floating plastic materials in a seawater–sediment interface;
- ISO 23977–1: aerobic biodegradation of plastic materials exposed to seawater (part 1).

The operating principle of volumetric instruments for aerobic processes is the specular image of the anaerobic assay: instead of measuring the gas produced by the fermentation and leaving the reactor, in the aerobic process, the instrument measures the quantity of O_2 entering the reactor. Figure 9.3 shows the experimental setup. The bottle is loaded with the inoculum, e.g., a sample of soil, and the sample we want to analyse. The headspace contains air, which is a mixture of 78% N_2 and 22% O_2. The microbes will consume O_2 and produce CO_2. An *in situ* CO_2 absorption cell filled with NaOH solution and attached to the bottleneck (called gas absorption attachment, GAA) will then capture the CO_2, causing a drop in the partial pressure inside the bottle. The bottle is connected to the exit of a flow measurement cell fed from a bag containing pure O_2. When the pressure difference between the bottle and the free surface of the water in the measurement cell is higher than that of the water head, O_2 will flow through the cell. Note that the cell's volume is either 2 or 9 ml, which is very small compared to the headspace volume of the bottle, which is usually larger than

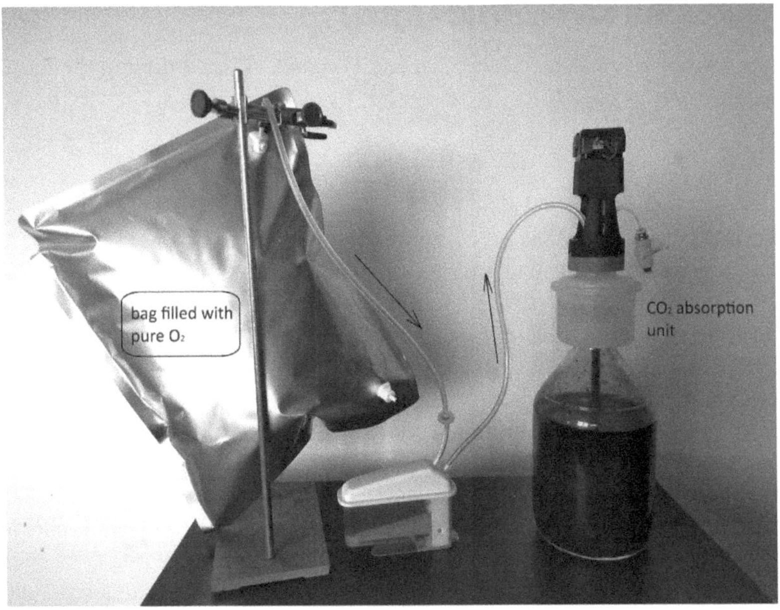

FIGURE 9.3 Experimental Setup for Aerobic Tests in the Liquid Phase. Bacteria consume O_2 in the reactor and produce CO_2, immediately captured in the NaOH solution contained in the GAA on the reactor. The pressure in the reactor's headspace drops, causing O_2 in the gas bag to pass through the FCU.

700 ml. This means that the concentration of O_2 inside the bottle remains constant, and equal to the atmosphere's, for the duration of the test. Since the respiration rate of the microbes is proportional to the partial pressure of O_2, i.e., to its concentration, the results obtained with this method represent the composting process much better than those obtained with barometric methods, in which the partial pressure of O_2 tends to zero (anaerobiosis!) unless air is periodically admitted to the reactor's headspace.

The setup shown in Figure 9.3 is that of the BPC Blue unit in use in the author's laboratory. BPC Blue is an instrument specifically designed for aerobic and anaerobic biodegradability tests. It allows the expression of the test results either as O_2 consumed or as a percentage of C degraded. Gas Endeavour II and III are suitable instruments for aerobic tests, too, but the data they produce are the flow and volume of consumed O_2. This means that the user must calculate the percentage of C degraded using an external spreadsheet or directly in the report generated by the instrument. It is theoretically possible to employ AMPTS III for aerobic tests. However, since its software is specifically conceived for BMP tests, the user must do all calculations manually with an external spreadsheet and the probability of human errors grows, so the author advises against it. AMPTS II is unsuitable for aerobic tests because it has open cells.

9.7 HYDROGENOTROPHIC ACTIVITY

Hydrogenotrophic archaea convert CO_2 and H_2 into CH_4 according to the following overall reaction:

$$CO_2 + 4H_2 \rightarrow CH_4 + 2H_2O$$

Despite the inefficiency of the process (already described in Chapter 7, Section 7.6.3), there is a growing interest in this technique from researchers and industries in Europe. The reason is that the EU Green Deal subsidises the production of 'clean hydrogen' to overcome the unbalance between electric demand and the production of wind, photovoltaic, and nuclear energy. Such a solution could perhaps make sense in Germany, Denmark, or Spain during a few hours of the year when there is overproduction of wind power but is not necessarily convenient elsewhere. Whatever the political reasons, right or wrong as they may be, the practical outcome is that measuring the hydrogenotrophic activity requires the development of a specific protocol. Till date, no standard exists on how to measure the specific hydrogenotrophic activity (SHA), but the procedure described here could be a departure point. The general idea is the same as the BMP assay, replacing the solid substrate with a gaseous substrate consisting of a 1:4 v/v mixture of CO_2 and H_2. The experimental setup is shown in Figure 9.4, which is basically the same one of the aerobic tests but without the CO_2 absorption unit, and the gas bag filled with the stoichiometric $CO_2 + H_2$ mixture instead of pure O_2. The headspace should be flushed with the same $CO_2 + H_2$ mixture to eliminate air before starting the test. The headspace should be two to three times

FIGURE 9.4 Measuring the Hydrogenotrophic Activity Requires the Same Experimental Setup of Aerobic Tests. The only difference is that the gas bag must contain a 5:1 (v/v) mixture of H_2 and CO_2.

the volume of inoculum to allow measuring as much gas mixture as possible before it is completely replaced with CH_4.

Operating principle: hydrogenotrophic archea will consume five volumes of the $CO_2 + H_2$ mixture per volume of CH_4 produced. This will create a difference in pressure between the headspace and the gas bag, causing the gas mixture to flow through the flow measurement cell. The process will continue until all the headspace is filled with CH_4. To keep coherence with the SMA test (see Chapter 5), it has been proposed that the hydrogenotrophic activity is expressed as the quotient between the maximum daily (or hourly) flow of the gas mixture and the total VS of the inoculum. In the author's opinion, this definition is incorrect because it does not take into account a peculiarity of the hydrogenotrophic process: its speed depends on both the activity of the inoculum and the gas–liquid transfer ratio, which in turn is a function of the partial pressures of the gases, the temperature of the aqueous medium, and the exchange coefficient of the gas–liquid interface, i.e., it depends on the test conditions. Hence, even if a norm may define the quantity of inoculum, the temperature, and the average test pressure, the test results will be different depending on the shape and dimensions of the reactor and probably with the stirring rate too. In any case, an engineer needing to design a hypothetic hydrogenotrophic reactor – no matter which type – would find the test results more useful if the SHA was defined including the gas exchange area.

9.7.1 Proposal of a New Definition of SHA

Including the gas exchange area in the formula of the SHA would allow the direct comparison of test results, regardless of the size and shape of the reactor employed.

$$SHA = \frac{4Q_{max}}{\pi \times d_i^2 \times VS_{inoc}},$$

where

SHA = specific hydrogenotrophic activity [(Ncm³/h of $CO_2 + H_2$ mixture) / (g_{inoc}· cm² $_{exchange\ area}$)];

Q_{max} = maximum hourly flow of the $CO_2 + H_2$ mixture [Ncm³/h];

VS_{inoc} = volatile solids of the inoculum [g];

d_i = internal diameter of the reactor, assumed to be a bottle standing in the usual vertical position [cm].

Since the flow of CH_4 produced is 1/5 of the flow of $H_2 + CO_2$ mixture consumed, it is also possible to express the SHA as

$$SHA = \frac{4Q_{max}}{5\pi \times d_i^2 \times VS_{inoc}},$$

where

SHA = specific hydrogenotrophic activity [(Ncm³/h of CH_4 produced) / (g_{inoc}· cm² $_{exchange\ area}$)];

Q_{max} = maximum hourly flow of the $CO_2 + H_2$ mixture [Ncm³/h];

VS_{inoc} = volatile solids of the inoculum [g];
d_i = internal diameter of the reactor, assumed to be a bottle standing in the usual vertical position [cm].

The incubation temperature must be equal to the intended operational temperature of the plant under design, and the atmospheric pressure at the moment of performing the test must be recorded in the test report. It is possible to read the latter value directly on the core unit's display (only in GE III, BPC Blue, or AMPTS III). Designing a hypothetical full-size hydrogenotrophic reactor will then be straightforward:

1. The gaseous exchange area results from the reactor's design, size, and geometry.
2. Operating pressures higher or lower than the atmospheric pressure will result in higher or lower SHA because the gas exchange rate is proportional to the partial pressure of H_2. Remember that the limiting factor of this process is the solubility of H_2 in water, so it is possible to correct the SHA measured in the laboratory by applying Henry's law to the actual partial pressure inside the reactor.
3. If the design requires that the operating temperature can vary within a given range, then it will be necessary to perform at least two tests: one at each end of the temperature range. Henry's law cannot be employed for calculating the response to temperature variations because the solubility of H_2 will drop with rising temperatures, but the biological activity of the hydrogenotrophic archaea will probably rise, so it is not easy to predict which of both phenomena will prevail.

9.8 FERMENTATION IN THE GASEOUS PHASE

Fermentation in the gaseous phase is a process gaining growing interest from the industry for a variety of biotechnological applications, such as the production of renewable liquid fuels and industrial chemicals from syngas (e.g.,14 –17). The direct application of gaseous-phase fermentation to biogas plants was already discussed in Chapter 8, Section 8.3. It consists of producing additional CH_4 by bubbling syngas into the digestate storage tank. Syngas could be obtained by the gasification of the solid fraction of digestate or any other lignin-rich feedstock. Evaluating the efficacy of such a process in the laboratory requires the same setup already described in Section 9.6, Figure 9.4, with the only difference being the gas composition inside the gas bag: a sample of syngas instead of the stoichiometric $H_2 + CO_2$ mixture.

9.9 DARK FERMENTATION

Dark fermentation is a special case of anaerobic digestion, in which the methanogenic archaea are inhibited or absent and the digestion products are H_2, CO_2, and VFA. The measured parameter is called biochemical hydrogen potential (BHP), and its test protocol is similar to the one for BMP. As of March 2024, the author only knows one proposal of a standardised protocol using AMPTS II [18], which of course is perfectly possible with third-generation instruments too. When using 500 ml reactors, the author suggests employing 2 ml cells because the total gas production may be very small. According to the author's limited experience with this particular biological process, the results may depend strongly on the inoculum's source and preparation. This could explain in part the high variability of values in the literature, but we must also consider the omnipresent problem already pointed out in Chapter 1: the instrumental error analysis is always missing because researchers continue to believe that the only uncertainty parameter to assess is the dispersion between replicates and that '$p > 5\%$' is a universal criterion for assessing the quality of a test's outcome. Readers researching this peculiar fermentation process should keep an open mind and a healthy dose of scepticism when comparing their results with those from the literature.

9.10 *IN SITU* AND *EX SITU* CO_2 ABSORPTION: ARE THEY EQUIVALENT?

The AMPTS family of instruments has employed *ex situ* CO_2 absorption filters (Figure 9.5) since the launch of AMPTS I in 2005.

In situ CO_2 absorption filters, aka GAA (Figure 9.6) were launched in 2023 and are meant for aerobic tests.

Comparing both photos, intuition indicates that GAA provides a more compact and leaner setup, with fewer tubes and a lower probability of gas leaks. Users are then tempted to employ the GAA for both aerobic and anaerobic tests. Figure 9.7 shows

FIGURE 9.5 Classical ***Ex Situ*** CO_2 Absorption Filters Employed in BMP Tests.

FIGURE 9.6 GAA or ***In Situ*** CO_2 Absorption Filters.

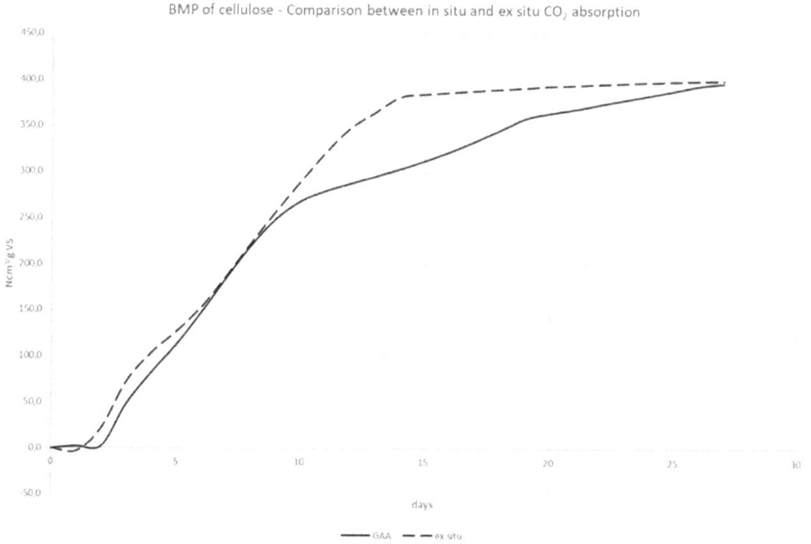

FIGURE 9.7 Comparison of BMP Tests of Cellulose Performed with GAA and ***Ex Situ*** CO_2 Absorption Units. The final BMP value is the same in both (difference within the error margin of the test), but the SMP curves have different shapes.

the outcomes of two simple BMP tests of cellulose (single blank and two replicates) conducted simultaneously with the same inoculum and cellulose sample, with GAA and *ex situ* CO_2 absorption units.

Note that the difference between the BMP calculated with both setups was 1% of the theoretical value (412 Ncm^3/g VS), the maximum uncertainty of the tests (metrological error) was equal to 1%, and the dispersion between the replicates was only 0.04% and 0.08%, respectively. So, we are tempted to conclude that both setups are equivalent because the difference between the calculated BMPs resulting from each one is within the error margin of the instrument. On the other hand, there is a remarkable difference in the shape of the curves. The possible explanations are two – or perhaps a combination of both:

a. The setup with GAA has a bigger headspace volume. In our case (2-l bottles), the headspace is 900 ml with the *in situ* setup and just 600 ml with the *ex situ*. Since the AMPTS III is a volumetric instrument, the headspace should be kept as small as possible when performing anaerobic tests, while the opposite is valid for aerobic tests. For this reason, the manufacturer recommends employing *ex situ* absorption units for performing BMP tests and GAA for aerobic tests. If the headspace is the same in all reactors, the induced error is systematic and will compensate when calculating the BMP. In our example, each group of reactors had the same headspace, but the difference between both groups is remarkable,300 ml.
b. The cellulose sample employed for this test was tissue paper. All reactors were loaded with the same sample quantity, which was weighed with a precision scale, ensuring a high repeatability of results. The paper sheets were shredded by hand, so the uneven granulometry led to the formation of paper lumps in the reactor and ultimately to different degradation rates in each sample reactor. Figure 9.8 shows that despite the ultimate dispersion between gas production of each single reactor being low, the individual gas production rates differed significantly over time. Such instantaneous differences result in BMP curves diverging from the classical sigmoid shape but converging all to the same BMP value (asymptote) when the sample is fully degradable and the digestion time is long enough.

Finally, it is worth noting a phenomenon that did not cause any problem in this particular case but could potentially yield overestimated results when testing substrates that yield more CO_2. At the end of the test described above, it was found that the GAA contained a thick layer of crystals (probably a mixture of sodium carbonate and bicarbonate). The reason is simple: the GAA contains half the amount of NaOH solution as the *ex situ* bottles, so all the NaOH was converted to carbonate and eventually to bicarbonate, presenting lower solubility and precipitating as crystals. If a lower ISR had been adopted for this test, the quantity of CO_2 produced would have been higher, risking then the complete saturation of the GAA and eventually clogging it with crystals. Under such a condition, the instrument would probably measure a mixture of CO_2 and CH_4 during the last days of the test, yielding an overestimated BMP and a curve with a slight ramp-up instead of the usually flat plateau.

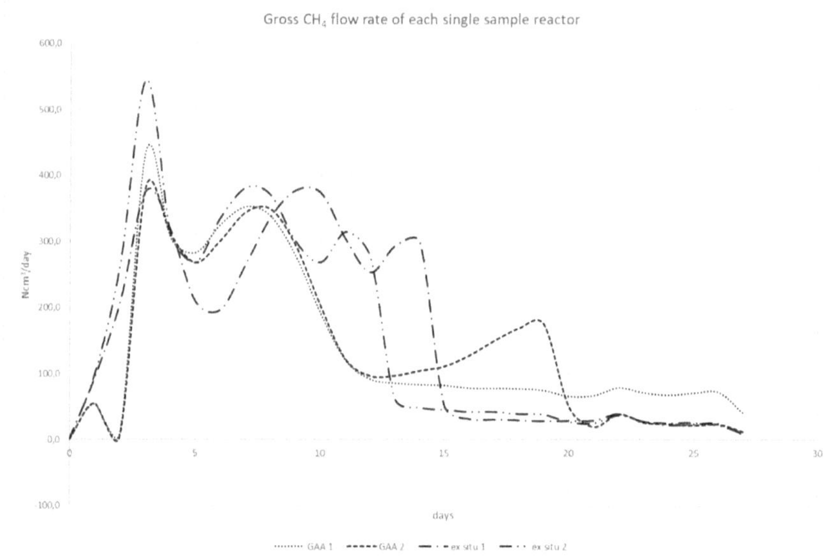

FIGURE 9.8 Comparison of Individual Methane Flows of Reactors with ***Ex Situ*** and GAA CO_2 Capture Units.

Although the GAA seems to work correctly in the test described above, it is safer to adhere to the manufacturer's recommendation and employ the GAA only for aerobic tests and the *ex situ* bottles for anaerobic ones.

ACKNOWLEDGEMENT

The author thanks Dr Mihaela Nistor of BPC Instruments AB for the hints and advice provided during the review of this chapter.

BIBLIOGRAPHY

[1] MA Rosato, *Managing biogas plants – A practical guide*, Chapter 3, Section 3.11, CRC Press (2017).
[2] MA Rosato, *Managing biogas plants – A practical guide*, Chapter 2, Section 2.1.2, CRC Press (2017).
[3] RH Perry & DW Green, *Perry's chemical engineer's handbook*, McGraw Hill (1999).
[4] ISO 9963–1:1994(en), *Water quality – Determination of alkalinity – Part 1: Determination of total and composite alkalinity*. https://www.iso.org/obp/ui/#iso:std:iso:9963:-1:ed-1:v1:en
[5] Jasmina Nikodinovic-Runic, Maciej Guzik, Shane T Kenny, Ramesh Babu, Alan Werker & Kevin E O Connor, "Chapter four – Carbon-rich wastes as feedstocks for biodegradable polymer (polyhydroxyalkanoate) production using bacteria," in Sima Sariaslani & Geoffrey M Gadd (eds), *Advances in applied microbiology*, Vol. 84 (2013), 139–200. Academic Press. https://doi.org/10.1016/B978-0-12-407673-0.00004-7. https://www.sciencedirect.com/science/article/pii/B9780124076730000047. ISSN 0065–2164; ISBN 9780124076730.

[6] Data from internal research, personal communication between the Author and Dr. Mihaela Nistor, BPC Instruments AB, September 2024.
[7] Personal communication between the Author and Dr. Mihaela Nistor, BPC Instruments AB, September 2024.
[8] The following are some of the said pertinent norms:

- BIPM (Bureau International des Poids et Mesures), Evaluation of measurement data. Guide to the Expression of Uncertainty in Measurement, JCGM 100, First edition, 2008
- ISO/IEC Guide 98–1:2009 – Uncertainty of measurement – Part 1: Introduction to the expression of uncertainty in measurement
- ISO/IEC Guide 98–3:2008 – Uncertainty of measurement – Part 3: Guide to the expression of uncertainty in measurement (GUM:1995)

[9] PH Robinson & G Getachew, *A practical gas production technique to determine the nutritive value of forages: The UC Davis approach*, California Chapter of the American Society of Agronomy (2005). https://animalscience.ucdavis.edu/sites/g/files/dgvnsk446/files/inline-files/Web200206.pdf
[10] Harinder PS Makkar & Philip E Vercoe, *Measuring methane production from ruminants*, Springer (2007). https://link.springer.com/book/10.1007/978-1-4020-6133-2
[11] A Quarantelli, M Renzi, M Simoni, MA Rosato & F Righi, Use of a new instrument for the in vitro evaluation of ruminants feeds through gas production: Preliminary study, *Proceedings of the 22nd ASPA Congress Perugia*, June 13th–16th, *Italian Journal of Animal Science*, 16:s1 (2017). ISSN 1594–4077.
[12] Klára Cverenkárová, Martina Valachovičová, Tomáš Mackuľak, Lukáš Žemlička & Lucia Bírošová, Microplastics in the food chain, *Life*, 11:12, 1349 (2021). https://doi.org/10.3390/life11121349
[13] OECD 307 – Guideline for testing of chemicals. Ready Biodegradability. Adopted by the Council on 17th July 1992. https://www.oecd.org/chemicalsafety/risk-assessment/1948209.pdf
[14] Basma Omar, Maie El-Gammal, Reda Abou-Shanab, Ioannis A Fotidis, Irini Angelidaki & Yifeng Zhang, Biogas upgrading and biochemical production from gas fermentation: Impact of microbial community and gas composition, *Bioresource Technology*, 286, 121413 (2019). https://doi.org/10.1016/j.biortech.2019.121413. ISSN 0960–8524.
[15] Wouter Van Hecke, Richard Bockrath & Heleen De Wever, Effects of moderately elevated pressure on gas fermentation processes, *Bioresource Technology*, 293, 122129 (2019). https://doi.org/10.1016/j.biortech.2019.122129. ISSN 0960–8524.
[16] IK Stoll, N Boukis & J Sauer, Syngas fermentation to alcohols: Reactor technology and application perspective, *Chemie Ingenieur Technik*, 92, 125–136 (2020). https://doi.org/10.1002/cite.201900118
[17] FE Liew et al., Carbon-negative production of acetone and isopropanol by gas fermentation at industrial pilot scale. *Nature Biotechnology*, 40, 335–344 (2022). https://doi.org/10.1038/s41587-021-01195-w
[18] Julián Carrillo-Reyes, Germán Buitrón, Iván Moreno-Andrade, Aida Cecilia Tapia-Rodríguez, Rodolfo Palomo-Briones, Elías Razo-Flores, Oscar Aguilar-Juárez, Jorge Arreola-Vargas, Nicolas Bernet, Adriana Ferreira Maluf Braga, Lucia Braga, Elena Castelló, Lucile Chatellard, Claudia Etchebehere, Laura Fuentes, Elizabeth León-Becerril, Hugo Oscar Méndez-Acosta, Gonzalo Ruiz-Filippi, Estela Tapia-Venegas, Eric Trably, Jorge Wenzel & Marcelo Zaiat, Standardized protocol for determination of biohydrogen potential, *MethodsX*, 7, 100754 (2020). https://doi.org/10.1016/j.mex.2019.11.027. ISSN 2215–0161.

10 The Technical Due Diligence of Biogas Projects

10.1 INTRODUCTION

In principle, a biogas plant should always be profitable regardless of tariff incentives; otherwise, how can we explain that over 40 million very small plants in developing countries, mostly devoid of any technology, provide energy at low cost despite not receiving any government subsidy? On the other hand, in Western countries, many industrial biogas projects have been closed or are operating at reduced capacity, and the entrepreneurs complain that they have not achieved the expected profitability.

Several factors condition the profitability of an industrial AD project. Some are location-dependent: local feed-in tariffs, local regulations, bureaucratic complications, capital expenditure (CAPEX), and operation expenditure (OPEX). The reader should then assess the pertinent local aspects or engage a local consultant since it is impossible to provide universally valid insights in this book. Therefore, this chapter will focus only on the potential technical issues that may lead to a biogas project's failure or, at least, hamper its profitability. For obvious privacy reasons, the content of this chapter will be generic, without references to people or companies involved in the cases or situations that the author encountered in the field. All presented data come from project reviews made before the plant's construction or from case studies in which the plants were already built and underperforming when the author was engaged to optimise them. Based on the said experiences, the aim of this chapter is to introduce a checklist of the most recurrent technical flaws to identify when reviewing biogas projects.

10.2 FINANCIAL DD, GOVERNMENTAL EIA, AND EU TAXONOMY: NECESSARY BUT NOT SUFFICIENT CONDITIONS

In the financial world, due diligence (DD) is the evaluation protocol that investors, business angels, and banks apply to decide whether to invest in or fund a project. In most European countries, environmental impact assessment (EIA) is used to evaluate all the risks that a work may pose for the environment and people at the local scale, a necessary step to obtain construction authorisations. Since the said analytical procedures were deemed insufficient, the European Commission (EC) published the so-called Taxonomy Regulation [1] in the *Official Journal of the European Union* on 22 June 2020, entering into force on 12 July 2020. The European Union (EU) Taxonomy is a classification system containing a list of environmentally sustainable economic activities. It provides appropriate definitions to companies, investors,

DOI: 10.1201/9781003609278-10

and policymakers on which economic activities can be considered environmentally sustainable. In the EC's intention, this should create security for investors, protect private investors from greenwashing, help companies plan the transition, mitigate market fragmentation, and eventually help shift investments from traditional markets to renewable energy and environmental protection projects. The taxonomy's practical application is explained in the Final Report on EU Taxonomy and Technical Annex [2]. The summary tables of the Taxonomy Report's Annex classify anaerobic digestion (AD) of biowaste as an economic activity that substantially contributes to short- and long-term climate change mitigation. Investing in this activity is recognised as not creating any lock-in effects in the mid-term, meaning the activity is intrinsically climate-friendly even in the long term. A few conditions must be satisfied to ensure that no significant harm is done (known as do-no-significant-harm (DNSH) criteria in the bureaucratic jargon): attention to methane leakage, emissions reduction or avoidance, and the subsequent agronomical use of the digestate. EUregulated biogas plants are supposed to ensure compliance with the said criteria.

Despite the CE's zealous overregulation approach, some AD projects finish by failing or do not achieve the expected profitability. Such a problem is not exclusively European because many biogas plants outside of the EU are designed and built using European criteria as a model, and consequently with all its inherent flaws. The failure's reason is that the said procedures contain a logical fallacy: the financial, administrative, and EU Taxonomy analysis procedures assume that the design choices and input data are correct, which, at least in the author's professional experience, rarely happens.

Passing the standard DD and Taxonomy procedures is then a 'necessary but not sufficient condition' to ensure the profitability of a biogas project. On the other hand, conclusions based on false assumptions can sometimes turn out to be true, but by pure chance. This explains why most biogas plants are reasonably profitable, despite having major design flaws and being managed with coarsely approximate methods. But can we accept that the profitability of a major investment like an anaerobic digestion plant relies on chance?

The following sections reflect the most frequent flaws that the author observed in Italian and Spanish anaerobic digestion projects, leading to low profitability and in a few cases even failure. Many of them are likely to occur in other countries, so analysing them is a good starting point for elaborating a checklist useful for the technical DD of a biogas project. Some flaws arise from cognitive biases, a few already discussed in Chapter 1. Other errors or weak points observed in business plans arise from the technical staff, who usually have cultural biases in favour of given solutions or design models. Others are just examples of superficiality and naivety. Hence, knowing the team members behind an AD project and their professional backgrounds helps the reviewer search for the more likely flaws and design errors that could hamper the project's profitability.

10.3 COGNITIVE BIASES OF INVESTORS

Cognitive biases in agricultural biogas business plans are more frequent when the promoters or investors are not agricultural entrepreneurs and, therefore, take for granted facts that a real farmer would not rely upon or at least would evaluate very carefully before making a decision. From a philosophical point of view, planning to cultivate with the scope of using the harvest to produce energy is a European cultural bias in itself. It is just a shortcut to building plants with 1 MWe power (the size that best suits European constructors) when the available quantity of waste would not be enough. Such a business model has been partly phased out in Europe, but many promoters exported it to developing countries. The ecological and strategic value of AD is the possibility of converting organic waste into energy and recycling the nutrients back into the soil. In any case, growing plants and using the harvest as AD feedstock is not forbidden, so if you are analysing a project of this type, be aware of the following potential pitfalls.

10.3.1 Extrapolation of Present Situations to the Future

The most common mistake in biogas projects carried out by capital groups is assuming that any quantity of by-products can be purchased at any time and for the price assumed in the business plan (oversimplification bias). Some examples are as follows:

a. A famous real estate entrepreneur decided to enter the renewable energies market. Such a decision is an example of a cognitive bias known as 'self-confidence bias': since the entrepreneur was very successful in the real estate business, he believed he would be successful trading renewable energies too. So, he founded a company and bought low-cost land in a rural area with a high density of pig farms. He imagined that all farmers would give him their slurry for free because pig slurry is a nuisance (oversimplification bias), so he built a biogas plant meant to digest only pig slurry. He didn't care about signing supply contracts (overconfidence bias again), so when the plant was complete and he tried collecting the slurry to start, the farmers in the area formed a cartel and demanded €2/m^3. Initially, the entrepreneur paid for the slurry, but the energy production was far below expected (see Section 10.3.2 for the cognitive bias that led to this). No other feedstock was available in the area, so the plant shut down a few months later.
b. An engineering company created an agricultural company for the sole purpose of building a biogas plant. Not having any residues or by-products of their own to feed the digester, they based their business model on buying corn silage and cereal meals in the animal fodder market at a given price, supposing it would remain stable for the next 20 years because it had been stable till then (a combination of oversimplification bias and gambler's fallacy). Other investors had the same idea, so the demand for silage grew fast and the price with it, lowering the project's profitability to nearly the breakeven point.

10.3.2 Basing the Business Plan on Pseudo-scientific or Hyped Information

According to the author's experience, many biogas plant projects failed to reach the expected productivity, and in a few cases were not productive at all, because the assumptions employed to redact their business plans were hyped or false. Such situations arise when the promoter or investor has a strong financial background but little or no knowledge of agronomy, chemistry, and physics. The present section intends to provide the reader with a method for critically reviewing such business plans. The said method is based on assessing the mass and energy balance. In general, all biogas projects contain a mass and energy balance, but it is often based on hypotheses that are either wrong, incomplete, hyped, and contradictory or have not been tested in the lab. Sometimes, it is possible to discover wrong assumptions just by *reductio ad absurdum*. Reduction to absurd is a technique of proofing the syllogism: if assuming the claims or hypotheses to be true one comes to an undoubtfully false or absurd conclusion, then at least one of the hypotheses must be false. Since the following examples derive directly from the author's personal experience, the information about geographical and brand references has been deliberately omitted to protect the privacy of the involved people.

10.3.2.1 Example 1: Biogas Plant Fed with Cactus (*Opuntia sp.*) Cladodes

The business plan relied on several assumptions, one of them being that '1 ton of cladodes produce 100 m^3 of biogas with 60% methane' and another one being that 'the moisture of cladodes is 91%'. The claim that biomass composed of carbohydrates can produce biogas with 60% methane violates the stoichiometry rules: carbohydrates always produce biogas with 50% methane. Such round figures are suspicious too. This should be enough to dismiss the whole business plan because if only one of the hypotheses is false, the conclusions cannot be true; hence, it is unlikely that the plant fed with the foreseen quantity of cladodes reaches the expected productivity and profit. The proponent claimed he had got the cladodes tested in a university lab. Hanlon's razor warns us about a case of carelessness: the man honestly believes in the quality of the 'scientific data' he purchased, but we cannot be sure of the quality of the test. We all know that batch tests tend to produce biogas with higher methane content than expected from stoichiometry (see Chapter 9 for more details). So, let's apply the method of reduction to the absurd: assume that it is true that 1 tonne of cladodes produces 100 m^3 of biogas and that the moisture of the cladodes is 91%. As usual in many business plans, the proponent did not specify whether they are Nm^3 or Sm^3 or other reference conditions. Since the project's proponent claims that the figure comes from a laboratory test at a university, it must be Nm^3 because the academic community works only with SI units. One mole of any gas under normal conditions occupies 22.4 dm^3. Hence, 1000 Ndm^3 contains 44.6428 moles of gas. If it were true that the said gas is composed of 60% CH_4 and 40% CO_2, then the density would be

$$\rho = 44.6428 \frac{\text{mol}}{\text{Nm}^3} \times (0.6 \times 16 + 0.4 \times 44) \frac{\text{g}}{\text{mol}} = 1.214 \text{kg/Nm}^3.$$

On the other hand, the productivity calculation assumes 91% moisture, i.e., 9% dry mass as feedstock to the digester. Incidentally, the 'university laboratory' did not specify the percentage of volatile solids (aka organic dry mass), or the person who redacted the business plan overlooked it. So let's assume, very optimistically, that the cladodes have no ash at all, and hence, 1 tonne of cladodes equals 90 kg of organic matter. Furthermore, suppose (again optimistically!) that the said organic matter is 100% biodegradable. Since the density of biogas with 60% CH_4 and 40% CO_2 is 1.214 kg/m^3, *how could 90 kg organic matter produce 121.4 kg of biogas?* Regardless of which one of the assumptions is false, we can drive a true conclusion: the productivity calculations in this business plan are hyped.

At this point, an investor could either stop the dealings and search for other opportunities or review the project with reliable data. In the second case, it is mandatory to find out the real biochemical methane potential (BMP) of the cladodes and their percentage of volatile solids (VS). Then the options are to recalculate the mass of cladodes needed to reach the desired energy production or, conversely, calculate how much energy can be produced with the available cladodes harvest.

Just to provide complete insights on the much-hyped potential of cactus, the author includes the results of some tests on *Opuntia ficus-indica* cladodes, performed many years before this case study, for another project in another location. Of course, it is impossible to grant the validity of these data for the case study discussed above, but at least the reader will have a reliable idea of the potential of cactus biomass as a feedstock for biogas plants. The sample of fresh cladodes tested in the author's laboratory had 8% VS, and their maximum BMP was 350 Nm^3/ton VS, with a maximum uncertainty of 5%. This means that 1 tonne of cladodes will produce a maximum of 28 Nm^3 of CH_4, i.e., 56 Nm^3 of biogas having 50% CO_2 because cladodes contain neither fats nor proteins. The density of such biogas is 1.339 kg/Nm^3. If we apply the mass balance analysis to these figures, 80 kg of net organic matter would produce 75 kg of biogas, i.e., the biodegradability of the sample was 93.75 %. The remaining 6.25% was probably lignin or became living bacterial biomass. In any case, this fraction would leave the digester as sludge together with the initial moisture and ash of the feedstock; hence, the business plan should account for the cost and benefits of digestate handling too, something that was missing in the business plan under revision.

10.3.2.2 Example 2: High-density Cultivation of Cactus

Another entrepreneur intended to cultivate *Opuntia sp.* at 'high density', claiming it would be possible to reach an annual productivity of 21,187 tonnes/ha. Such a huge estimation sounds unrealistic, but, according to the entrepreneur, it was the outcome of a study conducted by 'a team of expert agronomists', assuming no limitation in N, P, K, and water supply. The N application rate, for instance, resulted to be 28.4 tonnes/ha. Such a high concentration of N is several orders of magnitude the maximum allowable annual dosage for crops in the EU (0.35 tonnes/ha). Although it could be argued that European directives are meant for conventional crops, 81 times the maximum legal N dosage sounds like it would burn any plant, no matter if in Europe or abroad. Nevertheless, no literature on the toxicity limit of N on *Opuntia* was found, so it was impossible to dismiss the said hypothesis with experimental proof. The 'technical report' did not estimate the water needs, so this aspect is crucial to check whether the project is feasible or just a set of blind extrapolations. The reduction to absurd technique will prove useful again.

Suppose it is true that under the assumed growing conditions, it is possible to produce 21,187 tonnes/ha of fresh cactus biomass per year. Since it was agreed that the moisture content of the fresh cladode biomass at harvest is 90%, the plants must absorb 21,187 tonnes × 0.9 = 19,068 $m^3$of water every year. Suppose – very optimistically! – that the water use efficiency is 100% (i.e., the plants will absorb all the rain and irrigation water, with no leaks or evaporation, which is already contradictory with the hypothesis of cultivating in the desert). Then, 19,068 m^3 of water/10,000 m^2 (one ha) = 1,906 mm of equivalent rainfall. Such a water input is typical of a tropical rainforest [3] and is in net contradiction with the idea of cultivating biomass in the desert.

What if the assumed production figure had been a typing mistake and the fresh biomass production was just 2,118 tonnes/ha? Such productivity is still very high, but the required hydric needs would be just 191 mm/year, compatible with rainfall in Northern Africa and other arid areas. It must be noted that the said figure results from the unrealistic assumption of 100% water use efficiency. In this case, we should perform our calculations more accurately. Fortunately, the literature provides at least one study on the water use efficiency of *Opuntia ficus-indica* [4]. The cultivation conditions in the said study consider a much higher planting density: 71,428 plants/ha (plants spaced 0.14-min rows spaced 1 m, meant for forage production), against 15,000 plants/ha assumed in the biogas business plan. In any case, the paper provides information measured directly on the field that we can assume as true for benchmarking purposes. In the best case, the water use efficiency index measured in the field was 10.58 kg DM/ha per mm of total input (irrigation + rainfall). Producing the expected 21,187 tonnes/ha of fresh biomass with 90% moisture would then yield 2,118.7 tonnes of DM/ha, i.e., 2,118,700 kg DM/ha. The necessary water input results to be 2,118,700 / 10.58 = 200,255 mm equivalent rainfall. A clearly absurd conclusion; hence, even 2,118 tonnes/ha is an unrealistic biomass yield.

As a double proof, the planting density in the Brazilian experiment was 4.76 times higher than the assumptions in the biogas business plan, but the highest biomass yield was 218.2 tonnes of fresh biomass/ha, equal to 10.57 tonnes DM/ha, i.e., barely 5% DM content.

10.3.2.3 The General Case of 'High-biomass-yielding Crops'

If you intend to cultivate a given crop to feed the digester, check with a competent agronomist the required input of water, fertilisers, fuel for the machinery, and their corresponding costs. Assess the reasonable average biomass productivity and avoid assuming record figures from 'the literature' or claims of seed suppliers or business promotors as the base for your business plan. A second opinion from an independent professional will probably provide less biased insights.

In case of doubt, use the following technique of *reductio ad absurdum* to assess if claims about the biomass yield of any vegetal can be true.

1. The average lower heating value (LHV) of vegetal biomass is 19 MJ/kg DM = 5.28 kWh/kg DM = 5,280 kWh/tonne DM.
2. It is very improbable that the efficiency of photosynthetic conversion of sunlight into biomass exceeds 4% (algae in photobioreactors, sugarcane [5]),

most commonly being in the range of 1%–2% (high yield crops, water hyacinth under favourable conditions). Efficiencies beyond 1.5% are possible but improbable; one should reasonably doubt and check the yield in the field.

3. Calculate the efficiency of conversion per hectare with the following formula:

$$\eta = \frac{Y \times DM \times 5{,}280}{10{,}000 \times GHI}, \text{ where}$$

Y = yield of fresh biomass (tonnes/ha);
DM = percentage of dry matter at harvest;
GHI = global solar irradiation on a horizontal plane of the site (for instance, from the World Bank's app Solar Global Atlas: https://globalsolaratlas.info/), in kWh/m^2.year;
10,000 = m^2/ha.

Example (the same in Section 10.2.3.2): Suppose it is true that the yield of cactus fresh biomass is 21,187 tonnes/ha, with 8% DM. The envisaged cultivation locality is in the Doukkala-Abda region in Morocco, and the harvest will be once a year. According to the Global Solar Atlas (GSA) app, the average GHI is 1,950 kWh/m^2.year).

$$\eta = \frac{21{,}187 \times 0.08 \times 5{,}280}{10{,}000 \times 1{,}950} = 0.4589 \equiv 45.9\%$$

No plant on earth has such a high photosynthetic efficiency!

Example 2 (as a proof of this method's validity): According to the author's experience, the fresh biomass yield of corn in the Veneto region can be up to 60 tonnes/ha in very favourable seasons and with the correct dosage of fertilisers and irrigation. DM at harvest is around 35%. According to the GSA app, the total solar energy on the horizontal plane in April–September (the growth period of corn) for a generic location in the province of Venice is 905 kWh/m^2. Hence, the energy efficiency of corn biomass production is

$$\eta = \frac{60 \times 0.35 \times 5{,}280}{10{,}000 \times 905} = 0.012 \equiv 1.2\%.$$

10.3.3 Linear Thinking and Oversimplification Leading to Unrealistic Business Plans

An investor redacted his business plan based on a linear model, taking as design parameters the yearly availability of different feedstocks and the individual BMP of each (Table 10.1).

TABLE 10.1
Dimensioning the Digester on Average Annual Values (Wrong Approach)

Feedstock	Availability (tonnes/year)	Input (tonnes/day)	BMP (Nm^3/tonne VS)	VS (% ww)	CH_4 Production (Nm^3/day)
Chicken dung	21,900	60	200	18%	2,160
Grape marc	800	2	160	41%	142
Whey	2,000	5	330	5%	90
Olive mill effluent	5,000	14	300	2%	95
Citric waste pulp*	20,000	55	*220*	*18%*	*2,170*
Cactus cladodes	6,000	16	350	7%	403
Total	55,700	153			5,060

**BMP and VS from a degree thesis*

Digester volume **	4,578	m^3	** Assuming 1 tonne = 1 m^3 and 30 days hydraulic retention time (HRT)

It is worth noting that Table 10.1 contains three false assumptions that hamper the business plan's validity:

1. Not all feedstocks are available year-round. For instance, prickly pear cladodes, olive mill effluents, citric waste pulp, and grape marc are seasonal feedstocks.
2. 43% of the expected income is based on the supposition that citric waste pulp has 18% VS and its BMP is 220 Nm^3/ton VS. The investor never got a sample tested in a laboratory: he extracted the data from a degree thesis. A successive check showed the thesis text was vague about the test protocol, instruments, and inoculum employed to determine the BMP.
3. The digester's volume is just 30 times the total daily feedstock volume, a design criterion resulting from two false assumptions:
   a. Assuming the same density as water for all feedstocks, which is fairly approximate for whey, olive mill effluents, and cladodes reduced to slurry, but not true for the rest of the biomasses;
   b. Assuming that all feedstocks require 30 days for their complete digestion.

We could also argue that the BMP of a mixture of substrates is not always equal to the proportional sum of the individual BMPs. Nevertheless, we can adopt this criterion for a feasibility analysis (an example of abductive reasoning), and if the project turns out to be feasible, then check the actual BMP of the mixture in the laboratory before proceeding with the actual design.

Table 10.2 shows the result of a quick analysis of each feedstock's seasonal availability in the area. Note that even supposing that hypotheses 3a and 3b are true, the monthly variation of the volumetric load leads to digestion times shorter than 30 days during the highlighted months. The resulting organic loading rate (OLR) is much higher than the conventional design limit for CSRT digesters, 3 kg VS/m^3.day, except in July and October.

The first outcome of this more thorough analysis is that the digestion volume should be at least 8,000 m^3, almost double the initial supposition. This means that both the CAPEX (construction costs) and the OPEX (energy for the stirrers and pumps) will almost double too. On the other hand, Table 10.2 also shows that in the period July–October, the OLR is much less than the ideal 3 kg VS/m^3.day. This means that the digester is underloaded, i.e., there is a deficit of organic matter.

The second step is to check, at least theoretically, the methane production over the year and the upgrading system's load factor. The investor intended to install an upgrading plant with 250 Nm^3/h nominal capacity. A high loading rate ensures maximum efficiency and the shortest return on investment. Table 10.3 shows that the load factor is lower than 90% (conventional acceptable limit) as a consequence of the OLR being lower than the optimum value.

The final steps to conclude the revision of the business plan were

a. to find local suppliers of additional feedstock to increase production during July–October and define the VS percentage, the BMP, and the cost of each one and
b. to take samples of each feedstock at each source and check in the laboratory if the VS percentages and the BMPs are the same ones assumed for the calculations; otherwise, redo the calculations with the correct values.

During the final assessment phase, the laboratory tests showed that:

1. 20% of the assumed chicken dung quantity was unusable because the farmer employs sawdust as bedding.
2. The remaining 80% of the chicken dung had higher average VS (24%) and BMP (260 Nm^3/ton VS) than expected.
3. The waste citric pulp had a lower concentration of VS (16%), but its BMP was higher (293 Nm^3/ton VS) than expected. This represents 18% more productivity than expected but with some risk: this feedstock may cause partial inhibition. Figure 10.1 shows *diauxia*, i.e., a plateau between the 1st and the 5th day, then the usual exponential production reaching the final plateau within the 15th day. This is not a big problem by itself, but if a component of a mixture shows an abnormal shape of the BMP curve, then testing the whole mixture and checking the shape of its methane production curve are mandatory. Otherwise, it is impossible to ensure the absence of antalgic effects between the different substrates, leading to lower methane production than the expected proportional addition of the single BMPs.

TABLE 10.2
Assessing the Digester's OLR and HRT, Taking into Account the Seasonal Variabilities of Feedstock Supplies

	Feeding Rate [tonnes/day]												Consumption	Availability
Feedstock	**Jan**	**Feb**	**Mar**	**Apr**	**May**	**Jun**	**Jul**	**Aug**	**Sep**	**Oct**	**Nov**	**Dec**	**tonnes/year**	**tonnes/ year**
Chicken dung	60	60	60	60	60	60	60	60	60	60	60	60	21,900	21,900
Grape marc								13.4	12.8				799	800
Whey	5.5	5.5	5.5	5.5	5.5	5.5	5.5	5.5	5.5	5.5	5.5	5.5	2,008	2,000
Olive mill effluent										54	59	50	4,994	5,000
Citric waste pulp	84	84	84	84	79	79					83	83	19,962	20,000
Cactus cladodes							51	50	47	47			5,998	6,000
Total feedstock input	150	150	150	150	145	145	117	129	125	167	208	199	**55,661**	**55,700**
HRT with $V = 4,578\ m^3$ [d]	31	31	31	31	32	32	39	36	37	27	22	23		
OLR with $V = 4,578\ m^3$ [kg VS/m³.d]	5.7	5.7	5.7	5.7	5.5	5.5	3.2	4.4	4.3	3.4	6.0	5.9		
HRT with $V = 8,000\ m^3$ [d]	54	54	54	54	55	55	69	62	64	48	39	40		
OLR with $V = 8,000\ m^3$ [kg VS/m³.d]	3.3	3.3	3.3	3.3	3.2	3.2	1.8	2.5	2.4	2.0	3.4	3.4		

TABLE 10.3
Expected Monthly Methane Productions and Digester Load Factors

	Methane Production [Nm^3/day]												Total Production
Feedstock	**Jan**	**Feb**	**Mar**	**Apr**	**May**	**Jun**	**Jul**	**Aug**	**Sep**	**Oct**	**Nov**	**Dec**	**Nm_3/year**
Chicken dung	2,160	2,160	2,160	2,160	2,160	2,160	2,160	2,160	2,160	2,160	2,160	2,160	788,400
Grape marc								879.0	839.7				52,441
Whey	90.8	90.8	90.8	90.8	90.8	90.8	90.8	90.8	90.8	90.8	90.8	90.8	33,124
Olive mill effluent										324	354	300	29,964
Citric waste pulp	3,326	3,326	3,326	3,326	3,128	3,128					3,287	3,287	790,495
Cactus cladodes							1,250	1,225	1,152	1,152			146,951
Total daily production	**5,577**	**5,577**	**5,577**	**5,577**	**5,379**	**5,379**	**3,500**	**4,355**	**4,242**	**3,726**	**5,892**	**5,838**	**1,841,375**
Production rate (Nm^3/h)	232	232	232	232	224	224	146	181	177	155	245	243	
Load factor (250 Nm^3/h nominal capacity)	93%	93%	93%	93%	90%	90%	58%	73%	71%	62%	98%	97%	

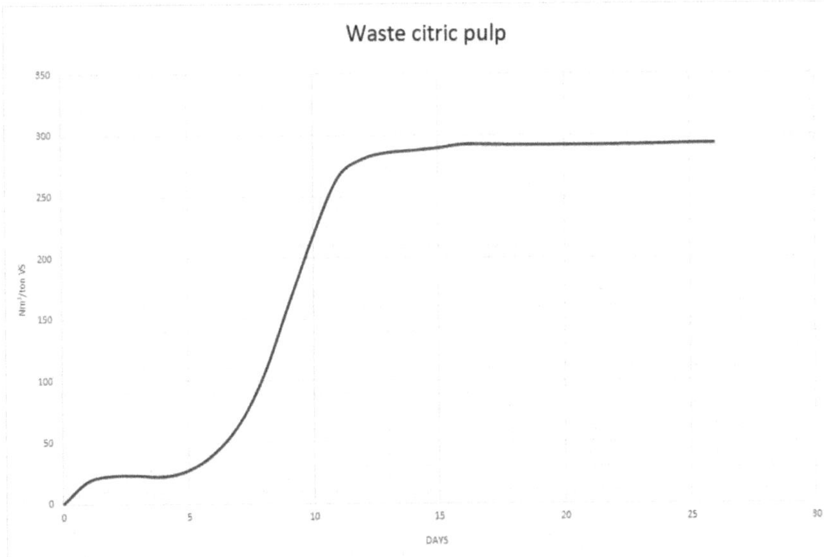

FIGURE 10.1 Cumulative SMP Curve of a Substrate Showing Diauxia.

10.4 COGNITIVE AND CULTURAL BIASES OF PLANT ENGINEERS AND PROJECT PROMOTERS

Sometimes, companies from the renewable energies or construction sectors without direct experience propose themselves as designers and builders of biogas plants, only to fail shortly thereafter. This happened in Italy at least five times in 2010–2024, so it is likely to happen in other countries experiencing a quick growth of the biogas economy. The fact that a subject has successfully designed and built a photovoltaic system or a concrete tank in the past, or that the company is a spin-off of a research group that published many research papers and degree theses, does not guarantee the capability to design and operate an industrial digestion plant.

On the other hand, reputed companies with a long track record in the AD market tend to have a standard plant design, often also some patented solution, that they try to adapt to any situation for purely commercial reasons. The result is usually a sub-optimal biogas plant: oversized digesters, undersized feeders, design data obtained by extrapolation, etc.

The following guidelines will help evaluate a provider's proposal. It is anyway a good idea to hire an independent consultant or to ask a competitor for an alternative quotation.

10.4.1 Cutting-edge AD Technology

Be wary of 'disruptive new digester technologies' especially if they have never been tested with feedstock available for your project. The following is a real case of such a situation: a start-up company proposed 'their special packed bed CSTR digesters' to several farms that intended to process cow dung with straw and other fibrous matter. Packed reactors may be effective –although not necessarily the most effective! – when treating very diluted wastewater carrying colloidal solids. High solids feedstock like cow farm waste clogged the digesters in a few months. The company had to face several lawsuits and failed.

10.4.2 The Right Digester for Each Feedstock

Check that the proposed digestion system suits your feedstock and local context. Most biogas plant constructors usually rely on a single digestion technology despite five different families of digesters being known, each one suitable for some specific feedstocks. It is common practice in the biogas industry to propose CSRT digesters for almost every substrate. CSRT digesters cannot work with mixtures containing more than 12%–15% total solids (TS), i.e., the process is wet digestion. However, some substrates like organic fraction of municipal solid waste (OFMSW) and cattle or chicken dung with bedding may contain 25%–40% TS. Most biogas plant builders will then prescribe diluting the feedstock with water. This facilitates transferring the feedstock from the feeding tank and between digestion stages. It is also commonly accepted that wet digestion ensures a higher BMP than dry digestion; however, it also has several drawbacks. Firstly, fresh water is not always available in large quantities everywhere. Recirculating part of the liquid fraction of digestate will help, but it is not possible to recirculate digestate forever because salts will accumulate and, in any case, fresh water will be necessary together with digestate to start the digester up. Furthermore, water occupies space, so the CSRT digester's volume will be three to four times bigger than a PFR for a given volumetric loading rate (VLR), which means three to four times more CAPEX. Finally, CSRT digesters require stirring – by definition – and stirrers may consume up to 10% of the electric energy produced. If the feedstock or mixture of biomasses has more than 15% TS, a semidry digester (PFR) may be more convenient. Most probably, the optimum solution will be a hybrid configuration: first stage with PFR and second stage with CSRT.

A rule for reviewing a biogas project: check the resulting TS of the envisaged feedstock mixture. If it is higher than 15%, then ask the engineering company to evaluate the biocell and/or PFR technologies, or ask for a quotation from an engineering or plant constructor specialised in such technologies.

An example common to three different case studies: after performing a market survey, the promoter planned to feed the digester with a given mixture of biomasses. Some were dry, others wet, and when putting them together in the resulting proportions, the overall percentage of solids was 23%. Since the chosen biogas plant builder specialises in CSTR digesters, the plant owner signed a contract obliging him to source enough water to dilute the feedstock. This may not be a problem in Central and Northern European countries with plenty of fresh water all year round. However, it is indeed a big problem when such companies propose CSRT digesters in countries that undergo long dry seasons or where water is brackish.

The following rules of thumb will allow the reader to quickly identify a potential bias in the digester technology choice:

a. UASB and EGSB digesters are suitable for high chemical oxygen demand (COD) wastewater with little suspended solids, i.e., the COD results mainly for diluted organic matter such as alcohols, sugars, and organic acids. Typical applications: sewage water, and effluents from beverage and food industries.
b. CSTR digesters are suitable for feedstock (or mixtures of feedstocks) with 2%–15% TS content (wet digestion). Typical applications: sludge from wastewater treatment, farm slurry, fresh fruit and vegetable biomass, and slaughterhouse waste.
c. Horizontal plug flow digesters and vertical downflow digesters (aka Dranco digesters) are suitable for feedstock with 15%–20% TS (semi-dry digestion). Typical applications: OFMSW and dairy dung with straw or other fibrous bedding.
d. Biocells (batch digesters with or without percolation) are suitable for high solids (20% or more) and coarse feedstock like urban garbage. The digestion is usually incomplete because of the low inoculum to substrate ratio (ISR) (usually <1). Typical applications: digestion of OFMSW and dairy dung with straw or other fibrous bedding, in all cases followed by composting (sometimes in the same biocell).
e. Covered lagoons are common in tropical countries because of their low cost and absence of temperature control. Lacking agitation, they can handle only swine slurry and thin sludge. The welded joints of their plastic membrane tend to fail after a few years of service, leading to slurry and biogas leaks.

10.4.3 Wrong Digester Size

Some biogas plant constructors tend to 'copy and paste' from former projects. The author has sometimes reviewed projects of biogas plants where the target rated power was relatively small (<250 kWe) but the digester size was the same as for a 1-MWe plant. This means higher power demand for agitation and thermal power for heating a bigger mass of slurry. In other cases, the designer assumed a higher specific methane production per volume unit of feedstock than the real one and calculated the digestion volume as 30 times the volumetric loading rate (VLR), resulting in an undersized digester for the expected organic load. The following guidelines will help in checking the digester volume.

1. Double-check the digester's design volume with two different criteria. The net digestion volume, V, must provide both a long enough HRT and an OLR smaller than the maximum acceptable for the type of digester foreseen. In the special case of CSTR digesters, the most frequent ones, the two conditions that must be satisfied are

   $HRT = V / VLR >$ digestion time

   and

OLR = daily quantity of feedstock × VS% / $V < 3$ kg VS/$m^3_{digester}$.day.

The digestion time results from the BMP curve measured in the laboratory. Assuming 30 days is only acceptable for a preliminary assessment of the investment's order of magnitude but not for deciding the size of the future digester because some substrates are readily digested in less than 2 weeks while others may require up to 60 days. Furthermore, the digestion time also depends on the inoculum employed to digest the sample.

2. As a guideline, the usual OLRs for the most common digester families are 3 kg VS/m^3.d for CSTR, 5 kg VS/m^3.d for PFR, 10 kg COD/m^3.d for UASB, and 1 kg VS/m^3.d for unstirred, unheated anaerobic covered lagoons. In the case of biocells, the volume of each one must be at least the daily volume of OFMSW (if the biocells will be loaded daily) plus the structuring material (between 20% and 35% of the OFMSW load) plus the inoculum (a quantity of digestate enough to ensure that the ISR is at least equal to 1).
3. In Europe, including a covered digestate storage tank with 90 (in some areas 120) days of HRT and recovery of the residual biogas production is mandatory. The reasons for such a prescription are multiple: the maturation time is said to allow an initial degree of humification of the undigestible lignin fraction, improving thus the digestate's agronomic quality; storing the digestate during the rainy season prevents nutrients from leaching into water bodies; exhausting the digestate completely prevents CH_4 emissions to the atmosphere; and eventual pathogens like coliforms will die after such a long treatment. Consider if any of the said factors apply to your project and always include an analysis of the digestate management's costs and benefits. If you include a digestate storage with biogas recovery, then it is admissible to calculate the digester's volume considering only the OLR criterion because the digestate storage is a post-digester at all effects.
4. Splitting the overall digestion volume into two or more stages and ensuring operation flexibility (e.g., including valves allowing to bypass the digesters for maintenance) minimise the probability of plant stops and profit loss during its lifetime.

A CASE STUDY OF 'COPY AND PASTE' ENGINEERING AND HOW TO IMPROVE THE ORIGINAL DESIGN

A large pig farm produces 70,000 m^3/year of slurry, i.e., 192 m^3/day. The plant constructor proposed a two-stage CSTR with four digesters of 30 m diameter and 6 m height each, connected in 'bridge configuration' (total V = 4 × 1,350 m^3 = 5,400 m^3). Such a configuration is typical for European agricultural biogas plants of 1 MWe rated power, fed mostly with corn silage. In the case of pig slurry, the choice of a CSRT is correct because the VS concentration is usually less than 5%. The necessary stirring power for 5,400 m^3 is about 48 kW. The constructor just assumed that the HRT would be enough for complete digestion:

$$HRT = \frac{5,400 \text{ m}^3}{192 \text{ m}^3 / \text{day}} = 28 \text{ days.}$$

They also assumed that the slurry had 10% DM and would produce 450 Nm^3/ton DM of biogas with 60% methane (data probably taken from some paper but 10% DM in pig slurry is far away from customary values found in the field). So, according to the constructor, the plant's rated power would be

$$P = 0.4 \times 9.94 \frac{\text{kWh}}{\text{Nm}^3} \times 0.6 \times 0.10 \times 450 \frac{\text{Nm}^3}{\text{tonne DM}} \times \frac{192 \frac{\text{tonne}}{\text{d}}}{24\ \text{h}} = 857\ \text{kW}.$$

Most probably, the constructor already had the drawings and calculations of a plant for 1 MW_{el} rated power and just proposed it with minimum or no adaptation work. The reality was quite different. The slurry sample had a 1.7% VS concentration, and the BMP measured in the author's laboratory with an adequate anaerobic inoculum was found to be 495 ± 10 Nm^3/tonne VS. The only correct assumption in the constructor's proposal was the digestion time equal to 28 days, as can be deduced from Figure 10.2.

The rated power resulting from feeding the plant only with the pig slurry would be

$$P = 0.4 \times 9.94 \frac{\text{kWh}}{\text{Nm}^3} \times 495 \frac{\text{Nm}^3}{\text{tonne VS}} \times 0.017 \times \frac{192 \frac{\text{tonne}}{\text{d}}}{24\ \text{h}} = 267\ \text{kW}.$$

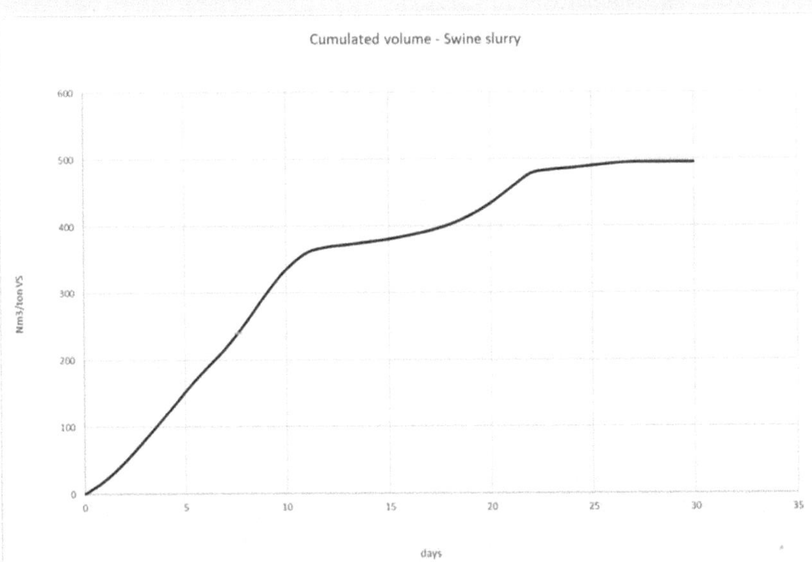

FIGURE 10.2 SMP Curve of Swine Slurry Reaching the Plateau in 25 Days.

Note that since the stirring power depends only on the digestion volume, the stirrers alone would consume 18% of the plant's produced energy. Also note that the OLR is

$$OLR = \frac{192{,}000\,\text{kg}/\text{d} \times 0.017}{5{,}400\,\text{m}^3} = 0.6\ \text{kg VS}/\text{m}^3\text{day}\,.$$

Such a low OLR means that the digester is working at just 20% of its potential. If built according to the constructor's proposal, the project would have failed. The options to improve the project in this case were two:

a. Install a centrifuge to concentrate the slurry to 10%–12%. The VLR would be reduced to just 27–33 m^3/day. The digestion volume calculated with the HRT criterion reduces to 30 times the VLR, i.e., 990 m^3. The volume calculated with the OLR criterion is

$$V = \frac{33{,}000\ \text{kg} \times 0{,}1}{3\ \text{kg VS}/\text{m}^3 \cdot \text{day}} = 1{,}100\ \text{m}^3.$$

   Hence, the digester's volume should be at least 1,100 m^3 to avoid organic overload. The rated electric power remains 267 kW because the organic load remains the same, but the necessary power for the stirrers reduces to roughly 10 kW, only 4% of the rated CHP power. This solution requires a storage volume for the water separated by the centrifuge and eventually a simple system to recover the nutrients, e.g., a struvite precipitation column.

b. Conduct a survey in the area to identify possible sources of digestible biomass to add to the slurry, reaching up to 10% – 12% VS concentration. Then take samples, measure the BMP of the resulting mixture, and recalculate both the CHP's rated power and the OLR for 5,400 m^3.

c. Regardless of the solution chosen, it is necessary to foresee a digestate storage tank to keep the digestate during the rainy season, a feature that the constructor did not foresee.

10.4.4 Assuming Data from the Literature as Universally Valid

Even reputed biogas plant builders tend to rely too much on productivity data 'from the literature'. The methanogenic potential (BMP) of biomass is not a 'universal constant' that can be taken from a table in some manual: it depends on both the substrate and the inoculum (differences up to 20% are common when measuring the BMP of a sample with inocula sampled from different plants). Average BMP data 'from tables' can be used for a preliminary feasibility study. If the feasibility study has a positive outcome, then such data must be checked by testing samples of the biomass available on site, and once the individual BMP has been ascertained, the BMP of the mixture must also be proven because it is not always equal to the proportional sum of the individual BMPs (see Chapter 1, Section 1.2.3, point 4: Information Quality Check).

10.4.5 Relying on Theoretical BMP Formulas

Some biogas plant builders rely on BMP tests performed on actual local samples but with non-biological methods. There is a strong cultural prejudice in favour of the German standard VDI 4630 by plant builders, business consultants, banks, and in some cases even public bodies that must evaluate the project (see the analysis of the flaws in the norm VDI 4630 and the derived IWA Protocol in [6]). One such logical flaw is accepting the Buswell formula for calculating the BMP. Some engineers rely on the Weende (alias Baserga) method, which was developed for assessing the metabolic energy of balanced feed for cattle, assuming that a digester is like a cow. Such methods do not account for the substrate's particle size distribution, the lignin content, the inoculum's specifical activity towards the substrate, and the presence of inhibitors. The following case study demonstrates, once again, why methods based on chemical analysis are not reliable.

A biogas plant was fed with sludge resulting from 'processing an organic food ingredient'. The plant owner sent a sample to a laboratory for an elementary analysis and calculated the BMP with the Buswell formula, obtaining a high value. When starting the plant's operation, the production was much below expected. The underperformance was attributed to defects in design or construction, so the owner filed a formal claim. The constructor, in turn, blamed the plant owner for adding the sludge to the digester's diet without prior agreement. The owner argued that the mixture contained a small percentage of the sludge and that it had been agreed that 'unforeseen substrates' should be tested according to VDI 4630, what he actually did. The author tested a sample of the sludge with two different inocula: sludge from the plant in question and digestate from former successful laboratory tests. The plant's inoculum had a low SMA, but the laboratory's one was within the norm. Nevertheless, the sludge caused the inhibition of both inocula, even when repeating the test with a high ISR. Further research showed that the sludge contained high percentages of anacardic acid, an irritating and antibiotic compound [7]. In this case, the plant owner and the engineering company had two cultural biases: assuming that a feedstock is digestible because it derives from 'organic food ingredients' (a pseudo-scientific belief that everything that is 'natural' or 'organic' is good) and that all methods in the VDI 4630 norm are equivalent and always yield reliable results.

10.4.6 Missing Start-up Strategy and Overlooking the Inoculum's SMA

Another common mistake of most biogas plant builders is omitting to perform a preliminary investigation to identify the most suitable anaerobic inoculum source for starting the plant up. The biological activity of the digestate used to start the plant directly affects the BMP of the biomass feedstock and, therefore, the energy productivity. The duration of the start-up phase and consequently the cost of fuel for heating the digester and the loss of profits because of the reduced productivity depend primarily on the specific methanogenic activity (SMA), which is not the same for all plants (see Chapters 5 and 6).

10.4.7 Extrapolation from Previous Experience in Different Contexts

We cannot take for granted that a solution that works or has worked fine somewhere can work equally well in a different context. Sometimes, the said extrapolation bias is amplified because a prestigious university or laboratory has been paid to 'validate' a given solution. Before adopting a given solution or technology in our biogas project, it is mandatory to check if the experimental conditions for such a validation are the same as expected in the plant and if the validation has considered the exergy balance (see the example of the sonicator technology in Chapter 2, Section 2.4.2).

10.4.8 Dismissing an Idea or Proposal without Experimental Proof

Conversely, assuming that a given solution or feedstock will not perform just because the plant constructor or the university 'validator' has never seen it may lead to losing market opportunities.

10.4.9 The Gambler's Fallacy

Assuming that the success obtained in the past in a given plant with a given strategy will also happen in the future is a bias known as the gambler's fallacy. There are many reasons why the plant's productivity can vary over time without any apparent change in the feedstock or the management strategy. Here are some examples of how Murphy's law will influence the biogas plant operation during its service life:

1. Plants 'age' because sediment deposits and/or floating mats form, reducing the digestion volume, which in turn reduces the HRT; therefore, digestion is not complete and energy production drops;
2. The quality of biomass varies over time because of abiotic or biotic stress to the crops and animals (drought, too much rain, contamination with aflatoxins, and health and diet of farm animals);
3. Different qualities of by-product batches from different suppliers (even cases of fraud: dishonest vendors selling adulterated cereal meal);
4. Seasonal variations in biomass quality (humidity, oxidation, and mould proliferation);
5. Human factor (more or less trained or diligent personnel);
6. Gradual reduction of the SMA of the bacterial ecosystem (e.g., due to too much recirculation of digestate and accumulation of salts, a deficit of some macro- or micronutrient, and excess dosage of some 'booster').

10.4.10 Error Propagation in the Business Plan's Spreadsheets

Sometimes, plant engineers and promoters employ spreadsheets for quick calculations. As the project evolves, they add interlinked sheets to the file. If a simple error, like a typo, is entered in a cell or a formula, and further calculations take that cell as one of the parameters, the result is a wrong business plan. The result can be either hyped or underestimated. In the first case, the risk is investing in a business that will never yield the expected profitability. In the second case, one risks losing the opportunity to invest in a profitable activity because the technical DD erroneously showed a low return on investment (ROI).

CASE STUDY 10.1

In Italy, the state pays an incentive bonus on the market value of biomethane injected into the grid. Such a bonus is the main reason why investors would endeavour the local tedious and never-ending bureaucratic procedure to obtain permission to build and operate the biomethane plant. In the case of a biomethane plant connected to the grid, the incentive paid on top of the market price of natural gas is the integral over time of the average higher heating value (HHV) of the biomethane, determined at least daily on the base of the chemical composition of the injected biomethane. The project promoter assumed this average HHV to be 38.5 MJ/Sm^3, which seemed too high. To assess the exact incentive, it is then necessary to consider the following:

1. The average composition of biomethane may vary between some limits defined by decree. In practice, the lowest percentage of methane acceptable for injection in the grid (in most EU countries) is 97%, with the remaining 3% being gases that do not contribute to the HVV. Pushing the upgrading system to obtain 99.9% purity is not necessarily the most economical option because the incentive will be higher but the energy consumption and maintenance costs of the upgrading system will grow more than proportionally. If we are to be conservative, as a good business plan should be, the average gas purity assumed for the calculations must be 97% of methane.
2. The HHV of pure methane (from any chemistry or engineering handbook) is 39.8 MJ/Nm^3. The biogas production was estimated from the BMP, expressed in Nm^3/tonne feedstock. So, all gas volumes in the business plan were expressed in Nm^3. However, the calculation of the incentive requires the HHV expressed in MJ/Sm^3, which is the usual reference condition employed in the natural gas industry.
3. The equivalence factor is 1 Nm^3 = 0,948 Sm^3 (rounded to 3 digits) or, conversely, 1 Sm^3 = 1,055 Nm^3. Both refer to the standard atmospheric pressure (101.32 kPa) but differ in the temperature: 0°C (273.15K) in the case of normal conditions and 15°C (288.15K) in the case of standard conditions.

   Demonstration of the equivalence
   According to the gas state equation:

   $$pV_{\text{normal conditions}} = nR(273.15+0)$$

   and

   $$pV_{\text{standard conditions}} = nR(273.15+15)$$

   dividing both equations member to member:

   $$V_{\text{normal conditions}} = \frac{273.15}{288.15} V_{\text{standard conditions}} = 0.948\ V_{\text{standard conditions}}$$

4. Hence, the HHV of pure methane is 39.8 MJ/Nm^3 x 0.948 Nm^3 / Sm^3 = 37.73 MJ/Sm^3. The conservative HHV to employ in calculating the incentive is 97% of the above value, i.e., 36.6 MJ/Sm^3 and not 38.5 MJ/Sm^3 as the promoter claimed, resulting in a 5% overestimation of the state incentive.
5. Corollary: why did the project promoter assume the HHV of biomethane to be 38.5 MJ/Sm^3? It turned out that the financial consulting company assumed methane to be equal to natural gas. This is an example of cultural bias because Italians commonly call natural gas 'metano'. Hence, the consultants took the average HHV of natural gas from the National Energy Authority's webpage because, in their view, 'it is the official value'. However, natural gas is a mixture of several gases. Its main constituent is methane, but depending on its origin, it may contain several other hydrocarbons that add to the HHV. Furthermore, natural gas in Italy is mostly imported from Libya and has a much higher content of secondary hydrocarbons, hence much higher HHV compared to gas from the Northern Sea available in other European countries. The average monthly value of natural gas HHV is published for transparency reasons because it varies over time depending on the import's mix and bills are calculated on the effective energy supplied to consumers. So, taking the published HHV in a given moment as equivalent to the HHV of methane is a coarse conceptual error.
6. Error propagation in the spreadsheet. On top of the conceptual error about the HHV, when converting the estimated methane production from Nm^3 to Sm^3 to calculate the income, the project promoter divided by 0.948 instead of multiplying. Hence, the business plan's income was inflated because they calculated a higher volume of gas than prospected, a higher methane percentage than the minimum required for an optimum operation of the upgrading system, and a higher HHV of methane than the actual figure.

10.4.11 Bias in Favour of Thermophilic Digestion

Some biogas plant builders tend to propose thermophilic digestion in every case. This is usually because 'the literature says that thermophilic digestion is more efficient' or 'the environmental impact will be lower because the digester will be smaller'. Such statements are only partially true (see Chapter 2, Section 2.2.1). The author has observed small differences between the BMP measured under mesophilic and thermophilic conditions, close to the error margins of the tests. Some of his customers repeated the tests, confirming that the differences were not significant, so they finally converted their plants from thermophilic to mesophilic without noticing any detrimental effects. Thermophilic digestion has some drawbacks: the bacterial biodiversity is reduced, reducing the biological stability in case of disruptions; the digester is indeed smaller but requires thicker insulation and higher heating power. Heat is not a problem for biogas plants with combined heat and power (CHP), but plants designed to produce biomethane will have to sacrifice part of their production just for heating the digester, so minimising the heat demand is crucial for maximising profits. Thermophilic digestion has two major advantages: the digestate is easier to process [8] and the process temperature above 50°C is an effective sanitation method for biowaste [9]. Unless mandatory for sanitation reasons, the convenience of a thermophilic process should be supported by thorough laboratory testing with samples of the inoculum and biomass available on site.

10.4.12 Branded Components

Purchasing the main components (CHP unit, pumps, and biogas upgrading system) from companies considered 'prestigious' certainly represents a higher initial cost but does not necessarily guarantee the success of the investment. In addition to the higher purchase cost, spare parts will also be more expensive and the system downtime while waiting for the visit of the 'authorised technician' could become too long. The best strategy to optimise the CAPEX is to search for components with an acceptable quality-to-price ratio and eventually invest in keeping some spares to reduce downtime in case of repairs.

10.4.13 Unnecessarily Oversizing the Digester

Some manufacturers tend to oversize the digestion volume, pumps, and stirrers in the belief that this makes the system more efficient. But they size the feeding system to the minimum, according to 'their' standard diet. The result is that when market contingencies or seasonal availability make it convenient to use lower quality, cheap, or even free biomass, the loading system is not able to feed the digester with the higher daily volume required. The loading system (automatic weighing system, auger or conveyor belt, mixing tank, etc.) becomes then the bottleneck that hampers increasing the OLR up to the optimum value for the digester. Furthermore, oversized pumps and stirrers will work at a fraction of their nominal power. Unless they are controlled through inverters, lightly loaded electric motors are less efficient. On the other hand, electric motors and transformers are designed to accept a permanent overload in the range of 5%–10%. So there is no point in oversizing motors. It is a good praxis to review the nominal power of the principal motors in a plant (stirrers, pumps, and compressors) and check if they are correctly dimensioned for the rated operational conditions.

10.5 EXTERNAL FAILURE CAUSES

An AD project may eventually fail or yield low profits despite having undergone a thorough financial and technical DD. Some causes like changes in regulations and policies, death of the promoter, litigation among investors, etc., are called *force majeure* in the legal jargon, and a small level of risk always exists. All business plans should include the cost of a good insurance covering as many risks as possible.

A well-designed AD plant may also underperform because of coarse and approximative management methods. Considering that an industrial AD plant costs several million euros, investing €30,000–40,000 in laboratory instruments to monitor the feedstock quality and the biological process will pay itself in less than 1 year. Investing in basic training of the plant personnel is fundamental.

10.6 CONCLUSIONS: A CHECKLIST FOR A SUCCESSFUL BIOGAS PROJECT

The following checklist will help the reader pinpoint the most frequent flaws in the redaction of biogas or biomethane business plans.

10.6.1 Feedstock Supply

a. Do you control the supply chain or is the feedstock a residue of your production? If not, you will have to purchase the feedstock for your digester, but prices are always subject to volatility and biomass may be subject to shortage (imagine a year with a severe drought or some disease drastically reducing the availability of agricultural by-products or residues). If you do not control the logistic chain but decide to proceed anyway, then sign contracts with your feedstock suppliers to secure the price, availability, and quality for the next years. The price should be calculated on the effective BMP and not on the bulk quantity supplied or its dry mass.
b. Check the feedstock's seasonal availability.
c. If you intend to cultivate some crops for feeding the digester, check the actual yield per hectare. In the case of 'exotic' crops, check if the claims in the literature are biologically possible (Sections 10.2.3.2 and 10.2.3.3).
d. Check the quality of samples taken from the envisaged feedstock suppliers. Do they contain – or are suspected to contain – sawdust, inhibitory substances like antibiotics, or more than 90% moisture?
e. Check the individual BMP of each feedstock composing the digester's diet. Data from 'the literature' are only acceptable for a preliminary rough assessment of the project's potential, but their uncertainty is too high to rely on for a business plan.
f. BMP tests must always include a control reactor (cellulose, acetate, or any other reference substrate) to assess the suitability and biological activity of the inoculum. The calculation of the test's metrological uncertainty and dispersion between replicates is mandatory for a sensitivity analysis.
g. When possible, choose the quantity of each feedstock so that the mixture's carbon-to-nitrogen ratio is $C/N \approx 30$.
h. Measure in the laboratory the actual BMP of the feedstock mixture. Assuming that the mixture's BMP is the proportional sum of the individual components is acceptable only for a feasibility study. In real life, the resulting BMP could be higher (if $C/N \approx 30$ and the inoculum is healthy enough) or lower (if one of the mixture's components is partially inhibitory, or if the resulting $C/N \ll 30$, or if there are antalgic interactions between substrates, or if the inoculum is not adequate). Check the density of each feedstock, and calculate the density of the resulting mixture. Then check both the OLR and VLR as explained in Section 10.4.3.

10.6.2 Plant Engineering

1. Correct choice of the digester type (see Section 10.4.2).
2. Digester size (see Section 10.4.3).
3. Mass balance. Never forget to perform a mass balance of the following:

   a. Each of the BMP tests. Checking both CH_4 and CO_2 productions (see Chapter 9) or at least using stoichiometric formulas (as in the example of Section 1.4.1) is a powerful tool to assess the quality of BMP tests (or claims that a given feedstock has been tested). In case of inconsistencies – like the biogas's mass bigger than the feedstock's VS in the example in Section 10.3.2.1 – engage another laboratory and repeat all the BMP tests. Caveat: a biogas mass of less than 95% of the VS mass of the substrate is acceptable. Note that biomass is seldom 100% digestible, and even when it is, a small percentage of the digestible fraction does not become biogas but living bacterial biomass. Vegetal biomass may contain up to 40% lignin and usually produces biogas with 50% CH_4 in volume, i.e., 26.7% in mass. So, one could expect the net mass yield of CH_4 to be as low as 0.5 × 0.267 = 13% of the VS mass.

   b. Always consider the error margin of the test because it will propagate into mass balance. In case of doubt, assume the lowest BMP value.

   c. The mass flows into and out of the digester. Knowing exactly the seasonal availability of feedstock, the feedstock input, and the CH_4 and CO_2 output, calculate the mass of digestate and define its management strategy.

4. Start-up strategy. Sourcing an adequate inoculum to fill the digester and starting the AD process are as important as securing feedstock supply contracts (see Chapter 6).
5. Process stability and energy balance

   a. If you are including a pretreatment, check that its energy demand is much lower than the increase in BMP expected. N.B.: Use the correct thermodynamical equivalences. If the pretreatment consumes electricity, the energy balance must consider how much electricity it is possible to generate with the expected additional CH_4 production:

   $$\Delta E = \Delta BMP \times 9.94 \frac{\text{kWh}}{\text{Nm}^3} \times 0.4$$

   If ΔE is smaller than the energy consumed in the pretreatment, then there is no point in pretreating the feedstock.

   b. Consider the self-consumption of energy. Rules of thumb:

      - 8–9 W/m^3 for the stirrers in the case of CSRT digesters. The overall energy consumption (stirrers + pumps + mills or shredders + conveyors) should be less than 10% of the rated electrical power of the generator.

- For biomethane production plants (injection into the grid), the necessary energy for upgrading (with selective membranes) and compression to 12 bars is about 0.25 kWh/Nm³ (Chapter 7, Section 7.4.1.1).
- For biomethane production plants (liquid biomethane aka LRNG), see Chapter 7, Table 7.3 because there are several different solutions.

6. Errors in the formulas and wrong conversion coefficients of the measurement units (see example in Section 10.4.10).

10.6.3 Income Sources in the Business Plan

a. In Europe, it is common practice to rely on state subsidies, carbon credits, and tariff incentives. Always remember that policies may change as the governments change political colour, so it is better to check if the biogas project is profitable with current energy prices and without any public subvention.
b. Assuming to get some income from selling the digestate as fertiliser is a common pitfall in which many investors get caught. If your plant is of the agricultural type, in the best case you will save some money by (partly) replacing commercial fertilisers with digestate. Composting the digestate often requires the addition of structuring material (woodchips, bark, straw, chopped reeds, and pruning). Straw is digestible, so it is more suitable as a substrate to produce energy. The other materials are undigestible, but at the same time they may have a market value as fuel. If their value as fuel is higher than the compost's value, then there is no convenience in composting the digestate. From an agronomical point of view, dry digestate is almost equivalent to compost (10 and 11), so there is little convenience in composting digestate unless legally required by local regulations.
c. If your plant is in Europe, check if the Nitrates Directive [12] applies to the site. If you do not possess enough land for distributing the digestate, you may need to install a denitrification system or any other posttreatment system, which means higher CAPEX and OPEX and no income.
d. Check if there is any industry requiring CO_2 in the neighbourhood. Selling CO_2 could provide a marginal income (see Chapter 8, Section 8.2).

10.7 FINAL ADVICE

Nullius in Verba!

BIBLIOGRAPHY

[1] Regulation (EU) 2020/852 of the European Parliament and of the Council of 18 June 2020 on the establishment of a framework to facilitate sustainable investment, and amending Regulation (EU) 2019/2088. Texts in all EU official languages here: https://eur-lex.europa.eu/legal-content/EN/TXT/?uri=CELEX:32020R0852 ii *Taxonomy: Final report of the technical expert group on sustainable finance*. https://finance.ec.europa.eu/document/download/9101b639-ebe5-413e-9c86-3e291722899e_en

[2] ii *Taxonomy: Final report of the technical expert group on sustainable finance*. https://finance.ec.europa.eu/document/download/9101b639-ebe5-413e-9c86-3e291722899e_en

[3] Encyclopaedia Britannica. https://www.britannica.com/science/tropical-rainforest/Environment
[4] V Andrade Fonseca et al., Morpho-physiology, yield, and water-use efficiency of Opuntia ficus-indica irrigated with saline water, *Crop Production; Acta Scientiarum. Agronomy*, 41 (2019). https://doi.org/10.4025/actasciagron.v41i1.42631. Open access text https://www.scielo.br/j/asagr/a/zhjyJdMzKrBmcHZPjd7sVCF/?lang=en#
[5] Alexandrius de Moraes Barbosa, Rafael Rebes Zilliani, Carlos Sérgio Tiritan, Gustavo Maia Souza & Marcelo de Almeida Silva, Energy conversion efficiency in sugarcane cultivars as a function of production environments in Brazil, *Renewable and Sustainable Energy Reviews*, 150, 111500 (2021). https://doi.org/10.1016/j.rser.2021.111500. ISSN 1364–0321.
[6] MA Rosato, *Managing biogas plants: A practical guide*, Chapter 6, CRC Press (2018). https://www.routledge.com/Managing-Biogas-Plants-A-Practical-Guide/Rosato/p/book/9781138626614. ISBN 9781138626614.
[7] National Library of Medicine, PubChem. https://pubchem.ncbi.nlm.nih.gov/compound/Anacardic-acid
[8] GD Gebreeyessus & P Jenicek, Thermophilic versus mesophilic anaerobic digestion of sewage sludge: A comparative review, *Bioengineering (Basel)*, 3:2, 15 (June 18, 2016). https://doi.org/10.3390/bioengineering3020015. PMID: 28952577; PMCID: PMC5597139.
[9] P Seruga, M Krzywonos, Z Paluszak, A Urbanowska, H Pawlak-Kruczek, Ł Niedźwiecki & H Pińkowska, Pathogen reduction potential in anaerobic digestion of organic fraction of municipal solid waste and food waste, *Molecules*, 25:2, 275 (January 9, 2020). https://doi.org/10.3390/molecules25020275. PMID: 31936589; PMCID: PMC7024283.
[10] CE Egene et al., Solid fraction of separated digestate as soil improver: Implications for soil fertility and carbon sequestration, *Journal of Soils and Sediments*, 21, 678–688 (2021). https://link.springer.com/article/10.1007/s11368-020-02792-z
[11] Vania Scarlet Chavez-Rico, Paul LE Bodelier, Miriam van Eekert, Valentina Sechi, Adrie Veeken & Cees Buisman, Producing organic amendments: Physicochemical changes in biowaste used in anaerobic digestion, composting, and fermentation, *Waste Management*, 149, 177–185 (2022). https://www.sciencedirect.com/science/article/pii/S0956053X22003063?via%3Dihub. ISSN 0956–053X.
[12] Council Directive of 12 December 1991 concerning the protection of waters against pollution caused by nitrates from agricultural sources (91/676/EEC). https://eur-lex.europa.eu/legal-content/EN/TXT/?qid=1561542776070&uri=CELEX:01991L0676-20081211

Abbreviations

a.a.	acetic acid
AD	anaerobic digestion
AMPTS	automatic methane potential test system
BMP	biochemical methane potential (Nm^3/tonne VS or Nm^3/tonne COD)
BHP	biochemical hydrogen potential (Nm^3/tonne VS or Nm^3/tonne COD)
CAPEX	capital expenditure = investment in a project
CHP	combined heat and power, aka 'cogeneration' in European texts
CSRT	continuously stirred reactor tank. This technology represents 90% of the digesters in the market
DD	due diligence
DF	dark fermentation
DM	dry matter (% of fresh mass)
E	absolute error (the same unit of the magnitude it refers to)
e	relative error (% of the measured value)
EBITDA	Earnings Before Interest, Taxes, Depreciation, and Amortisation. It is an alternate measure of profitability to net income. It is used to assess a company's profitability and financial performance.
EU	European Union, sometimes employed as a synonym of European
EC	European Commission
FCU	flow cell unit
FOS/TAC (German)	VFA/TA (English) = ratio of the total VFA, expressed as a.a., to total alkalinity, expressed as CO_3Ca
GAA	gas absorption attachment
GHG	greenhouse gas
HHV	higher heating value (usually MJ/kg or kWh/kg)
HRT	Hydraulic retention time (usually days)
ISR	inoculum to substrate ratio = $VS_{inoculum}$ / $VS_{substrate}$
LHV	lower heating value (usually MJ/kg or kWh/kg)
LRNG	liquid renewable natural gas = liquefied biomethane
NO/NC	normally open/normally closed contacts of a relay or switch
OFMSW	organic fraction of municipal solid waste
OLR	organic loading rate (kg VS/$m^3_{digester}$.day)
OPEX	operation expenditure = operation cost (energy, personnel, feedstock, etc.)
ORP	oxidation–reduction potential
PFR	plug flow reactor
Q	flow (usually m^3/day)

SMA	specific methanogenic activity (Ncm^3 CH_4/g $VS_{inoculum}$.day)
SOFC	solid oxide fuel cell
SWOT	Strengths, Weaknesses, Opportunities, Threats
TS	total solids (see DM)
UP	ratio of stirring power (at the shaft) to V (W/m^3)
V	volume of the digester (m^3)
VFA	volatile fatty acids, usually expressed as equivalent of a.a.
VLR	volumetric loading rate (m^3 of feedstock/day). N.B.: If water is required to prepare the slurry to feed the digester, as is usually the case with CSRT digesters, then the daily volume is the sum of the feedstock and the dilution water volumes.
VS	volatile solids = TS – Ash. Usually expressed as % of fresh mass, sometimes as % of DM. In this book, the first convention applies.
WW	wastewater
WWTP	wastewater treatment plant
η	efficiency. It is usually a quotient expressed in %. For example, the efficiency of digestion is the percentage of the feedstock's VS converted into methane.

Index

Q

R

S

T

U

FSC
www.fsc.org
MIX
Papier | Fördert gute Waldnutzung
FSC® C083411

Kolibri 360 GmbH
Ferdinand-Jühlke-Straße 7
99095 Erfurt, Deutschland
produktsicherheit@kolibri360.de